Natural Resource Degradation and Human-Nature Wellbeing

Rashed Al Mahmud Titumir · Tanjila Afrin ·
Mohammad Saeed Islam

Natural Resource Degradation and Human-Nature Wellbeing

Cases of Biodiversity Resources, Water Resources, and Climate Change

Springer

Rashed Al Mahmud Titumir
Department of Development Studies
University of Dhaka
Dhaka, Bangladesh

Tanjila Afrin
Department of Development Studies
Bangladesh University of Professionals
Dhaka, Bangladesh

Mohammad Saeed Islam
Department of Development Studies
Bangladesh University of Professionals
Dhaka, Bangladesh

ISBN 978-981-19-8660-4 ISBN 978-981-19-8661-1 (eBook)
https://doi.org/10.1007/978-981-19-8661-1

This Springer imprint is published by the registered company Springer Nature Singapore Pte Ltd.
The registered company address is: 152 Beach Road, #21-01/04 Gateway East, Singapore 189721, Singapore

Preface and Acknowledgements

The natural resource bank is gradually being drawn down. This book scrutinises the underlying causes of the degradation of natural resources. In this regard, it develops a new framework of sustainability in the case of the usage and management of natural resources by incorporating the idea of human sociality. Human sociality implies that humans are social beings, who behave in a reciprocal manner. As nature provides numerous benefits to human beings, they are naturally inclined to conserve and protect nature in return. Therefore, nature and human beings have a mutually beneficial relationship. This relationship, however, becomes distorted in a market economy.

The alternative framework emphasises the revitalisation of the symbiotic relations in order to progress toward a sustainable transformative pathway that can ensure the well-being of both nature and human beings by ensuring sustainable governance of natural resources. The framework thus entails that the well-being of human beings and nature must go side by side because one without the other is certainly not a viable option. Simultaneously, it considers the necessity of recognising political economy factors to pinpoint the root causes of natural resource degradation. Overall, the necessary condition is the revitalisation of mutual relations between human beings and nature while the sufficient condition is captured in terms of the nature of institutions and political settlement.

Specifically, the book examines biodiversity resources, water resources and climate change—the three most pressing issues in the realm of the current natural resource governance regime in the context of a developing country. It demonstrates, through theoretical and empirical analyses, that natural resources have been exploited beyond sustainable limits due to the commodification process, the existence of fragile institutions and unequal power-sharing arrangements. This results in the dissipation of natural resource rent, leading to the degradation of biodiversity resources, depletion and inequitable distribution of water resources, climate change as well as unsustainable development of the economy. Such degradation also steers intergenerational injustice because the younger generation of today is likely to suffer more in the future than their older counterparts.

The book relates to the COVID-19 pandemic, as there is a direct link between the pandemic and the destruction of nature. The 2030 Agenda for Sustainable Development with its biodiversity, water, climate change and other environment-related goals and targets is even more important today in the face of the pandemic, which has reinforced the need to protect biodiversity, ensure water security and halt climate change. In the light of the pandemic and its effects, accordingly, the book scrutinises the challenges of governing biodiversity, water and climate change.

Overall, the book offers important insights for academics and researchers specifically interested in the field of development studies who wish to gain a deeper understanding of sustainable natural resource governance, specifically in the context of developing countries. For policymakers and policy advocates, the book serves as a groundwork by providing pertinent contents in outlining the justification for policy objectives concerning biodiversity, water resources and climate change.

The authors would like to thank *Unnayan Onneshan*—an independent research think tank in Bangladesh—for providing support to conduct several research projects down the years to help build the foundation to plan and write this book. A significant amount of data for empirical analysis in this book was also collected under several research projects of this organisation. In line with that, the authors express gratitude to the researchers, Jayanta Kumar Basak, Mohammed Abdul Baten, Md Humayain Kabir, Tahera Akter, among others as well as the respondents who participated in those research. In addition, the authors would like to thank Azmol Hossain, Coordinator, and Mamun-ur-Rashid, Community Researcher, who also provided their support in conducting the studies.

Furthermore, the authors also express their gratitude to the anonymous reviewer(s) for their valuable insights for the development of the book. The authors' special thanks go to Assistant Prof. Md. Zahidur Rahman of Bangladesh University of Professionals, Dhaka, Bangladesh, for going through the first draft of the manuscript, which helped them in revising the chapters thoroughly. Last, but not least, the authors are grateful to their family members for their unwavering support throughout the journey in writing this book.

The authors are solely responsible for any errors, inadequacies and omissions that may still remain in the book even after a thorough review and proofreading. The authors will certainly address those in the next edition.

Dhaka, Bangladesh
December 2022

Rashed Al Mahmud Titumir
Tanjila Afrin
Mohammad Saeed Islam

Contents

About the Authors

Rashed Al Mahmud Titumir is a Professor of Economics at the Department of Development Studies, University of Dhaka, Bangladesh and is currently holding the charge of the Chairman of the Department. He is the Founder-Chairperson of *Unnayan Onneshan*—a Dhaka based multidisciplinary think-tank, Vice Chairperson of IUCN Asia Regional Members Committee and Chairperson of IUCN National Committee of Bangladesh. His latest books are: *Fiscal and Monetary Policies in Developing Countries: State, Citizenship and Transformation* (Routledge); *State Building and Social Policies in Developing Countries: The Political Economy of Development* (Routledge); *Why Agriculture Productivity Falls: The Political Economy of Agrarian Transition in Developing Countries* (Purdue University Press); and *Numbers and Narratives in Bangladesh's Economic Development* (Palgrave Macmillan). He edited *Sundarbans and Its Ecosystem Services: Traditional Knowledge, Customary Sustainable Use and Community Based Innovation* (Palgrave Macmillan) and co-edited *COVID-19 and Bangladesh: Response, Rights and Resilience* (University Press Ltd).

Tanjila Afrin is an Assistant Professor at the Department of Development Studies, Bangladesh University of Professionals, Bangladesh. Her research focuses on the political economy of natural resource governance, environment and development, biodiversity conservation, climate change, poverty and inequality, livelihoods and governance.

Mohammad Saeed Islam is an Assistant Professor at the Department of Development Studies, Bangladesh University of Professionals, Bangladesh. His areas of research interest are environmental and resource economics, water economics and policy, environmental protest, agriculture and rural development, poverty and inequality and welfare economics.

Abbreviations

AEZs	Agro-Ecological Zones
AR	Assessment Report
ATC	Average Total Cost
BADC	Bangladesh Agricultural Development Corporation
BARC	Bangladesh Agricultural Research Council
BBS	Bangladesh Bureau of Statistics
BCCSAP	Bangladesh Climate Change Strategy and Action Plan
BCM	Billion Cubic Meter
BCR	Benefit Cost Ratio
BFD	Bangladesh Forest Department
BIGD	BRAC Institute of Governance and Development
BLC	Boat License Certificate
BMD	Bangladesh Meteorological Department
BoB	Bay of Bengal
BPDB	Bangladesh Power Development Board
BWP	Bangladesh Water Partnership
CBA	Cost Benefit Analysis
CBD	Convention on Biological Diversity
CDC	Centre for Disease Controls and Prevention
CDM	Comprehensive Disaster Management
CDMP II	Comprehensive Disaster Management Programme II
CEDMHA	Center for Excellence in Disaster Management and Humanitarian Assistance
CEGIS	Centre for Environmental and Geographic Information Services
CMAAS	Community-based Mangrove Agro Aqua Silvi
COP	Conference of the Parties
CPI	Climate Policy Initiative
CRI	Climate Risk Index
CS	Commercial Shrimp
DDM	Department of Disaster Management
DMB	Disaster Management Bureau

DoE	Department of Environment
DoF	Department of Fisheries
DPHE	Department of Public Health and Engineering
DPSIR	Driving forces-Pressures-State-Impacts-and-Response
DSSAT	Decision Support System for Agro-technology Transfer
DWASA	Dhaka Water Supply and Sewerage Authority
ECA	Ecologically Critical Area
EEZ	Exclusive Economic Zone
EIA	Environmental Impact Assessment
EJF	Environmental Justice Foundation
EU	European Union
FAO	Food and Agriculture Organization of the United Nations
FD	Forest Department
FPP	Forest People's Programme
FSF	First Start Finance
GBF	Global Biodiversity Framework
GCF	Green Climate Fund
GCM	General Circulation Model
GCRI	Global Climate Risk Index
GDP	Gross Domestic Product
GED	General Economic Division
GHGs	Green House Gases
GI	Geographical Indication
GMB	Ganges-Brahmaputra-Meghna
GoB	Government of Bangladesh
GoN	Government of Netherlands
HBS	Heinrich Boll Stiftung
HRAs	High-Risk Areas
IBNET	International Benchmarking Network for Water and Sanitation Utilities
IBRD	International Bank for Reconstruction and Development
ICZM	Integrated Coastal Zone Management
IDMC	Internal Displacement Monitoring Centre
IPBES	Intergovernmental Science-Policy Platform on Biodiversity and Ecosystem Services
IPCC	Intergovernmental Panel on Climate Change
IPLCs	Indigenous Peoples and Local Communities
IRP	International Resource Panel
IRSWR	Internal Renewable Surface Water Resources
IUCN	International Union for Conservation of Nature
IWM	Institute of Water Modelling
IWRM	Integrated Water Resources Management
JMP	Joint Monitoring Programme
LPI	Living Planet Index
LPL	Lower Poverty Line
LUC	Land Use Change

LUCF	Land Use Change and Forestry
MC	Marginal Cost
MCM	Million Cubic Meter
MEA	Millennium Ecosystem Assessment
MEB	Multiple Evidence Base
MoDMR	Ministry of Disaster Management and Relief
MoEF	Ministry of Environment and Forest
MoF	Ministry of Finance
MoFA	Ministry of Foreign Affairs
MoUs	Memoranda of Understanding
MoWR	Ministry of Water Resources
MR	Marginal Revenue
NAPA	National Adaptation Programmes of Action
NB	Net Benefit
NIE	New Institutional Economics
NOAA	National Ocean and Atmospheric Administration
NPV	Net Present Value
NTFPs	Non-Timber Forest Products
ODA	Official Development Assistance
OECD	Organisation for Economic Cooperation and Development
PA	Protected Area
PES	Payments for Ecosystem Services
PPGIS	Public Participation Geographic Information System
PVB	Present Value of Benefits
PVC	Present Value of Costs
RF	Reserve Forest
RGR	Rechargeable Groundwater Resources
RSLR	Relative Sea Level Rise
SCBD	Secretariat of the Convention on Biological Diversity
SDGs	Sustainable Development Goals
SLR	Sea Level Rise
SMRC	SAARC Meteorological Research Centre
SRDI	Soil Resource Development Institute
SRF	Sundarbans Reserve Forest
TEV	Total Economic Value
TIB	Transparency International Bangladesh
TK	Traditional Knowledge
TRWR	Total Renewable Water Resources
TWAP	Transboundary Waters Assessment Programme
UN	United Nations
UNDP	United Nations Development Programme
UNDRR	United Nations Office for Disaster Risk Reduction
UNEP	United Nations Environmental Programme
UNESCO	United Nations Educational, Scientific and Cultural Organization
UNFCCC	United Nations Framework Convention on Climate Change

UNHCR	United Nations High Commissioner for Refugees
UNICEF	United Nations Children's Fund
USAID	United States Agency for International Development
WARPO	Water Resources Planning Organisation
WASH	Water, Sanitation and Hygiene
WB	World Bank
WC	Working Circle
WHO	World Health Organization
WMO	World Meteorological Organisation
WRIS	Water Resources Information System
WWAP	World Water Assessment Programme
WWF	World Wide Fund for Nature
YPSA	Youth Power in Social Action

List of Figures

List of Tables

List of Maps

Chapter 1
Setting the Context

1.1 Introduction

The worldwide concerns for deforestation, degraded soil, polluted air, declined water, endangered species and climate change are all well-meaning. Yet the causes, effects and solutions warrant meticulous scrutiny. The concerns have heightened in the face of the COVID-19 pandemic, as there is a direct linkage between nature's destruction and the outbreak of disease. It is imperative to protect nature by halting natural resource degradation to prevent future pandemics. In this backdrop, the book focuses on natural resource governance mechanisms illustrated with cases of biodiversity, water resources and climate change to identify the underlying causes of resource degradation, especially in the context of developing countries. It also attempts to bring forth a new understanding regarding the usage, management and conservation of natural resources in a sustainable manner.

The broad concerns that arise in the contemporary era are to (a) conserve nature in all its forms and functions and (b) create an equitable home and achieve a high standard of living for people on this planet without destabilising critical planetary processes (O'Neill et al., 2018; WWF, 2016). These dual challenges are also delineated in the United Nations' 2030 agenda for sustainable development. The Sustainable Development Goals (SDGs)—17 in total—under the agenda include several goals (Goals 6, 13, 14 and 15) which focus on nature and the environment directly (Elder & Olsen, 2019; Scharlemann et al., 2020), while other goals have indirect environmental implications as well.

However, specific targets to protect climate and biodiversity (under Goals 13, 14 and 15) are completely off track (Nature, 2020). Despite a temporary drop in carbon dioxide emissions due to the pandemic, the world is still heading for a temperature rise of more than 3 °C this century. This goes far beyond the Paris Agreement goals of keeping global warming below 2 °C and aiming for 1.5 °C (UNEP, 2020). Simultaneously, around 25% of biodiverse species on an average are threatened indicating that around 1 million species have already faced extinction (IPBES, 2019). Ensuring access to clean water (under Goal 6) also remains a major challenge. The

R. A. M. Titumir et al., *Natural Resource Degradation and Human-Nature Wellbeing*,
https://doi.org/10.1007/978-981-19-8661-1_1

goal itself recognises that managing water sustainably goes beyond simply providing clean water to address broader issues such as water quality and watershed management, water scarcity and usage efficiency and restoration of water-related ecosystems (Dahan & Kashiwase, 2016). Essentially, the problems in the realm of biodiversity, water or climate are more acute in the context of developing countries due to the context-specific dynamics of resource governance. Moreover, there is also a paucity of studies in understanding the dynamics in that specific context. The book, accordingly, draws on the empirical evidence of a developing country—in this case Bangladesh—to highlight the complexity of issues.

The mainstream theories and policy options under the neoliberal regime primarily follow the market-centric approach to deal with the problems of natural resource degradation. The approach clearly suggests that the market is the best possible mechanism to allocate and distribute resources efficiently through a voluntary exchange, based on Pareto optimality arguments (Arrow et al., 1996; Freeman III et al., 2014; Pearce, 1991; Pearce et al., 1989; Stavins, 1989, 2011; Tietenberg, 1990; Zhang, 2013).

The assumption of voluntary exchange, however, does not always fit well with the real world, and the arguments on efficiency are not viable. The heterogeneity of ownership of natural resources causes the scope of inefficient consumption and overexploitation, which results in 'market failure' (see Chap. 2 for details). Thus, the continuous degradation of natural resources could not be stopped. Moreover, the market-centric approach ignores the factors related to political economy, which are the key forces that cause the failure of market mechanisms in transitional economies (Broad, 1995; Clark & York, 2012) and politics, in particular, can affect the exploitation of natural resources (Collier, 2010; López & Toman, 2006). Recent literatures recognise that a solution can be arrived at through the revitalisation of mutual relationships between human beings and nature (Díaz et al., 2015; Gu & Subramanian, 2012; Ichikawa, 2012). It is, however, yet to examine the actual implications of this understanding in resource governance regime in general and in developing countries in particular.

The book tries to contribute to this arena of resources governance regime by formulating a new framework pivoted around the concept of 'human sociality'. The framework puts together variables such as institutions, power, political settlement along with human sociality into the alternative resource governance framework that helps understand the pitfalls in the mainstream governance framework to ensure a sustained relationship between human beings and nature. Overall, the book firstly scrutinises the market-centric perspectives and, secondly, combines political economy questions and human sociality that are usually overlooked in the discussions of the current governance framework. It ultimately develops an alternative framework to examine the reasons behind the degradation of natural resources and to offer viable and sustainable solutions to the problem.

The empirical analysis of the book demonstrates that in developing countries—as evident in Bangladesh—natural resources have been exploited beyond the sustainable limit due to structural rigidities, embedded in and reproduced by fragile institutions and unequal power-sharing arrangements under a market-centric economy. These countries are also bearing a larger share of the burden emerging from climate change as compared to developed countries due to the factors pertaining to the international political economy. Therefore, in order to reverse the condition, it is important to recognise the inherent values of nature going beyond its narrow conceptualisation through a market-centric lens. This would ultimately ensure the well-being of nature and human beings simultaneously.

1.2 Biodiversity Resources

The loss of biodiversity, both at the regional and global levels, is quite evident. The Living Planet Index (LPI) of 2020 shows that between 1970 and 2016, the planet lost 68% of its biodiversity (WWF, 2020). Previous assessments also showed a persistent downward trend across different periods of time (e.g., LPI of 2018: 60% between 1970 and 2014, LPI of 2016: 58% between 1970 and 2012 and LPI of 2014: 52% between 1970 and 2010) (WWF, 2014, 2016, 2018). It implies that the rate of loss has also increased over the years. Moreover, the trend of biodiversity populations under various ecosystems exhibits a decline, and the greatest losses can be observed in the freshwater environment, followed by the terrestrial ecosystem (Table 1.1).

Additionally, the degrading trend in biodiversity varies from one region to the other (Table 1.2). The highest rate of degradation (94%) has been observed in Latin America and the Caribbean. The Asia Pacific region is in the third position, losing almost half of its total biodiversity over the same timeframe. Thus, differences are acute in the abundance of biodiversity resources across regions, with the largest decline being in tropical areas.

Table 1.1 Trend in biodiversity based on LPI 2016

Types of LPI	Trend in population (between 1970 and 2012) (%)	Annual decline (%)
Terrestrial	– 38	1.1
Freshwater	– 81	3.9
Marine	– 36	1.0

Source WWF (2016)

Table 1.2 Declining trend in biodiversity across different regions from 1970 to 2016

Species name	Percentage
North America	– 33
Europe and Central Asia	– 24
Latin America and Caribbean	– 94
Africa	– 65
Asia Pacific	– 45

Source WWF (2020)

Table 1.3 Percentage of threatened species

Species name	Percentage
Amphibians	41
Mammals	26
Conifers	34
Birds	14
Sharks and rays	37
Reef Corals	33
Selected crustaceans	28

Source IUCN (2021)

In terms of diversity among species, more than 38,500 are threatened with extinction (IUCN, 2021). Specifically, amphibians are the most threatened species followed by conifers, according to the IUCN Red List of 2020 (Table 1.3). The share of threatened species of other types is also quite significant.

The degradation of different ecosystems (e.g., forest, wetlands and marine) also indicates a threat to the volume of biodiversity on the planet as they occupy a vast amount of such resources. FAO's Global Forest Resources Assessment illustrates that while 129 million hectares (million ha) of forest have been lost since 1990 on a net basis, 10 million ha a year of forest cover have been lost since 2015 (FAO, 2015, 2020a).

The assessment also reveals that the global forest area is shrinking, though the rate of loss has slowed down between 1990 and 2020 (Fig. 1.1). In the period 2010–2020, Africa experienced the highest annual rate of net forest loss, with 3.9 million ha, followed by South America with 2.6 million ha. Since 1990, the rate of net forest loss in Africa has increased in each of the three decades. However, it has dropped significantly in South America, to nearly half the rate in 2010–2020 compared to 2000–2010. Asia had the highest net gain in forest area, followed by Oceania and Europe in the same period. Nonetheless, net gains in both Europe and Asia were substantially lower in 2010–2020 than in 2000–2010. Oceania also observed net losses in forest area in the decades of 1990–2000 and 2000–2010. Tropical forests are becoming degraded primarily because of massive deforestation, limiting their ability to provide the ecological services that they formerly provided in abundance (McNeely, 2021). The degradation of tropical forests due to ongoing deforestation is reducing the ability to deliver their former abundance of ecosystem services (McNeely, 2021). For instance, approximately 8,000 of the world's estimated 60,000 tree species are already considered globally threatened, while 8% of forest plants, 5% of forest animals and 5% of fungi are critically endangered (IUCN, 2020; One Tree Planted, 2021).

Mangrove forests—a unique combination of terrestrial and aquatic ecosystems—support diverse groups of flora and fauna. The total area of mangroves was estimated to be 15.6 million ha in 2010, down from 16.1 million ha in 1990 (FAO, 2010), which has decreased by 1.04 million ha and now stands at 14.56 million ha (FAO, 2020b).

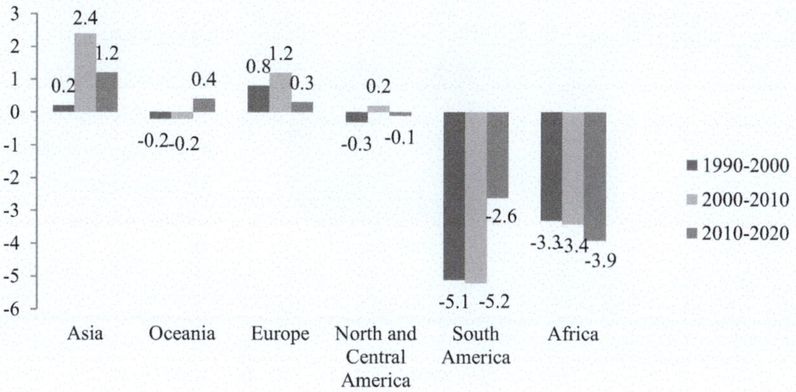

Fig. 1.1 Annual forest area net change, by decade and region, 1990–2020. *Source* FAO (2020a)

In Asia alone, 92,135 ha of mangroves were deforested and 80,461 ha reforested between 2000 and 2012, resulting in a net loss of 11,673 ha (Giri et al., 2015). In this process, nearly 50% of the mangrove biome has been lost since the 1950s, 16% critically endangered, 10% near-threatened and 15% facing the risk of extinction, out of 70 mangrove species in total (Sala, 2020; Sarker et al., 2016).

The wetland ecosystem also supports a diverse source of biodiversity resources. Freshwater ecosystems support approximately one in every ten known species of plants, mammals, fish, reptiles, insects and molluscs, totalling over 126,000 species, despite covering approximately 0.8% of the earth's surface and 0.01% of the world's water (UN Environment/UN-Water, 2018; UNESCO/UN-Water, 2020). According to the freshwater LPI, roughly 880 of those species have declined by 83% with reptiles, amphibians and fish being the most vulnerable, and regions most at risk are the Neotropics (– 94%), the Indo-Pacific (– 82%) and the Afrotropics (– 75%) (WWF, 2018). In particular, fish had the highest extinction rate worldwide in the twentieth century (IPBES, 2019; UNESCO/UN-Water, 2020). Since 1970, there has been a 70% increase in the number of invasive alien species in wetlands (IPBES, 2019), posing a major threat to the survival of native fish species.

Knowledge of the marine ecosystem in terms of the number of species it supports is incomplete, with only 11% described as such (Luypaert et al., 2020). It is clear, nonetheless, that marine biodiversity is under threat. The abundance of marine fish species has decreased by 38% compared to the level of 1970 (Hutchings et al., 2010). It is obvious as 63% of fish stocks in the ocean are currently experiencing overfishing (Sala, 2020). Of the total coastal and marine species, 36% have already declined (Ramsar Convention on Wetlands, 2018), whereas the LPI for the marine population showed a 49% drop between 1970 and 2012 (WWF, 2015). The risk of extinction appears to be increasing, with projections indicating that by 2100, more than half the marine species may be on the verge of extinction unless dramatic adjustments are made (UNESCO, 2017).

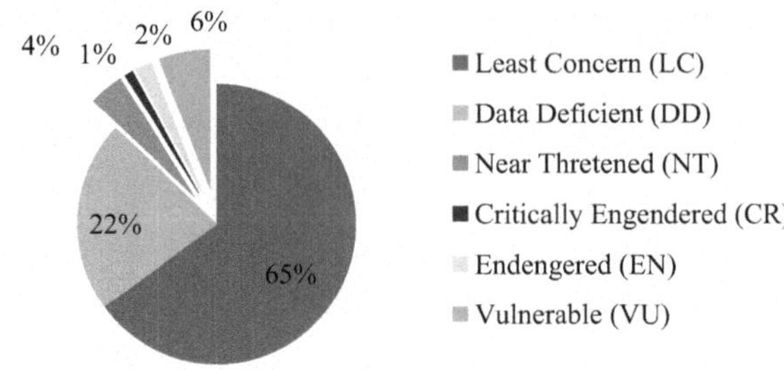

Fig. 1.2 Status of marine species. *Source* IUCN (2018)

The status of marine species in terms of the IUCN Red List reveals that a significant percentage of the 12,924 species assessed (only 4.2% of those currently defined) fall into the NT, CR, EN and VU categories (Fig. 1.2). Overall, 13% of species are identified as threatened with extinction. However, the assessment rate of marine species for extinction risk is not satisfactory (Luypaert et al., 2020) and therefore, it might not be possible to fully comprehend the extent of the loss of resources in this particular ecosystem.

The overall trend in the extent of biodiversity resources across the world thus signifies an increasing trend of degradation of such resources. A vast number of species have recently become extinct or are threatened with extinction at the most expeditious rate ever recorded. The loss of biodiversity over the last century has been so severe that many biologists believe that the planet is on the brink of ‘the sixth extinction’, coming 65 million years after the fifth extinction, which saw the disappearance of dinosaurs (Ceballos et al., 2020). In the past, the extinction of certain species was mainly caused by natural events such as tectonic movements leading to continental interchange, but today this has been caused primarily by human activities.

The developing countries, however, are suffering a disproportionately greater loss of biodiversity compared to the developed ones. While there is an increase of 10% in the extent of biodiversity in high-income countries and a loss of 18% in middle-income countries, the astounding figure of loss for low-income countries is 58% (WWF, 2014). Notably, the losses in the last two cases cancelled out the gains of the first category countries. Overall, the overwhelming evidence regarding the loss of biodiversity worldwide indicates that it is necessary to act promptly to conserve this valuable resource base for the sake of humanity’s survival, which has also been recognised in the recently adopted post-2020 Global Biodiversity Framework (SCBD, 2022).

1.3 Water Resources

Water, another vital component of nature, is under serious pressure due to severe unequal distribution and degradation. Since the 1980s, global water consumption has increased by about 1% per year, owing to a combination of population growth, socio-economic development and changes in consumption patterns (UNESCO/UN-Water, 2019). The sustainable management of water resources includes multiple challenges, which are indeed difficult to classify, since there are diverse types of water resources.

Access to and provisioning of safe water for drinking, household activities and agricultural production has remained a prime concern in the realm of water governance. There are already two billion people who do not have daily access to a safe source of drinking water (WHO & UNICEF, 2019). Climate change has brought about additional hardships for them, highlighting the distinction between health and sickness, thriving or barely surviving (Water Aid, 2020; UNESCO/UN-Water, 2020). The UN data shows that about 1.5 billion people are affected by acute water scarcity (Harvey, 2020) and the number is likely to climb to 5 billion by 2050 (Water Aid, 2019). In addition, by 2050, over 50% of the world's population will struggle to get water for at least a month each year (Water Aid, 2020).

The issue has emerged as a major challenge particularly in developing countries (UNESCO/UN-Water, 2019). Per capita water usage in the majority of these countries remains far below than that of developed countries. None of the top 10 countries, which lack access to safe water, is a developed country (Table 1.4). Eritrea is right at the top where about 80.7% of the total population do not have access to basic water services, whereas Mozambique occupies the tenth position with more than half the population suffering in this regard.

The disparity among countries in terms of access to safe water is also evident based on their economic status (Table 1.5). The share of the population, who got access to water free from contamination, had increased across all countries from 2000 to 2020. Nevertheless, the coverage reached 60.9% in lower-middle-income economies, 76.8% in upper-middle-income economies and 99.2% in high-income economies in 2020. On the contrary, low-income economies could manage to provide contamination-free water to only 35.8% of the people in the same year.

Table 1.4 Top 10 countries with the greatest percentage of people living without access to basic water services

Rank	Country	Percentage (%)
1	Eritrea	80.7
2	Papua New Guinea	63.4
3	Uganda	61.1
4	Ethiopia	60.9
5	Somalia	60.0
6	Angola	59.0
7	Democratic Republic of the Congo	58.2
8	Chad	57.5
9	Niger	54.2
10	Mozambique	52.7

Source Water Vision (2020)

Table 1.5 Access to water (free from contamination) among countries based on economic status

Year	Low-income economies (%)	Lower-middle-income economies (%)	Upper-middle-income economies (%)	High income economies (%)
2000	26.6	47.6	68.8	97.6
2005	27.9	48.6	71.0	97.8
2010	29.8	51.1	73.1	98.2
2015	32.3	56.1	75.0	98.9
2020	35.8	60.9	76.8	99.2

Source WHO/UNICEF (2021)

Moreover, the consumption of water has increased across different countries over the years, while the disparity has widened among the high-income and low-income countries (Table 1.6). Per capita consumption of water in developing countries has increased between 2002 and 2016, but the amount is still too low compared to their rich counterparts.

The inequality in getting water access in terms of the socio-economic condition is also evident at the national level. The wealthy groups generally receive high-quality service at a modest cost, whereas the poor often pay much higher prices for services of equivalent or lesser quality (UNESCO//UN-Water, 2019). The situation is worse in developing countries where the majority of the poor have to pay a higher price to access water through informal channels. The price of water through the house or piped connections (formal channel) is almost always lower than informal connections (World Bank, 2019). The amount of cost that the poor in developing countries burden actually comprises a significant share of their household income.

Table 1.6 Unequal consumption of water in some selected countries

Country	Per capita domestic water use (litre/day) In 2002	Per capita domestic water use (litre/day) In 2016	Per capita total water use (m^3/year) In 2002	Per capita total water use (m^3/year) In 2016
USA	575	609	1923	1570*
Australia	493	570	1118	854
India	135	205	586	631**
Bangladesh	46	92	–	242***
Kenya	46	42	68	79****
Ethiopia	15	32	79	103
Mozambique	4	54	47	53

Source Authors' calculation based on IBNET and FAO (2018)
*For USA, the latest available data is up to 2010
**For India, the latest available data is up to 2010
***For Bangladesh, the latest available data is up to 2008
****For Kenya, the latest available data is up to 2010

For instance, in Cambodia, the cost of 50 L of water is equivalent to 108% of a typical poor person's salary; it is 54% in Papua New Guinea; 45% in Madagascar; 25% in Ghana; 17% in India; 15% in Ethiopia; and 13% in Mozambique (Water Aid, 2016). On the contrary, in the case of rich countries, the price of water is only about 0.1% of a person's minimum wage. Thus, there are issues with the provision of and access to water, especially in terms of overall consumption and household usage across and within countries.

Concerns over water resources are also increasing due to unilateral developments in the spheres of *transboundary water*. The number of Transboundary Rivers in the world is 276. In Asia, there are about 60 Transboundary Rivers (Table 1.7). More than 31,340 large dams (65.3% of the total) have been commissioned on the rivers of Asia to harness hydropower and irrigate agricultural land.

Upstream countries are interested in building large dams and barrages in order to secure the availability of water for their lands. Downstream countries, therefore, suffered largely from an inadequate inflow of water available for consumption and irrigation. Consequently, the rivers in downstream are drying up and facing a shortage of water. The total number of large dams was 4270 in 1950, but the trend of building new dams increased manifold since 1950 and it peaked in the 1970s (Fig. 1.3). However, the number slowed down in the 2000s.

Table 1.7 Region-wise large dams and their capacities

Category	World	Europe	Asia	North and Central America	South America	Africa	Austral-Asia
Transboundary River basin	276	68	60	46	38	64	–
Total number of large dams	~ 48,000	5480	31,340	8010	979	1269	577
Average height (m)	31	33	33	28	37	28	33
Average reservoir capacity (million m^3)	269	70	268	998	1011	883	205
Technically feasible hydroelectric potential (TWhr/yr)	14,370	1225	6800	1660	2665	1750	270
Annual hydroelectric production (TWhr/yr)	2643	552	753	700	534	62	42

Source Authors' compilation from Gleick (2014), UN Water (2014)

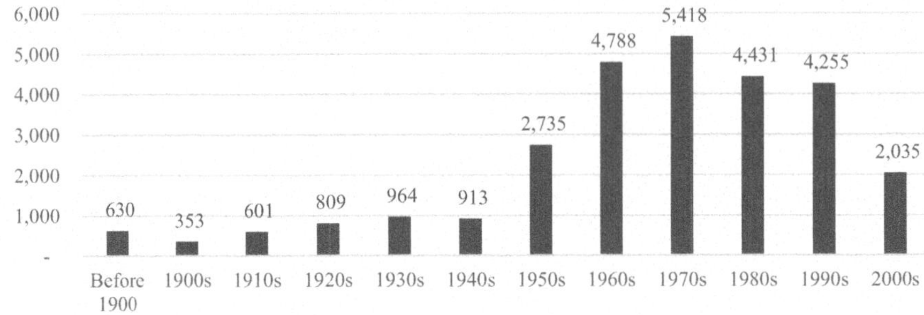

Fig. 1.3 Number of large dams commissioned each decade. *Source* Compiled by the authors from Gleick (2014), McCully (2007)

The construction of dams, incidentally, has posed a serious threat to regional water security. It is revealed that the risk of severe political conflict is likely to occur or escalate in the context of transboundary water governance (Baranyai, 2020). Despite the fact that multiple policy options have already been initiated, particularly under the Water Convention (United Nations, 2022), there is still an unequal sharing of the benefits of Transboundary Rivers. It has, therefore, become a major concern to find out a realistic solution to the problem that may help all the stakeholders in the long run as well as ensure the sustainability of the existing river system.

Water-related ecosystems, also known as *wetlands*, are among the most ecologically diverse environments on the planet and provide numerous benefits to society (UNESCO/UN-Water, 2018). This ecosystem continues to decline globally. Global wetland outlook exhibits 35% of total wetlands, where data is available, have been lost since 1970 (Ramsar Convention on Wetlands, 2018), while another estimate, covering a longer timeframe, shows that 85% have lost during 1700–2000 (UNESCO/UN-Water, 2020). Both of the estimates agree that the rate of declination is three times greater than that of the forests. Furthermore, the rate of declination appears to have reached a critical level in recent years as during 1970–2008 about 25–30% of the wetlands disappeared worldwide (IPBES, 2015) whereas during 1970–2015, the extent of wetlands sharply declined by 35% (Crump, 2017; Ramsar Convention on Wetlands, 2018).

The trend in declination, however, varies across regions. The largest declines (60%) of natural wetlands have taken place in Europe and Asia, whereas the rate is 20% each in North America and Oceania and 30% in Africa (IPBES, 2015). Another estimation signifies that the Neotropic and the Asian region are facing a quick rate of declination estimated at − 1.956% and − 1.515% annually (Table 1.8). Moreover, the rates of depletion of the wetlands (both inland and coastal) in the Asian region are higher compared to that in Europe and North America.

Such scenarios are a cause of concern for the ecological system of both the inland and coastal wetlands since the freshwater and marine species are also dwindling with the degradation of wetlands. The sustainable usage and management of wetlands and

Table 1.8 Rate of change in the area of natural wetlands in different regions during the twentieth and early twenty-first centuries

Region		*N**	Average rate of change (% per year)	s.d.**
Africa				
	All wetlands	10	− 0.927	1.260
Asia				
	Inland wetlands	16	− 1.885	1.872
	Coastal wetlands	12	− 1.102	0.488
	All wetlands	28	− 1.515	1.494
Europe				
	Inland wetlands	26	− 1.027	0.912
	Coastal wetlands	17	− 0.986	0.625
	All wetlands	50	− 1.057	0.767
Neotropics (Caribbean, Central and South America)				
	All wetlands	7	− 1.956	1.639
North America				
	Inland wetlands	7	− 0.347	0.504
	Coastal wetlands	7	− 0.508	0.368
	All wetlands	14	− 0.428	0.432
Oceania (incl. Australasia)				
	All wetlands	4	− 1.062	1.799

Source Davidson (2014)
*‘*N*’ denotes the number of records in each period
**‘s.d.’ denotes the standard deviation of the estimation

the resources provided by them, therefore, have also emerged as a key challenge in the realm of water resources governance.

The book also focuses on *marine resources*. The marine ecosystem is currently considered to be the most vital form of water resource that can largely contribute to accelerating the national and global economy. It is also separately termed as ‘blue economy’ considering the value of resources that can be harvested from the oceans. The ecosystem is full of highly productive and valued resources that include hydrocarbons, precious minerals, large stocks of fisheries and unique marine biodiversity. However, the striking fact is that some of the resources remain underutilised on the one hand as there is a shortage of data regarding the extent of marine resources in both the global and national contexts, while some resources are being overexploited and degraded on the other. Under the consideration of the global concerns (Table 1.9), marine resources become a contested resource largely stemming from the demarcation problem and security perspective.

Table 1.9 List of major global concerns related to the marine resources

Concerns	Status
Fishing	• 53% of the world's ocean fisheries are fully exploited and 32% overexploited, depleted or recovering from depletion • Most of the top 10 marine fisheries, accounting for about 30% of all capture fisheries production, are fully exploited or overexploited, and several important commercial fish populations have dwindled to the point that their survival is threatened • Each year, billions of unwanted fish and other mammals such as dolphins, marine turtle, seabirds, sharks and corals die due to inefficient, illegal and destructive fishing practices
Climate change	• Marine species affected by climate change include plankton which forms the basis of marine food chains—corals, fish, polar bear, walrus, seals, sea lions, penguins and seabirds • The current increase of 0.7 °C in global temperature since pre-industrial times is disrupting life in the oceans, from the tropics to the poles. The prediction of a further rise in temperature between 1.4 and 5.8 °C by the end of the century can be the cause of extinction for many species
Inadequate protection	• Whereas 70% of the planet's surface is covered by marine ecosystem, only 7% of the ocean is designated as protected and only 2.4% is fully protected from fishing and other activities
Pollution	• Over 80% of the waste, dumped in the ocean, come from land-based activities. Moreover, oil spill and other anthropogenic activities like ship-breaking are a major threat to the destruction of the marine ecosystem

Source Adapted from WWF (2017), Sala (2020)

In particular, the utilisation of marine resources efficiently and sustainably has emerged as a major challenge for the developing countries as they lack in technological advancement and in playing an appropriate role in the global platform in the context of a complex international political economy.

As regards both freshwater and marine ecosystems, the *fisheries* sector can also be recognised as one of the most valuable resources in the water resources regime. The sector provides natural nutrition, generates employment and contributes as a livelihood option for thousands of people. The fisheries and aquaculture sector has grown significantly in recent decades, with total production, trade and consumption reaching an all-time record in 2018 (FAO, 2020c). Sustainable management of fisheries, however, remains a challenge. The percentage of fish stocks, which was 90% within a sustainable level in 1990, dropped to 65.8% in 2017 (Fig. 1.4), which is a cause for concern.

Furthermore, a large portion of fisheries and aquaculture production is either lost or squandered—approximately 35% of global harvest—because of inefficient capture methods, a lack of an appropriate regulatory framework, infrastructure and market access (FAO, 2020c). As the problems are more evident in developing countries, developed countries still dominate the fish markets in trade worldwide. The

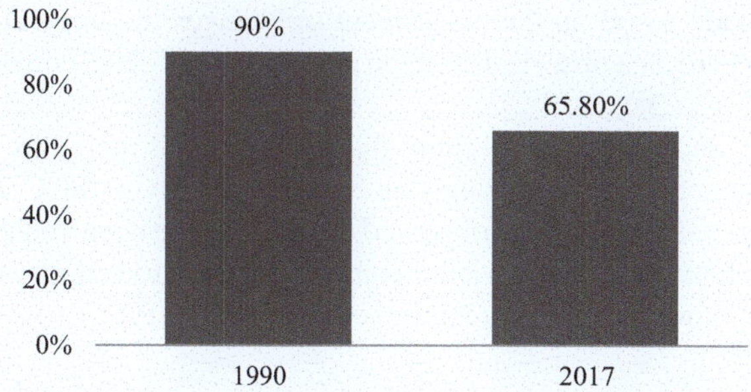

Fig. 1.4 Percentage of fish stocks within biologically sustainable levels. *Source* FAO (2020c)

importance of developing countries, however, as consumers as well as producers, has been steadily increasing.

Overall, the challenges connected to the water resources regime can be seen as the growing demands and anthropogenic interventions in the water system leading to negative distributional effects. In particular, water resources are degrading on one hand, while the resources are distributed unevenly on the other. The rich countries are well ahead in providing access to water, whereas the developing and poor countries are hamstrung by lack of infrastructure and poor governance. Against this backdrop, an alternative multi-level analysis of the governance of water regimes is important to understand the welfare of resources. The crucial fact to note is that the issues relating to water resources governance are quite complex to categorise, analyse and find a uniform solution.

1.4 Climate Change

Climate change has appeared as one of the major challenges in the path of ensuring sustainable development. It has had a significant impact on the earth and its systems over the last few decades, and it is likely to continue doing so in the intermediate and long-term future (Nichols et al., 2011). There is also evidence that climate change brings stark intergenerational injustice (Carrington, 2021).

Changes in climate are primarily evident from the increase in global average air and ocean temperatures, widespread melting of snow and ice and rising global average sea level (IPCC, 1990, 1995, 2001, 2007, 2014, 2018, 2021). The IPCC's Fifth Assessment Report (AR5) presented a detailed scenario of the increasing trends in temperature, sea level and atmospheric concentrations of major greenhouse gases (GHGs)—carbon dioxide (CO_2), methane (CH_4) and nitrous oxide (N_2O) (IPCC, 2014). The IPCC later issued a special report warning that if the current trend sustains, global warming would likely hit 1.5 °C between 2030 and 2052 (IPCC, 2018). The Sixth Assessment Report (AR6) recently confirmed that temperatures would rise by more than 1.5 °C above pre-industrial levels in the next two decades, breaching the ambition of the Paris climate agreement (IPCC, 2021). The projection has been made based on the assessment of changes in temperature covering a longer period where data observed over the period of 1850–2020 shows an unprecedented warming (Fig. 1.5).

The assessments claim that the observed increases are unequivocally caused by human activities through the emission of GHGs (IPCC, 2018, 2021). The comparative scenario among the observed changes, those occurred due to natural factors only and others due to human and natural factors, clearly signify that the natural factors played a minor role in increasing the temperature (Fig. 1.6).

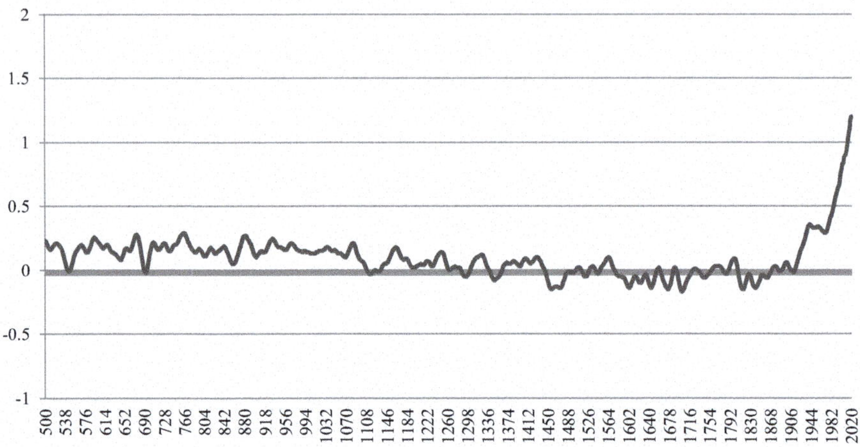

Fig. 1.5 Change in global surface temperature (decadal average) relative to 1850–1900. *Data Source* IPCC (2021)

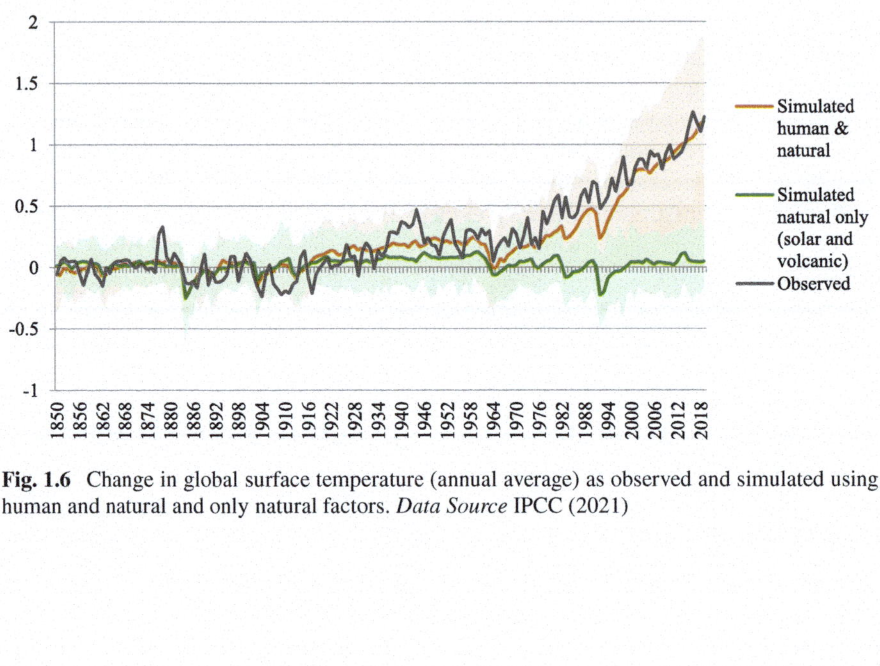

Fig. 1.6 Change in global surface temperature (annual average) as observed and simulated using human and natural and only natural factors. *Data Source* IPCC (2021)

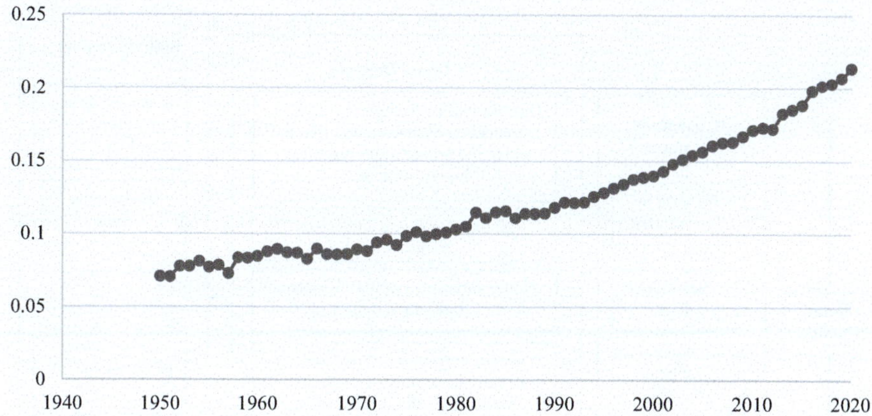

Fig. 1.7 Global mean sea-level change relative to 1900. *Data Source* IPCC (2021)

The Sixth Assessment Report also points out that the sea level has been rising precipitously (Fig. 1.7). Global mean sea level increased by 20 cm between 1901 and 2018. While temperature and sea-level rise have clear trends, accurate annual precipitation volumes are unknown in many regions. Such uncertainties, however, do not preclude that extreme precipitation events have increased over the years and are likely to intensify more in future (UNESCO/UN-Water, 2020). The IPCC projection indicates that precipitation would increase at high latitudes in the equatorial Pacific and parts of the monsoon regions, but decrease in the subtropics and limited areas of the tropics (IPCC, 2021). Overall, there are significant changes in climate parameters resulting from different anthropogenic pressures.

Assessment of the impacts of climate change, however, is scientifically challenging and has, until now, been fragmented. The challenges of climate change are, in fact, multi-dimensional, multi-sectoral and have both immediate as well as long-term effects (Islam, 2016). Multiple direct and indirect impacts of climate change range from economic, social, environmental as well as political parameters (Fig. 1.8). The changes in climate parameters have a direct impact on the hydrological system resulting in the reduction of surface water, groundwater and loss of wetlands. Extreme weather events, salinity intrusion, and hotter summer have also increased due to the direct impact of climate change. Most significantly, it has resulted in the displacement of a large number of people. Climate change has also had an effect on income status, agriculture and food security, biodiversity, human health and social stability indirectly. All these have clearly affected the lives and livelihoods of millions of people. The Sixth Assessment Report, in assessing the impacts of climate change in particular, has also reported recently that climate change has severely affected both ecosystems and human systems worldwide in diverse ways (IPCC, 2022).

The impacts, however, are unevenly distributed (IPCC, 2022) and generally greater for the disadvantaged communities and for the low-lying developing and underdeveloped countries. Incidentally, there is no end to the vulnerability of people residing

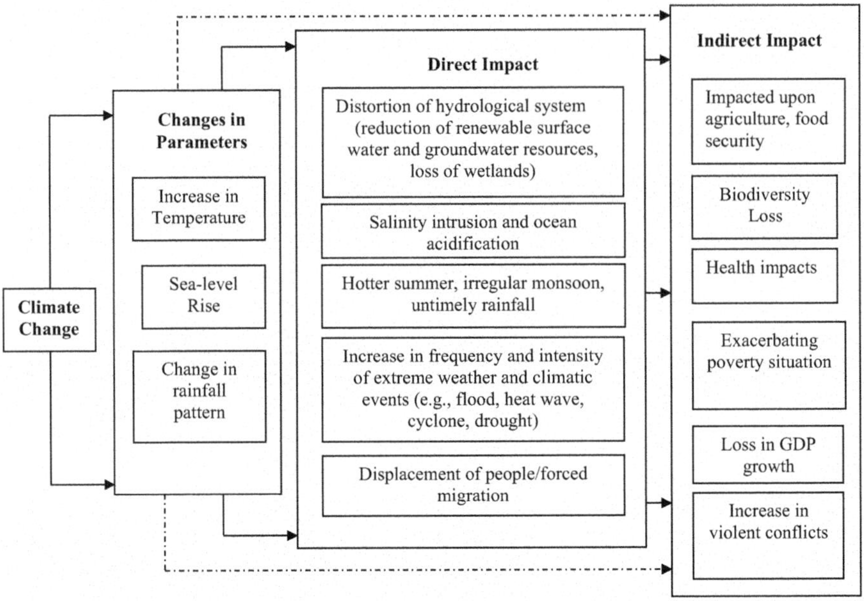

Fig. 1.8 Impacts of climate change. *Source* Prepared by the authors

in the coastal regions of developing countries. According to the Global Climate Risk Index (GCRI)-2020, seven of the ten most affected countries and territories from 1999 to 2018 were low-income or lower-middle-income developing countries, two were upper-middle-income countries (Thailand and Dominica), and one was a high-income country (Puerto Rico) (Eckstein et al., 2020). It confirms that the less developed countries are generally more affected by climate change than the developed ones. The striking reality is that the former ones are the least contributors to the emission of GHGs. Climate change thus reflects and reinforces inequalities. It is yet to come up with a globally acknowledged solution to this environmental crisis of climate change.

Overall, contemporary natural resource governance problems and the subsequent environmental degradation do not respect state borders and, thereby, pose challenges that are local, regional and global in nature.

1.5 Biodiversity, Water and Climate Change: The Case of Bangladesh

The growing global concerns regarding biodiversity, water resources and climate change as discussed above have implications in the Bangladesh context, too. Bangladesh—a developing country in South Asia—is endowed with exceptionally rich biodiversity. The extinction and gradual decrease in the country's flora and fauna, however, signify that this *biodiversity resource* base is under potential threat.

The total number of faunal species (Table 1.10) does not necessarily imply that the number is increasing in Bangladesh, as newer ones were included in the assessment list of some studies. For instance, crustaceans and butterflies were added to the list for the first time by IUCN in 2015.

Concentration on a specific group rather indicates a note of caution that many species have already become extinct, while some others are facing a threat of extinction (Table 1.11). As many as 13 species in all were marked as extinct from the country on the Red List of 2000, whereas the number increased to 31 in 2015. The total number of threatened species also increased from 147 to 414 during that time, including the number of crustaceans, butterflies and freshwater fish. Though there has been a decrease in the number of assessed species, an increasing threat to the faunal species, on the whole, has become obvious.

Many of the floral species of Bangladesh are also reportedly facing a threat in different assessment reports. The first volume of the Red Data Book of Vascular Plants of Bangladesh listed 106 species (Khan et al., 2001), whereas the second volume included 120 species as threatened following IUCN's Red List categories (Ara et al., 2013). On the other hand, the Encyclopaedia of Bangladesh on flora and fauna listed 486 vascular plants as threatened (Table 1.12). Thus, several species of flora and fauna in Bangladesh are facing the threat of extinction.

Such degrading trend in biodiverse resources is associated with the increasing vulnerability of different ecosystems of the country. The forest ecosystem has decreased significantly in both quantity and quality over the last few decades. The extent of forest cover in Bangladesh from 1990 to 2015 shows a gradual decline (Fig. 1.9). The forest cover has slipped down to 1429 thousand ha in 2015 from 1494 thousand ha in 1990. Marine and wetlands-based biodiversity resources in Bangladesh are also degrading, and they are often undervalued in terms of assessment of the biodiversity portfolio of the country.

The Sundarbans is the richest reservoir of Bangladesh's biodiversity resources, which is importantly a combination of forest, wetland and coastal ecosystems. It alone supports 53% birds, 43% animals, 42% reptiles, 36% amphibians, 29% plants and 17% fish species of the country's total biodiversity resources (Baten & Kumar, 2010). Over 40 species of amphibians, reptilians, aves and mammalians in this forest, however, are listed as critically endangered or vulnerable (Aziz & Paul, 2015). The decadal changes in forest cover depicted through Public Participation Geographic Information System (PPGIS) in the Sundarbans reveal that the number of trees is declining at a tremendous pace, resulting in an increasing amount of fallow land in the forest (*Unnayan Onneshan*, 2020). In two decades, the total area of dense forests in the case study sites has come down to half. In addition, the value of various waterways of the forest such as rivers, canals and creeks has decreased in the last two decades (2000–2020) with respect to the availability of big fish and crabs, which indicates that aquatic resources are also depleting drastically (*Unnayan Onneshan*, 2020).

Table 1.10 Status of wildlife faunal species in Bangladesh

Group	Khan (1982)	IUCN Bangladesh (2000)	IUCN Bangladesh (2007)	Khan (2008)	Ahmed et al. (2007, 2008, 2009)*	Khan et al. (2019)	Khan (2015)	IUCN Bangladesh (2015)
Butterflies	–	–	–	–	–	–	–	305
Crustaceans	–	–	–	–	185	–	–	141
Freshwater—fish	–	263	–	–	270	–	–	253
Amphibians	19	22	22	53	34	42	64	49
Reptiles	124	126	109	158	147	157	174	167
Birds	578	628	388	690	650	718	711	566
Mammals	119	113	110	121	120	124	133	138
Total	840	889	629	1022	951	1041	1082	1619

Source Authors' compilation from different sources
*Different volumes of Encyclopedia of flora and fauna of Bangladesh

Table 1.11 Extinct and threatened faunal species in Bangladesh

Group	Total no. of living species		Extinct		Threatened							
					Critically endangered		Endangered		Vulnerable		Total threatened	
	2007	2015	2007	2015	2007	2015	2007	2015	2007	2015	2007	2015
Amphibians	22	49	0	0	0	2	3	3	5	5	8	10
Reptiles	109	167	1	1	12	17	24	10	22	11	58	38
Birds*	388	566	2	19	19	10	18	12	4	17	41	63
Mammals	110	138	10	11	21	17	13	12	6	9	40	38
Freshwater fishes	–	253	–	0	–	9	–	30	–	25	–	64
Crustaceans[a]	–	141	–	0	–	0	–	1	–	10	–	11
Butterfly[b]	–	305	–	0	–	1	–	112	–	75	–	188
Total	629	1619	13	31	52	56	48	180	37	152	147	414

Source Prepared by the authors based on Red List report of IUCN Bangladesh (2007, 2015)

*Excluding migratory birds; *a* and *b* new groups included in the national Red List of Bangladesh for the first time in 2015

Table 1.12 Number of threatened species in major vascular plant groups according to the 'Encyclopedia of flora and fauna of Bangladesh'

Group	Total no. of species	Critically endangered (CR)	Endangered (EN)	Vulnerable (VU)	Total no. of threatened species (% of total species)
Pteridophytes	195	0	0	36	36 (18.46)
Gymnosperms	7	0	1	0	1 (14.29)
Angiosperms	3611	30	126	293	449 (12.43)
Dicotyledons	2623	8	80	179	267 (10.18)
Monocotyledons	988	22	46	114	182 (18.42)
Total	3813	30	127	329	486 (12.75)

Source Irfanullah (2011)

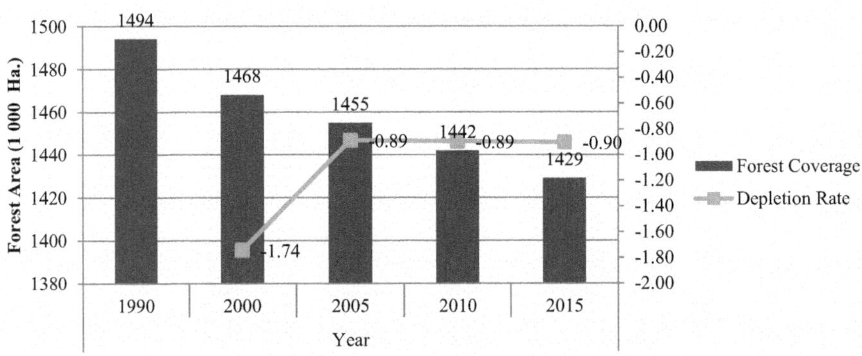

Fig. 1.9 Extent of forest area and rate of depletion in Bangladesh (1990–2015). *Source* Authors' calculation based on data from FAO (2015)

Regarding *water resources*, first, the reduced amount of surface water available for consumption has been putting pressure on the groundwater table. The groundwater is now considered to be the most important source of water for domestic consumption, irrigation and industrial use in the country. Groundwater levels, however, continue to decline due to excessive extraction. The growing pressures, because of the increasing population and extensive groundwater withdrawals for domestic consumption and irrigation, have lowered the water table in many areas below the effective reach of hand tube wells.

The estimate shows that Bangladesh's water table dropped drastically by 32% during 2003–2013 at an average of 8.73 mm per year (Khaki et al., 2018). Northern districts in particular have been facing an acute problem. A study revealed that in 15% of the monitoring wells, located in Bogura, Rajshahi, Naogaon, Joypurhat and Chapai Nawabganj districts, groundwater tables were found to remain below six metres throughout the year (Mojid et al., 2019).

Table 1.13 Average rate of annual groundwater decline in Dhaka during 1970–2000 as per the estimate of Bangladesh Water Development Board

Timeline	Average rate of annual groundwater decline (in metre)
1970–1980	0.17–0.6
1980–1990	0.15–0.69
1990–2000	1.24–3.00

Source Alam (2018)

Urban areas are also experiencing a dwindling water table as a result of intensive groundwater extraction, with Dhaka—the capital city—being the most vulnerable. Dhaka has a daily water demand of 2.45 billion litres, with groundwater exploitation from underground aquifers meeting 78% of that demand (DWASA, 2020). With the increase in demand, the rate of groundwater declination has increased significantly over the years (Table 1.13). According to the 2030 Water Resources Group, the city's annual water decline rate is 3 m per year, and if no preventive measures are implemented, the groundwater level will drop to 100–150 m by 2050 (Islam, 2021).

Apart from excessive extraction associated with rising demand, the filling up of lowlands, canals, water bodies, pollution of rivers adjacent to the city and unplanned urbanisation are some other reasons responsible for the depletion of the groundwater table in Dhaka. Against this backdrop, access to this water source and consequential inequality in terms of provision has become a contested issue.

Bangladesh has also been suffering from the unilateral withdrawal of trans-boundary water by the upstream countries, along with seasonal variations in water flow causing floods as well as shortages. The country, being located in the lower part of the Ganges–Brahmaputra–Meghna (GBM) basin, depends largely on the water from these river systems. Additionally, all the major rivers in Bangladesh are connected to the GBM basin. As a result, any intervention in the flow of the GBM basin hampers Bangladesh severely.

The river system of Bangladesh shows that around 1500 rivers were available in Bangladesh during the eleventh century, which decreased to 800–700 rivers in 1947 and then to 310–230 rivers in 2017 (Table 1.14) of which 25 are already dead. This is mainly due to the obstacles built over the natural flow in the upstream. The dams and the barrages established by the upstream tend to upset the ecological balance of the region. In particular, Bangladesh and India share some of the most intricate river systems. The water, therefore, is now subject to an unequal exchange between these co-riparian countries.

Being among the low-lying river basin countries in South Asia, Bangladesh is also a favourable region for the formation of wetlands. According to the Ramsar

Table 1.14 Decreasing number of inland rivers of Bangladesh

Year	1000–1100	1947–1971	2017
Number of inland rivers of Bangladesh	1500–1400	800–700	310–230

Source Matin (2017)

Convention, wetlands comprise two-thirds of the country's total land (Byomkesh et al., 2009) and almost 50% of the population depend on wetland resources (Islam, 2010). The wetlands of Bangladesh, however, have been declining and their resources degrading gradually (Byomkesh et al., 2009; Haque & Basak, 2017; Islam, 2010; Shopan et al., 2013). Approximately, 2.1 million ha of wetlands have been lost in the Ganges–Brahmaputra floodplains because of flood control, drainage and irrigation development (Shopan et al., 2013).

Several anthropogenic pressures such as overexploitation, illegal encroachment, construction of flood embankments and industrialisation have increased the vulnerability of the ecosystem. This is evident from the degradation of the available biodiversity resources and the deterioration of the livelihood condition of the communities dependent on wetlands. In particular, the contribution of the fishery resources of the harbour region to the national economy is on the decline. Overall, Bangladesh is facing the loss of its wetland areas and their resources.

The marine ecosystem is also a vital source of water resources for Bangladesh. Bangladesh gained an entitlement of 118,813 sq. km in the Bay of Bengal (BoB), comprising the territorial sea and Exclusive Economic Zone (EEZ), following the final settlement of border disputes with neighbours Myanmar and India, in 2012 and 2014, respectively (MoFA, 2014). Bangladesh's total marine water area is now 121,110 sq. km. The resources of the ecosystem have enormous potential to contribute to the country's socio-economic development. However, challenges remain in terms of equal distribution, technological capacity and sustainable fisheries production.

Finally, Bangladesh, as a low-lying developing country, is one of the most vulnerable to climate change. The Global Climate Risk Index (GCRI)-2010 also identified the country as the most exposed to extreme climate events during the period 1990–2008 (Hermeling, 2010). The GCRI-2020, however, indicated that Bangladesh is one of the top 10 nations (7th) facing the adversities of climate change (Eckstein et al., 2020). Several scientific assessments, using various models, have shown changes in the climatic variables of Bangladesh, including temperature, rainfall pattern and relative Sea-Level Rise (SLR) (e.g., Basak et al., 2013; Choudhury et al., 1997; Khan et al., 1999; Karmakar & Shrestha, 2000; Khan et al., 2019) and also have projected for changes in future (e.g., Agarwal et al., 2003; Becker et al., 2020; BUP-CEARS-CRU, 1994).

One such estimate shows that Bangladesh's temperature has increased predominantly over the last 21 years (1990–2010) compared to the last 63 years (1948–2010) (Hasan & Rahman, 2013). Moreover, fluctuations in the rainfall pattern and events related to extreme precipitation have increased (Basak et al., 2013; Mondol et al., 2018). Finally, there is an increasing trend in the SLR indicator, which is the most significant cause of concern for the country. It results in the submerging of land that is likely to have a massive impact on the lives and livelihoods of thousands.

An overall projection scenario (Table 1.15) exhibits that temperature would rise by 1.0 °C, 1.4 °C and 2.4 °C by 2030, 2050 and 2100, respectively, in Bangladesh. The rising trend is evident not only on an annual basis but also during winter and monsoon. On the other hand, the percentage of change in rainfall is expected to rise during monsoon and decrease in winter. Above all, SLR is expected to rise by 14 cm,

Table 1.15 Projections for changes in temperature, precipitation and sea-level rise for Bangladesh based on GCM (General Circulation Model)

Year	Temperature change (°C) mean			Rainfall change (%) mean			Sea-level rise (SLR) (cm)
	Annual	DJF	JJA	Annual	DJF	JJA	
2030	1.0	1.1	0.8	+ 5	− 2	+ 6	14
2050	1.4	1.6	1.1	+ 6	− 5	+ 8	32
2100	2.4	2.7	1.9	+ 10	− 10	+ 11	88

Source Adapted from Agarwal et al. (2003), MoEF (2005)
Note DJF represents December, January and February, usually the winter months. JJA represents June, July and August, the monsoon months

32 cm and 88 cm, respectively, in 2030, 2050 and 2100. It has also been warned that a one-metre sea-level rise will inundate 15–20% of the coastal Bangladesh in the upcoming decades (MoEF, 2012).

These changes in climate parameters have had a huge impact on nature and the people of Bangladesh, which have a complex relationship with each other (Fig. 1.10). Rise in temperature, along with other changes, causes scarcity of water, impacts food security and degrades biodiversity resources. In addition, erratic rainfall worsens water scarcity in the dry season and threatens food security while creating higher river flow during monsoons. This results in drainage congestion and floods particularly in the city areas. Several climate hazards in the form of natural disasters have also emerged as major threats for Bangladesh.

There is an increased intensity and frequency of cyclones and tidal surges, especially in coastal areas. Increased magnitude of floods, flash floods and river erosion becomes a common factor in the rainy season throughout the country. Such hazards have a negative impact on agriculture, biodiversity and livelihood, ultimately causing the displacement of people and the increase in poverty rate. Last but not the least, the rise in sea level also threatens food security by affecting agricultural production and biodiverse regions. Most importantly, it is the prime cause of the climate-induced displacement of marginalised communities. Overall, the impact of climate change has become a challenging task to address for Bangladesh in ensuring sustainable development.

The cumulative effect of the past and present mismanagement of natural resources in this country has brought about a significant reduction in the economic, social and environmental benefits to society. To halt and reverse the degradation of natural resources, thereby, it is required to ensure sustainable governance of those resources.

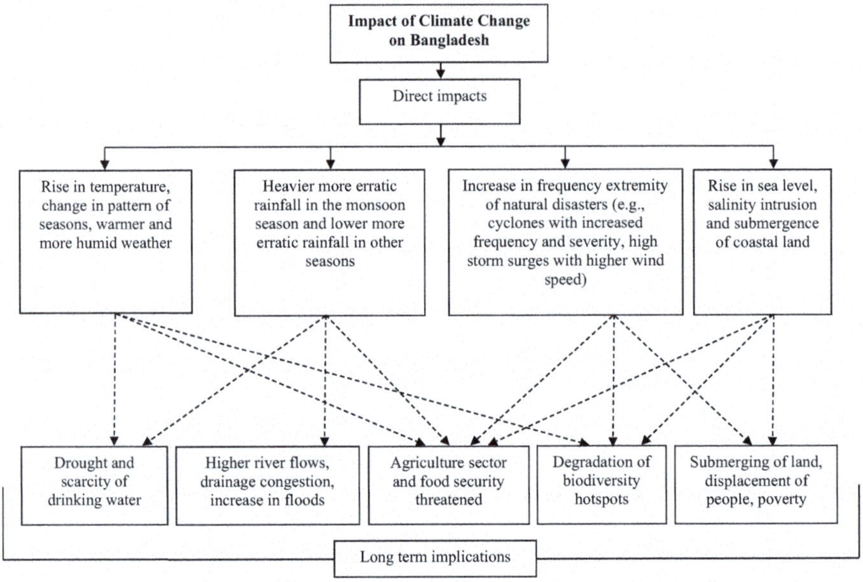

Fig. 1.10 Direct impact and long-term implications of climate change in Bangladesh. *Source* Prepared by the authors

1.6 Scope and Approach of the Book

The book discusses the issue of natural resource degradation and conservation through the lens of development studies. Development studies attempt to find explanations to 'development-related problems' such as extreme poverty, social inequality, natural resource degradation as well as analyse the conflicting meanings of development from an interdisciplinary perspective. One of the fundamental challenges to development is the degradation of natural resources across the globe. The trend of degradation is evident across regions, but the extent is more acute in developing countries. Moreover, there is a lack of knowledge in the area of natural resource governance in developing countries or transitional economies, which calls for scrutiny. In this backdrop, the book explores the cases of decline in biodiversity resources, exploitation of water resources and climate change considering the context of Bangladesh.

There have been a number of significant scholarships available on natural resource governance issue, but the majority of them only touched on the topic in general terms. There is a paucity of studies on natural resource governance mechanisms across different contexts (e.g., Shivakoti et al., 2016 explored the dynamics of natural resource management in Asia), particularly in developing or transitional economies. Furthermore, most of the books on natural resources and environmental issues are related to economics and policy analysis (e.g., Chichilnisky & Rezai, 2020; Hackett,

2006; Helm & Hepburn, 2009; Kula, 1994). Some others have attempted to identify environmental problems while examining the interrelations between economic growth and sustainability (e.g., López & Toman, 2006). Studies are also available on specific issues such as on biodiversity (e.g., Adhikari and Lawson, 2018; Kirton et al., 2002; Laird, 2002; Laladhas & Oommen, 2017; ten Kate & Laird, 2000), or on water (e.g., Earle et al., 2010; Farolfi et al., 2013; Grigg, 2016; Islam & Smith, 2019), or on climate change (e.g., Betzold & Weiler, 2018; Bulkelly and Newell 2010; Filho, 2015; Helm & Hepburn, 2009; Huq et al., 2019). Most of these books focus on either theoretical understanding or offer some case studies in the context of different countries.

Considering the existing loopholes, this book aims to provide a thorough understanding of natural resource governance mechanisms in developing countries through theoretical as well as empirical analysis of three of the major concerns in the field of natural resources from an interdisciplinary perspective (Fig. 1.11). Overall, the scope of the book is broadly divided into two sections having the following functions. First, it attempts to add value in explaining the dynamics of natural resource governance in the context of a developing country as this area of knowledge is restricted by a lack of research and relevant literature. Second, it also attempts to fill up the knowledge gap existing in the field of natural resource governance and sustainability in general.

In doing so, it draws on variables such as institutions, power relations, political settlement and human sociality that are considered necessary to integrate into the framework of natural resource governance. Institutions are the humanly devised constraints that structure political, economic and social interaction (North, 1991), which are important to have sound economic fundamentals. If the institutions are found to be fractured in a country, there are problems in maintaining the pace of development. Therefore, institutional fragility is necessary to consider in the case of natural resource management, because excessive natural resource degradation is associated with institutional failures and market imperfections (López & Toman, 2006). As long as the institution remains ineffective, the causality of market failure eventually threatens the resources and questions the sustainability arguments. The effectiveness of institutions is, however, primarily based upon power dynamics and political settlement.

Political settlement is defined as the distribution of organisational power (Khan, 2010). Indeed, it can help understand why some institutions that worked well in some contexts appeared to achieve poorer results in others (Khan, 2017). Unequal sharing of resources also results from the imbalance of power relations among different countries. Accordingly, this book considers it essential to study the influence of power and political settlement on institutions and policies related to the governance of natural resources.

Another key aspect of the book is the complementary relationship between human beings and nature. Human beings play a crucial role in ensuring the sustainability of nature, and this relation remains at the core of sustainable governance of natural resources. The commodification of nature, however, results in the severance of this relation. The revitalisation of such mutual relations brings forward the idea of 'human sociality'. The book puts forth the idea that "human groups maintain a high level of

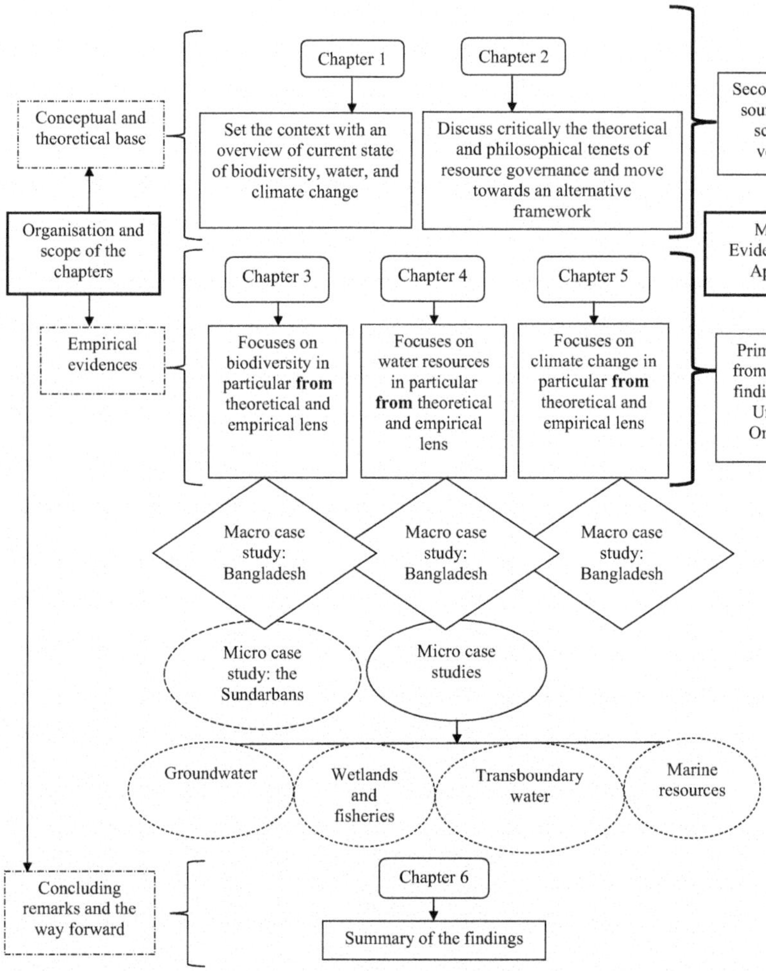

Fig. 1.11 Scope, approach and chapters of the book in summary. *Source* Prepared by the authors *Note* **a** Secondary and primary data have been used across different chapters. However, secondary data has been more used for the conceptual and theoretical part, and primary data has been more used in empirical evidence part as denoted by black-coloured bold right brace; **b** detail justification for the micro-case studies has been described in the relevant chapters

sociality despite a low level of relatedness among group members" (Gintis, 2000, p. 169). It signifies that pro-social behaviour exists in human beings called 'strong reciprocity', according to Gintis, which may partly explain sociality. To understand the current dilemma of 'sustainability', it is crucial to have a theoretical as well as a philosophical understanding on the relations between nature and human beings as argued in this book.

The recently developed conceptual framework by the Intergovernmental Science-Policy Platform on Biodiversity and Ecosystem Services (IPBES) has emphasised

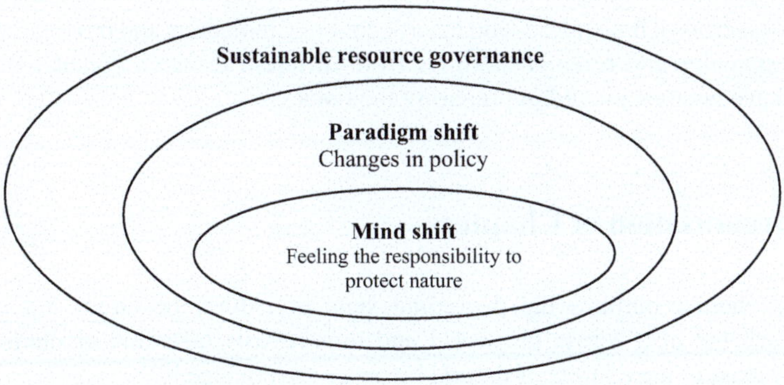

Fig. 1.12 Bringing 'mind shift' to ensure sustainable resource governance. *Source* Prepared by the authors. *Note* The structure of the diagram adapted from Silvius et al. (2012)

the connection between people and nature. It depicts the six primary socio-ecological constituents and the relationships among them, namely: (a) nature; (b) nature's benefit to people; (c) anthropogenic assets; (d) institutions and governance systems and other indirect drivers of change; (e) direct drivers of change and (f) good quality of life (Diaz et al., 2015). It should be noted that the IPBES framework is only relevant to the conservation and management of biological diversity. The book, however, argues that recognising and revitalising the connection are at the core of any natural resource-related problem from biodiversity conservation to climate change. It emphasises bringing out a 'mind shift' in the resource governance regime instead of only a 'paradigm shift', which is at the core of sustainable resource governance (Fig. 1.12). In this mind shift, the change in resource governance is no longer imposed; rather, the change will happen automatically based upon human sociality.

Finally, the book follows the Multiple Evidence Base (MEB) approach for collecting empirical data (Fig. 1.11). The MEB approach signifies that the studies are based on both primary and secondary data, bringing heterodox multi-disciplinary perspectives to a common platform. A significant amount of data (primary data) was gathered through participatory observations, questionnaire surveys, key person interviews, focus group discussions, Participatory Rural Appraisal (PRA) tools such as social mapping, impact assessment by the respondents, etc. under different research projects of the *Unnayan Onneshan* research institute—a progressive think tank in Bangladesh.

The institute has conducted several research on biodiversity, water and climate change over the years, and thus, the data reservoir of the institute is the key source of information for the book. In addition, the book also uses secondary data and statistics that have been collected from reputable national and international scholarly volumes, including journal articles, books and newspapers. Put differently, evidence collected from the field has been comprehensively rechecked and cross-examined with the available relevant literature from secondary sources. On the whole, empirical evidence in terms of both micro- and macroeconomic indicators and proxies, country-specific statistics and research findings from different research conducted by the authors have been employed in preparing the book.

1.7 Organisation of Chapters

The first chapter outlines the degrading state of natural resources not only in Bangladesh but also across the world and thus reveals facts and challenges that currently exist in the context of natural resource governance.

In response to the emerging facts and challenges, Chap. 2 of the book constructs a new framework by challenging the existing theoretical positions of natural resource governance. In this case, it critically scrutinises the mainstream theoretical positions and then uses the new framework to find solutions for a peaceful resolution of conflicts over natural resources and their sustainable, efficient and equitable governance.

The theoretical and empirical analysis of biodiversity resource degradation has been done in Chap. 3 through a case study of the Sundarbans, which is not only the largest mangrove forest in the world but also a World Heritage Site and a Ramsar site. The chapter first develops a conceptual framework to help identify the major drivers of biodiversity resource degradation of the Sundarbans. It also examines the alternative means and measures for the conservation, restoration and sustainable utilisation of those resources. Thereafter, the empirical evidence has been discussed under the conceptual framework to test the hypotheses by juxtaposing the mainstream thoughts and national-level data, collected from primary and secondary sources.

Chapter 4 tries to understand the complexities of water resource governance mechanisms in the context of Bangladesh as a macro-case study. As the water resources are diverse in nature, it is challenging to thoroughly examine the problems of governance of each resource. Simultaneously, it is difficult to generalise the reasons for governance failure by considering water as a single unit of analysis. Therefore, the chapter would take some specific micro-case studies such as groundwater, wetlands, marine resources, fisheries and transboundary water to examine the complexity of problems in each case. This chapter, however, also attempts to develop an alternative understanding based on the human–nature nexus in order to comprehend the sustainable governance of water and its resources.

Chapter 5 attempts to understand climate change from theoretical debates on the one hand and the implications of change on the ground, based on the case study of a developing country—such as Bangladesh—on the other. Moreover, it also develops an alternative understanding of the concept of 'sustainability'. In doing so, the chapter first offers a detailed discussion on climate change, first in the context of Bangladesh and then the world. The chapter, then, critically examines the existing theoretical underpinnings of climate change and sustainability and builds an alternative framework. The empirical sections put the whole sequence as stated in the propositions and concomitant hypotheses into practical terms, in order to test them in an empirical manner.

Finally, Chap. 6 wraps up the major findings and offers specific suggestions towards sustainable transformative pathways.

References

Adhikari, K., & Lawson, C. (Eds.). (2018). *Biodiversity, genetic resources and intellectual property: Developments in access and benefit sharing*. Routledge.

Agarwal, S., Ota, T., Ahmed, A. U., Smith, J., & van Aalst, M. (2003). *Development and climate change in Bangladesh: Focus on coastal flooding and the Sundarbans*. Organization for Economic Co-operation and Development (OECD).

Ahmed, Z. U., Begum, Z. N. T., Hassan, M. A., Khondker, M., Kabir, S. M. H., Ahmad, M., Ahmed, A. T. A., Rahman, A. K. A., & Haque, E. U. (Eds.). (2007). *Encyclopedia of flora and fauna of Bangladesh, volume 23: Freshwater fishes*. Asiatic Society of Bangladesh.

Ahmed, Z. U., Begum, Z. N. T., Hassan, M. A., Khondker, M., Kabir, S. M. H., Ahmad, M., Ahmed, A. T. A., Rahman, A. K. A., & Haque, E. U. (Eds.). (2008). *Encyclopedia of flora and fauna of Bangladesh, volume 26: Birds*. Asiatic Society of Bangladesh.

Ahmed, Z. U., Begum, Z. N. T., Hassan, M. A., Khondker, M., Kabir, S. M. H., Ahmad, M., Ahmed, A. T. A., Rahman, A. K. A., & Haque, E. U. (Eds.). (2009). *Encyclopedia of flora and fauna of Bangladesh, volume 25: Amphibians and reptiles and volume 27: Mammals*. Asiatic Society of Bangladesh.

Alam, H. (2018). Move to check groundwater depletion. *The Daily Star*. https://www.thedailystar.net/backpage/news/move-check-groundwater-depletion-1661962

Ara, H., Khan, B., & Uddin, S. N. (Eds.). (2013). *Red data book of vascular plants of Bangladesh*, (Vol. 2). Bangladesh National Herbarium.

Arrow, K. J., Cropper, M. L., Eads, G. C., Hahn, R. W., Lave, L. B., Noll, R. G., Portney, P. R., Russell, M., Schmalensee, R., Smith, V. K., & Stavins, R. N. (1996). Is there a role for benefit-cost analysis in environmental, health, and safety regulation? *Science, 272*(5259), 221–222.https://doi.org/10.1126/science.272.5259.221

Aziz, A., & Paul, A. R. (2015). Bangladesh Sundarbans: Present status of the environment and biota. *Diversity, 7*, 242–269. https://doi.org/10.3390/d7030242

Baranyai, G. (2020). Emerging challenges to transboundary water governance. In G. Baranyai (Ed.), *European water law and hydropolitics. Water governance—Concepts, methods and practice* (pp. 53–68). Springer. https://doi.org/10.1007/978-3-030-22541-4_6

Basak, J. K., Titumir, R. A. M., & Dey, N. C. (2013). Climate change in Bangladesh: A historical analysis of temperature and rainfall data. *Journal of Environment, 2*(2), 41–46.

Baten, M. A., & Kumar, U. (2010). *Responses to the changes in the Sundarbans*. Paper presented at the International Conference on Biological and Cultural Diversity, UNCBD, 3–5 May, Montreal, Canada.

Becker, M., Papa, F., Karpytchev, M., Delebecque, C., Krien, Y., Khan, J. U., Ballu, V., Durand, F., Le Cozannet, G., Islam, A. K. M. S., Calmant, S., & Shum, C. K. (2020). Water level changes, subsidence, and sea level rise in the Ganges–Brahmaputra–Meghna delta. *PNAS, 117*(4), 1867–1876. https://doi.org/10.1073/pnas.1912921117

Betzold, C., & Weiler, F. (2018). *Development aid and adaptation to climate change in developing countries*. Palgrave Macmillan.

Broad, R. (1995). The political economy of natural resources: Case studies of the Indonesian and Philippine forest. *The Journal of Developing Areas, 29*(3), 317–340.

Bulkeley, H., & Newell, P. (2010). *Governing climate change*. Routledge.

BUP-CEARS-CRU. (1994). *Bangladesh: Greenhouse effect and climate change*. Briefing Documents No. 1–7. Bangladesh Unnayan Parishad (BUP), Centre for Environmental and Resource

Studies (CEARS), University of Waikato, New Zealand and Climate Research Unit (CRU), University of East Anglia, Norwich, UK.
Byomkesh, T., Nakagoshi, N., & Shahedur, R. M. (2009). State and management of wetlands in Bangladesh. *Landscape and Ecological Engineering, 5*, 81–90. https://doi.org/10.1007/s11355-008-0052-5
Carrington, D. (2021). Children set for more climate disasters than their grandparents, research shows. *The Guardian*. https://www.theguardian.com/environment/2021/sep/27/children-set-for-more-climate-disasters-than-their-grandparents-research-shows
Ceballos, G., Ehrlich, P. R., & Raven, P. H. (2020). Vertebrates on the brink as indicators of biological annihilation and the sixth mass extinction. *PNAS, 117*(24), 13596–13602. https://doi.org/10.1073/pnas.1922686117
Chichilnisky, G., & Rezai, A. (2020). *Handbook on the economics of climate change*. Edward Elgar.
Choudhury, A. M., Haque, M. A., & Quadir, D. A. (1997). Consequences of global warming and sea level rise in Bangladesh. *Marine Geodesy, 20*(1), 13–31. https://doi.org/10.1080/01490419709388092
Clark, B., & York, R. (2012). Techno-fix: Ecological rifts and capital shifts. In A. Hornborg, B. Clark & K. Hermele (Eds.), *Ecology and power: Struggles over land and material resources in the past, present and future*. Routledge.
Crump, J. (Eds.). (2017). *Smoke on water—Countering global threats from peatland loss and degradation*. A UNEP Rapid Response Assessment. United Nations Environment Programme and GRID-Arendal.
Collier, P. (2010). The political economy of natural resources. *Social Research, 77*(4), 1105–1132.
Dahan, S., & Kashiwase, H. (2016). SDG 6 on water and sanitation is essential for sustainable development. *World Bank Blogs*. Retrieved June 03, 2021, from https://blogs.worldbank.org/opendata/sdg-6-water-and-sanitation-essential-sustainable-development
Davidson, N. C. (2014). How much wetland has the world lost? Long-term and recent trends in global wetland area. *Marine and Freshwater Research, 65*(10), 934–941. https://doi.org/10.1071/MF14173
Díaz, S., Demissew, S., Carabias, J., Joly, C., Lonsdale, M., et al. (2015). The IPBES conceptual framework—Connecting nature and people. *Current Opinion in Environmental Sustainability, 14*, 1–16.https://doi.org/10.1016/j.cosust.2014.11.002
DWASA. (2020). *Dhaka water supply and sewerage authority-annual report 2018–2019*. Dhaka Water Supply and Sewerage Authority.
Earle, A., Jägerskog, A., & Öjendal, J. (Eds.). (2010). *Transboundary water management: Principles and practice*. Routledge.
Eckstein, D., Künzel, V., Schäfer, L., & Winges, M. (2020). *Global climate risk index 2020*. Briefing Paper. Germanwatch.
Elder, M., & Olsen, S. H. (2019). The design of environmental priorities in the SDGs. *Global Policy, 10*(1), 70–82. https://doi.org/10.1111/1758-5899.12596
FAO. (2010). *Global forest resources assessment 2010—FAO Forestry Paper 163*. Food and Agriculture Organization of the United Nations.
FAO. (2015). *Global forest resources assessment 2015—Desk reference*. Food and Agriculture Organization of the United Nations.
FAO. (2020a). *Global forest resources assessment 2020—Key findings*. Food and Agriculture Organization of the United Nations.
FAO. (2020b). *The state of the world's forests: Forests, biodiversity and people*. Food and Agriculture Organization of the United Nations.
FAO. (2020c). *The state of world fisheries and aquaculture: Sustainability in action*. Food and Agriculture Organization of the United Nations.
Farolfi, S., Perret, S., & Hassan, S. (2013). *Water governance for sustainable development: Approaches and lessons from developing and transitional countries*. Routledge.
Filho, W. L. (2015). *Handbook of climate change adaptation*. Springer.

Freeman, A. M., III., Herriges, J. A., & Kling, C. L. (2014). *The measurement of environmental and resource values: Theory and methods*. RFF Press.

Gintis, H. (2000). Strong reciprocity and human sociality. *Journal of Theoretical Biology, 206*(2), 169–179. https://doi.org/10.1006/jtbi.2000.2111

Giri, C., Long, J., Abbas, S., Murali, R. M., Qamer, F. M., Pengra, B., & Thau, D. (2015). Distribution and dynamics of mangrove forests of South Asia. *Journal of Environmental Management, 148*, 101–111. https://doi.org/10.1016/j.jenvman.2014.01.020

Gleick, P. H. (Eds.). (2014). *The world's water-Volume 8: The* biennial *report on freshwater resources*. Pacific Institute for Studies in Development, Environment, and Security. Island Press.

Grigg, N. S. (2016). *Integrated water resource management: An interdisciplinary approach*. Palgrave Macmillan.

Gu, H., & Subramanian, S. M. (2012). *Socio-ecological production landscapes: Relevance to the green economy Agenda*. UNU-IAS Policy Report. The United Nations University Institute for the Advanced Study of Sustainability (UNU-IAS).

Hackett, S. C. (2006). *Environmental and natural resources economics: Theory, policy, and the sustainable society* (3rd ed.). M.E. Sharpe Inc.

Haque, M. I., & Basak, R. (2017). Land cover change detection using GIS and remote sensing techniques: A spatio-temporal study on Tanguar Haor, Sunamganj, Bangladesh. *The Egyptian Journal of Remote Sensing and Space Sciences, 20*(2), 251–263. https://doi.org/10.1016/j.ejrs.2016.12.003s

Harvey, F. (2020). More than 3 billion people affected by water shortages, data shows. *The Guardian*. https://www.theguardian.com/environment/2020/nov/26/more-than-3-billion-people-affected-by-water-shortages-data-shows

Hasan, A. B. M. S. U., & Rahman, M. Z. (2013). Change in temperature over Bangladesh associated with degrees of global warming. *Asian Journal of Applied Science and Engineering, 2*(2), 62–75.

Hermeling, S. (2010). *Global climate risk index 2010*. Briefing Paper. Germanwatch.

Helm, D., & Hepburn, C. (Eds.). (2009). *The economics and politics of climate Change*. Oxford University Press.

Huq, S., Chow, J., Fenton, A., Stott, C., Taub, J., & Wright, H. (Eds.). (2019). *Confronting climate change in Bangladesh: Policy strategies for adaptation and resilience*. Springer Nature.

Hutchings, J. A., Minto, C., Ricard, D., Baum, J. K., & Jensen, O. P. (2010). Trends in the abundance of marine fishes. *Canadian Journal of Fisheries and Aquatic Sciences, 67*(8), 1205–1210. https://doi.org/10.1139/F10-081

IBNET & FAO. (2018). *The international benchmarking network water supply and sanitation utilities database and FAO AQUASTATE database*. The International Benchmarking Network & Food and Agriculture Organization of the United Nations.

Ichikawa, K. (Ed.). (2012). *Socio-ecological production landscapes in Asia*. United Nations University Institute for the Advanced Study of Sustainability (UNU-IAS).

IPBES. (2015). *Wetland extent index*. Intergovernmental Platform on Biodiversity and Ecosystem Services. Retrieved April 10, 2020, from http://www.ipbes.net/sites/default/files/Metadata_UNEPWCMC_Wetland_Extent_Trend_Index.pdf

IPBES. (2019). Summary for policymakers of the global assessment report on biodiversity and ecosystem services. In S. Díaz et al. (Eds.), *Intergovernmental science-policy platform on biodiversity and ecosystem services*. IPBES secretariat.

IPCC. (1990). Climate change: The IPCC scientific assessment. In J. T. Houghton, G. J. Jenkins, & J. J. Ephraums (Eds.), *Contribution of working group I to the first assessment report of the intergovernmental panel on climate change*. Cambridge University Press.

IPCC. (1995). Climate change 1995: Impacts, adaptations and mitigation of climate change: Scientific-technical analyses. In R. T. Watson, M. C. Zinyowera, R. H. Moss (Eds.), *Contribution of working group II to the second assessment report of the intergovernmental panel on climate change*. Cambridge University Press.

IPCC. (2001). Climate change 2001: Impacts, adaptation, and vulnerability. In J. J. McCarthy, O. F. Canziani, N. A. Leary, D. J. Dokken, & K. S. White (Eds.), *Contribution of working group*

II to the third assessment report of the Intergovernmental Panel on Climate Change. Cambridge University Press.
IPCC. (2007). Climate change 2007: Synthesis report. In R. K. Pachauri, & A. Reisinger (Eds.), *Contribution of working groups I, II and III to the fourth assessment report of the intergovernmental panel on climate change.* IPCC.
IPCC. (2014). Climate change 2014: Synthesis report. In R. K. Pachauri & L. Meyer (Eds.), *Contribution of working groups I, II and III to the fifth assessment report of the intergovernmental panel on climate change.* IPCC.
IPCC. (2018). Summary for policymakers. In V. Masson-Delmotte, P. Zhai, H.-O. Pörtner, D. Roberts, J. Skea, P. R. Shukla, A. Pirani, W. Moufouma-Okia, C. Péan, R. Pidcock, S. Connors, J. B. R. Matthews, Y. Chen, X. Zhou, M. I. Gomis, E. Lonnoy, T. Maycock, M. Tignor & T. Waterfield (Eds.), *Global warming of 1.5°C. An IPCC Special Report on the impacts of global warming of 1.5°C above pre-industrial levels and related global greenhouse gas emission pathways, in the context of strengthening the global response to the threat of climate change, sustainable development, and efforts to eradicate poverty.* In Press.
IPCC. (2021). Summary for policymakers. In: Climate change 2021: The physical science basis. In V. Masson-Delmotte, P. Zhai, A. Pirani, S. L. Connors, C. Péan, S. Berger, N. Caud, Y. Chen, L. Goldfarb, M. I. Gomis, M. Huang, K. Leitzell, E. Lonnoy, J. B. R. Matthews, T. K. Maycock, T. Waterfield, O. Yelekçi, R. Yu & B. Zhou (Eds.), *Contribution of working group I to the sixth assessment report of the intergovernmental panel on climate change.* Cambridge University Press. In Press.
IPCC. (2022). Summary for policymakers. In: Climate change 2022: Impacts, adaptation and Vulnerability. In H.-O. Pörtner, D.C. Roberts, E.S. Poloczanska, K. Mintenbeck, M. Tignor, A. Alegría, M. Craig, S. Langsdorf, S. Löschke, V. Möller, A. Okem (Eds.), *Contribution of working group II to the sixth assessment report of the intergovernmental panel on climate change.* Cambridge University Press. https://doi.org/10.1017/9781009325844.001.
Irfanullah, H. M. (2011). Conserving threatened plants of Bangladesh: Miles to go before we start? *Bangladesh Journal of Plant Taxonomy, 18*(1), 81–91. https://doi.org/10.3329/bjpt.v18i1.7844
Islam, S. M. (2021). Groundwater depletion in Dhaka: Too little too late? *The Daily Star*. https://www.thedailystar.net/shout/news/groundwater-depletion-dhaka-too-little-too-late-2064933
Islam, S. N. (2010). Threatened wetlands and ecologically sensitive ecosystems management in Bangladesh. *Frontiers of Earth Science in China, 4*(4), 438–448. https://doi.org/10.1007/s11707-010-0127-0
Islam, M. R. (2016). *Bangladesh disaster-related statistics 2015: Climate change and natural disaster perspectives.* Planning Commission, Government of the People's Republic of Bangladesh.
Islam, S., & Smith, K. M. (2019). *Interdisciplinary collaboration for water diplomacy: A principled and pragmatic approach.* Routledge.
IUCN Bangladesh. (2000). *Red list of threatened animals of Bangladesh.* International Union for Conservation of Nature, Bangladesh Country Office.
IUCN Bangladesh. (2007). *Red list of Bangladesh, Volume 1: Summary.* International Union for Conservation of Nature, Bangladesh Country Office.
IUCN Bangladesh. (2015). *Red list of Bangladesh, Volume 1: Summary.* International Union for Conservation of Nature, Bangladesh Country Office.
IUCN. (2018). *The IUCN red list of threatened species, Version 2018-1.* International Union for Conservation of Nature.
IUCN. (2020). *The IUCN red list of threatened Species, Version 2020-1.* International Union for Conservation of Nature.
IUCN. (2021). *The IUCN red list of threatened species, Version 2021-1.* International Union for Conservation of Nature.
Karmakar, S., & Shrestha, M. L. (2000). *Recent climate change in Bangladesh.* SMRC Series No. 4. SAARC Meteorological Research Centre.

Khaki, M., Forootan, E., Kuhn, M., Awange, J., Papa, F., & Shum, C. K. (2018). A study of Bangladesh's sub-surface water storages using satellite products and data assimilation scheme. *Science of the Total Environment, 625*, 963–977. https://doi.org/10.1016/j.scitotenv.2017.12.289

Khan, M. A. R. (1982). *Wildlife of Bangladesh-A checklist*. University of Dhaka.

Khan, M. A. R. (2015). *Wildlife of Bangladesh: Checklist and guide*. Chayabithi.

Khan, M. H. (2010). *Political settlements and the governance of growth-enhancing institutions*. School of Oriental and African Studies, University of London.

Khan, M. H. (2017). Political settlements and the analysis of institutions. *African Affairs, 117*(469), 636–655. https://doi.org/10.1093/afraf/adx044

Khan, M. H. R., Rahman, A., Luo, C., Kumar, S., Islam, G. M. A., & Hossain, M. A. (2019). Detection of changes and trends in climatic variables in Bangladesh during 1988–2017. *Heliyon, 5*(3). https://doi.org/10.1016/j.heliyon.2019.e01268

Khan, M. M. H. (2008). *Protected areas of Bangladesh—A guide to wildlife*. Nishorgo Program. Bangladesh Forest Department, Government of the People's Republic of Bangladesh.

Khan, M. S., Rahman, M. M., & Ali, M. A. (Eds.). (2001). *Red data book of vascular plants of Bangladesh*. Bangladesh National Herbarium.

Kirton, J. J., von Moltke, K., & Le Prestre, P. G. (2002). *Governing global biodiversity: The evolution and implementation of the convention on biological diversity*. Routledge.

Khan, T. M. A., Quadir, D. A., Akter, F., & Sarker, M. A. (1999). Sea level changes and coastal erosion problems along Bangladesh coast. *Journal of Remote Sensing and Environment, 3*, 87–103.

Kula, E. (1994). *Economics of natural resources, the environment and policies* (2nd ed.). Chapman & Hall.

Laird, S. A. (2002). *Biodiversity and traditional knowledge: Equitable partnership in practice*. Earthscan Publications.

Laladhas, N. P., & Oommen, O.V. (2017). *Biodiversity for sustainable development*. Springer Nature.

López, R., & Toman, M. A. (Eds.). (2006). *Economic development and environmental sustainability: New policy options*. Oxford University Press.

Luypaert, T., Hagan, J. G., McCarthy, M. L. & Poti, M. (2020). State of marine biodiversity in the Anthropocene. In S. Jungblut, V. Liebich & M. Bode-Dalby (Eds.), *YOUMARES-9-The oceans: Our research, our future* (pp. 57–82). https://doi.org/10.1007/978-3-030-20389-4_4

Matin, M. A. (2017,). River pollution in Bangladesh: Unabated atrocity on environment. *The Daily Star*. http://www.thedailystar.net/environment-and-climate-action/unabated-atrocity-environment-1367095

McCully, P. (2007). *Silenced rivers: The ecology and politics of large dams*. Zed Book.

McNeely, J. A. (2021). Nature and COVID-19: The pandemic, the environment, and the way ahead. *Ambio, 50*, 767–781. https://doi.org/10.1007/s13280-020-01447-0

MoEF. (2005). *National adaptation programme of action (NAPA)*. Ministry of Environment and Forest, Government of the People's Republic of Bangladesh.

MoEF. (2012). *Rio + 20: National report on sustainable development*. Ministry of Environment and Forest, Government of the People's Republic of Bangladesh.

MoFA. (2014). *Press release: Press statement of the Hon'ble Foreign Minister on the verdict of the Arbitral Tribunal/PCA*. Ministry of Foreign Affairs, Government of the People's Republic of Bangladesh.

Mojid, M. A., Parvez, M. F., Mainuddin, M., & Hodgosn, G. (2019). Water table trend—A sustainability status of groundwater development in north-west Bangladesh. *Water, 11*(6), 1182. https://doi.org/10.3390/w11061182

Mondol, M. A. H., Mamun, A., Iqbal, M., & Jang, D. (2018). Precipitation concentration in Bangladesh over different temporal periods. *Advances in Meteorology, 28*.https://doi.org/10.1155/2018/1849050

Nature. (2020). Get the sustainable development goals back on track. *Nature: The International Journal of Science, 577*, 7–8. https://doi.org/10.1038/d41586-019-03907-4

Nichols, J. D., Koneff, M. D., Heglund, P. J., Knutson, M. G., Seamans, M. E., Lyons, J. E., Morton, J. M., Jones, M. T., Boomer, G. S., & Williams, B. K. (2011). Climate change, uncertainty, and natural resource management. *The Journal of Wildlife Management, 75*(1), 6–18. https://doi.org/10.1002/jwmg.33

North, D. C. (1991). Institutions. *The Journal of Economic Perspectives, 5*(1), 97–112. https://doi.org/10.1257/jep.5.1.97

O'Neill, D. W., Fanning, A. L., Lamb, W. F., & Steinberger, J. K. (2018). A good life for all within planetary boundaries. *Nature Sustainability, 1*, 88–95. https://doi.org/10.1038/s41893-018-0021-4

One Tree Planted. (2021). *The state of the world's forests in 2020.* Retrieved September 12, 2021, from https://onetreeplanted.org/blogs/stories/state-of-the-worlds-forests

Onneshan, U. (2020). *Biodiversity, climate change and traditional resource users in the Sundarbans: An exploration through public participation geographic information system (PPGIS).* Unnayan Onneshan.

Pearce, D. (Ed.). (1991). *Blueprint 2: Greening the world economy.* Earthscan.

Pearce, D., Markandya, A., & Barbier, E. (1989). *Blueprint for green economy.* Earthscan.

Ramsar Convention on Wetlands. (2018). Global wetland outlook: State of the world's wetlands and their services to people. Ramsar Convention Secretariat.

Sala, E. (2020). *These trees are crucial to our planet's survival—Why are we not protecting them?* Retrieved April 01, 2021 from https://www.weforum.org/agenda/2020/01/threat-to-biodiversity-risks-a-flood-of-economic-ruin-1e1591540c/

Sarker, S. K., Reeve, R., Thompson, J., Paul, N. K., & Matthiopoulos, J. (2016). Are we failing to protect threatened mangroves in the Sundarbans world heritage ecosystem? *Scientific Reports, 6*, 21234. https://doi.org/10.1038/srep21234

SCBD. (2022). Decision adopted by the Conference of the Parties to the Convention on Biological Diversity: Kunming-Montreal Global Biodiversity Framework. Secretariat of the Convention on Biological Diversity. https://www.cbd.int/doc/decisions/cop-15/cop-15-dec-04-en.pdf

Scharlemann, J. P. W., Brock, R. C., Balfour, N., Brown, C., Burgess, N. D., Guth, M. K., Ingram, D. J., Lane, R., Martin, J. G. C., Wicander, S., & Kapos, V. (2020). Towards understanding interactions between sustainable development goals: The role of environment–human linkages. *Sustainability Science, 15*, 1573–1584. https://doi.org/10.1007/s11625-020-00799-6

Shivakoti, G., Pradhan, U., & Helmi, H. (Eds.). (2016). *Redefining diversity and dynamics of natural resources management in Asia, Volume 1.* Elsevier.

Shopan, A. A., Islam, A. K. M. S., Dey, N. C., & Bala, S. K. (2013). *Estimation of the changes of wetlands in the northwest region of Bangladesh using Landsat images.* Paper for 4th International Conference on Water and Flood Management (ICWFM), 9–11 March, Dhaka, Bangladesh.

Silvius, A. J. G., Schipper, R., & Nedeski, S. (2012). *Sustainability in project management: Reality bites.* Proceedings of the 26th IPMA World Congress, Crete, 1053–1061.

Stavins, R. (1989). Harnessing market forces to protect the environment. *Environment: Science and Policy for Sustainable Development, 31*(1), 5–35. https://doi.org/10.1080/00139157.1989.9929926

Stavins, R. N. (2011). The problem of the commons: Still unsettled after 100 years. *The American Economic Review, 101*(1), 81–108. https://doi.org/10.1257/aer.101.1.81

ten Kate, K., & Laird, S. A. (2000). *The commercial use of biodiversity: Access to genetic resources and benefit-sharing.* Earthscan Publications.

Tietenberg, T. H. (1990). Using economic incentives to maintain our environment. *Challenge, 33*(2), 42–46. https://doi.org/10.1080/05775132.1990.11471412

UN Environment/UN Water. (2018). *Progress on water-related ecosystems—Piloting the monitoring methodology and initial findings for SDG indicator 6.6.1.* UN Environment.

UNEP. (2020). Emission gap report 2020. United Nations Environment Programme.

UNESCO. (2017). Facts and figures on marine biodiversity. United Nations Educational, Scientific and Cultural Organization. Retrieved November 22, 2020, from http://www.unesco.org/new/en/natural-sciences/ioc-oceans/focus-areas/rio-20-ocean/blueprint-for-the-future-we-want/marine-biodiversity/facts-and-figures-on-marine-biodiversity/

UNESCO/UN-Water. (2018). *The United Nations world water development report 2018: Nature-based solutions for water*. United Nations' World Water Assessment Programme (WWAP). United Nations Educational, Scientific and Cultural Organization.

UNESCO/UN-Water. (2019). *The United Nations world water development report 2019: Leaving no one behind*. United Nations' World Water Assessment Programme (WWAP). United Nations Educational, Scientific and Cultural Organization.

UNESCO/UN-Water. (2020). *The United Nations world water development report 2020: Water and climate change*. United Nations' World Water Assessment Programme (WWAP). United Nations Educational, Scientific and Cultural Organization.

United Nations. (2022). The water convention: 30 years of impact and achievements on the ground. Water Convention Secretariat, United Nations.

UN Water. (2014). *Transboundary waters: The factsheet*. Retrieved April 08, 2017, from http://www.unwater.org/statistics/thematic-factsheets/en/

Water Aid. (2016). *Water: At what cost? The state of the world's water*. Water Aid.

Water Aid. (2019). *Beneath the surface: The state of world's water*. Water Aid.

Water Aid. (2020). *On the frontline: The state of world's water*. Water Aid.

WHO & UNICEF. (2019). *Joint monitoring programme: Progress on household drinking water, sanitation and hygiene, 2000–2017*. World Health Organization & United Nations Children's Fund.

WHO & UNICEF. (2021). *Joint Monitoring Programme (JMP) for water supply and sanitation*. World Health Organization & United Nations Children's Fund.

World Vision. (2020). 10 worst countries for access to clean water. Retrieved August 10, 2021, from https://www.worldvision.org/clean-water-news-stories/10-worst-countries-access-clean-water

WWF. (2014). *Living planet report 2014: Species and spaces, people and places*. World Wide Fund for Nature.

WWF. (2015). *Living blue planet report 2015: Species, habitats, and human well-being*. World Wide Fund for Nature.

WWF (2016). *Living planet report 2016: Risk and resilience in a new era*. World Wide Fund for Nature.

WWF. (2017). *Threats to ocean and coasts*. Retrieved April 22, 2017, from http://wwf.panda.org/about_our_earth/blue_planet/problems/

WWF. (2018). *Living planet report 2018: Aiming higher*. World Wide Fund for Nature.

WWF. (2020). Living planet report 2020: Bending the curve of biodiversity loss. In R. E. A. Almond, M. Grooten & T. Petersen (Eds.), World Wide Fund for Nature.

World Bank. (2019). *Informal water markets in an urbanising world: Some unanswered questions*. Report No: AUS0000606. The World Bank.

Zhang, B. (2013). Market-based solutions: An appropriate approach to resolve environmental problems. *Chinese Journal of Population Resources and Environment, 11*(1), 87–91. https://doi.org/10.1080/10042857.2013.777526

Chapter 2
Managing Natural Resources Sustainably: Market and Non-market Approaches

2.1 Introduction

This chapter aims at constructing a conceptual framework to find the causality of natural resource degradation and suggesting an alternative understanding of sustainable resource governance based primarily upon the human–nature relationship. The discussion, in this regard, critically reviews two major theoretical approaches—(a) the market-centric approach and (b) the political economy approach. The chapter, at its core, interrogates the concepts of commodification, valuation and markets in addressing the issues of natural resource governance. It also argues that political economy factors and the complementary relations between human beings and nature need to be incorporated into the framework of resources governance to define the sustainable usage, management and distribution of resources, particularly in the context of a transitional economy.

Means and measures for natural resource governance are mostly drawn from the market-centric approach as a part of the neoliberal intellectual project (Gunningham & Holley, 2010; Titumir et al., 2020). Under this mainstream approach, 'neo-classical economics' defines natural resource degradation in terms of market failure by arguing that the market cannot always warrant an efficient allocation of natural resources. Neo-classical economics especially suggests two types of policy instruments (a) altering the prices of existing market-oriented functions by taxing environmental damage (e.g., pollution) or by subsidising environmental improvement and (b) creating markets for environmental goods and services. The argument is that the market has the power to allocate environmental resources efficiently and in a socially optimal manner (Beder, 2011; Zhang, 2013). However, instituting corrective measures such as taxes and subsidies necessitates the intervention of a regulatory agency. Therefore, this policy option can be criticised because it advocates for market mechanisms on the one hand while proposing certain remedies through government interventions on the other. Most importantly, it considers the value of nature (and its services) in monetary terms, which can yield a flow of services to human beings

R. A. M. Titumir et al., *Natural Resource Degradation and Human-Nature Wellbeing*,
https://doi.org/10.1007/978-981-19-8661-1_2

(Freeman III et al., 2014), and thus, breaks down natural resources into commodities and tends to disaggregate nature into discrete units (Sullivan, 2013). This way of valuing nature largely ignores its 'intrinsic value' (that nature is in itself morally considerable), and as a result, the resources are considered as extractable for the benefits of human beings as much as possible which leads to their destruction (Titumir et al., 2019). Therefore, the branch fails to offer any constructive solution to the problem, rather aggravates it.

Within the framework of institutional economics—which modifies market centrism with some new explanations—the overuse of natural resources is seen as the consequence of the lack of appropriate and stable institutional arrangements. The proposed solution, for example, is to introduce clearly defined property rights over the resources, arguing that under stable property rights "compensations change hands according to which party holds the natural resource, ensuring an efficient degree of economic activity" (Titumir et al., 2019, p. 101). This school of thought thus emphasises primarily formal institutional arrangement while focusing less on informal institutions and largely ignoring political-economy factors. However, informal institutions such as social norms, values and customs are also crucial for resource conservation and management, whereas political-economic aspects are required to determine the causes of resource degradation. Overall, the market-centric approach views the natural resources degradation problem through a narrow lens and contends that the problem can be resolved through technological development. The attempt to fix the problem only with technological innovation is faulty because it disregards the ecological contradictions associated with the technological fixes (Clark & York, 2012; Hornborg, 2010; Li, 2008; York & Clark, 2010). Accordingly, it omits the issues of equitable distribution, intergenerational effects and the sustainability of resources (Titumir et al., 2019). Moreover, such approach neglects humans and their behaviour as a critical certainty for sustainable governance of natural resources (Fulton et al., 2011).

On the contrary, the political economy approach, which is adopted here, can help in scrutinising the problems of the resource governance regime by incorporating some major factors that can explain how the natural resources are being accumulated for personal gains under the market economy. The book specifically takes the Marxian political economy, political ecology and ecological economics into consideration to construct a blended understanding of the political economy approach. The analysis contends that it is necessary to shed light on the political elements such as power structures or relations in the resource governance regime, as the unequal power distribution leads to overextraction and concomitantly the degradation of natural resources (Titumir et al., 2019, 2020). In particular, the political-economic factors are important to consider while examining the governance mechanisms in a transitional economy. The approach, however, ignores incorporating the human–nature relations aspect into its framework. Recent conservation frames such as 'people and nature', in which humans and ecosystems are not perceived as separate elements but as integrated socio-ecological systems, recognise the co-existence, but metrics and management models under the frame are still in the early stages of development (Garcia-Llorente et al., 2018). Indeed, integrating human behaviour into the model of natural resources governance is still a major challenge (Fulton et al., 2011; Jansen & Jager, 2000; Milner-Gulland, 2012; Schlüter et al., 2012).

The built framework of this chapter has regarded the interrelation as a key component and emphasised the 'human sociality' perspective concerning the conservation and sustainable management of resources. The concept of human sociality entails that there is a long-standing embeddedness of human beings into nature, which has created reciprocal relations between the two entities. This unexplored or unrecognised relation in the resources governance framework necessities to be revitalised. The overall argument is that to understand the current dilemma of 'sustainability' in resource governance, it is crucial to establish a theoretical and philosophical understanding of the relations between nature and human beings.

To sum up, the new framework underscores three significant departures. First, the framework challenges the market-based approach in valuing natural resources. Second, it brings the institutional vulnerability aspect and power-sharing arrangement into the analysis of resource governance. Third, it formulates a new understanding of sustainability by incorporating the idea of human sociality regarding the governance of natural resources.

2.2 Market-Centric Approach: Neo-classical Economics and New Institutional Economics

An essential contribution of economics is the development of market-based tools to tackle problems relevant to natural resources and the environment (Arrow et al., 1996; Freeman III et al., 2014; Stavins, 1989, 2003, 2011). Adam Smith—the progenitor of classical economic theory—described how in a free market condition where competition prevails among the buyers and sellers, the 'invisible hand' can bring the most efficient outcomes. According to Smith (1776) every individual:

> …neither intends to promote the public interest, nor knows how much he is promoting it. …he intends only his own gain, and he is in this, as in many other cases, led by an invisible hand to promote an end which was no part of his intention. Nor is it always the worse for the society that it was no part of it. By pursuing his own interest he frequently promotes that of society more effectually than when he really intends to promote it (Book IV, Chapter II, pp. 593-594).

In later periods, neo-classical economics initiated the model of perfect competition. Austrian economics and new institutional economics also emerged as refinements of the market-centric approach, and at present, the distillation of which is known as 'neoliberalism'. The fundamental essence of market centrism remains, however, unchanged.

The model of perfect competition implies that market exchange based on price is the best mechanism to ensure the most efficient outcome at the 'Pareto optimal' position. 'Market failure'—the deviation from the 'Pareto optimal' position—occurs when existing markets are unable to allocate resources optimally because their full costs or benefits are not reflected in pricing. The outcome is suboptimal to society's well-being. Environmental economics—as a distinct branch concerned with nature and environment—applies these neo-classical principles to the study of how natural resources and the environment are managed (Field, 1997; Pearce, 1991, 2002; Pearce et al., 1989; Sagoff, 2012; Stavins, 2003; Stavins & Whitehead, 1992; Tietenberg, 1990).

Externality and public goods characteristics are two of the market failure cases, which are relevant to natural resources. Valuation of nature based on price mechanism (Adler & Posner, 2006; Arrow et al., 1996; Daily et al., 2000; Stavins, 2011) is also a key attribute of this school of thought. The focus of neo-classical economics applied to natural resources is, therefore, to look at issues of (a) externality, (b) public goods characteristics and (c) valuation of nature.

Externalities arise in the market for natural resources because of the disparity between the private and social costs and benefits of the use and conservation of natural resources (Dixon & Sherman, 1990). Pigou ([1920] 2013) proposed, in this regard, imposing a tax on negative externality generators while subsidising positive externality generators to maximise social welfare. Pigou's solution to the problem of resource depletion is thus essentially authoritarian (Kula, 1994).

Public goods are the type of goods the consumption of which is non-excludable and non-rival. Such goods are difficult to allocate efficiently based on the model of perfect competition. Though natural resources are usually considered public goods, it is sometimes difficult to identify them as pure public goods. The diverse classification of such goods, therefore, emerged as an issue of concern in the case of natural resource governance (see discussion on 'property rights' in later part for details).

The failure of the market to allocate natural resources efficiently due to the externality and public good characteristics problem creates the necessity of economic measures of values to guide policymaking (Freeman III et al., 2014). A significant concern, thus, arises in determining how nature or the environment should be valued. 'Value' is a word with all the complexity of life itself (Foster, [1997] 2003). The classical theory of value argued that it creates wealth. Wealth was initially associated with what John Locke referred to as 'intrinsic value' in classical political economy, and what later political economists were to call 'use-value' (Brown, 1984). Material use-values have always existed and are the foundation of human existence, but commodities produced for market sale embodied something else that has 'exchange value' (Foster & Clark, 2009). The production that needs to be exchanged exhibits the divergence between the use-value and the exchange value. Use-value only can be translated into the price if it can be expressed into the exchange value.

Neo-classical or environmental economics focuses particularly on the exchange value and puts a price tag on environmental goods by arguing that nature's value can be measured in monetary terms. Simply, the environment has value to the extent that it is useful to human beings. It also devises some valuation methods (e.g., contingent valuation, choice experiment, cost–benefit analysis, hedonistic pricing, revealed preference) to construct the hypothetical demand curves for environmental goods and services (Kula, 1994; Tietenberg & Lewis, 2012). The argument is that markets can better reflect environmental impacts (externalities) and resource scarcity in prices, allowing producers and consumers to respond appropriately (Rademaekers et al., 2011) and ensure the most efficient allocation of those resources. Overall, environmental economics treats natural resources as commodities that can be bought, sold, traded, saved and invested like any other commodity (Nadeau & Cleveland, 2007). Such valuation also considers natural capital as substitutable with produced capital (e.g., factories, roads) and human capital (e.g., knowledge, productive skills) (Beder, 2011; Solow, 1974). Accordingly, it argues that technological advancement can substitute for the loss caused by the degradation of natural resources. This proposition relies on the heart of the capitalist ideology, which implies that technological innovations can address each problem related to nature, allowing society to overcome natural limits and socio-ecological challenges without changing the economic system (Clark & York, 2012; York & Clark, 2010).

Overall, under this orthodox paradigm, nature's degradation is viewed as a result of 'market failure' to properly value the resources (Cutting & Cahoon, 2005; Roma, 2006). The suggestion is, accordingly, to create market signals that would charge appropriately for the use of the resources (Holley et al., 2012). The understanding is that placing an economic value on nature would encourage a more ecologically sensitive and sustainable model of development (Bresnihan, 2017). Thus, nature beyond human beings is understood and enacted as calculable, exchangeable, substitutable and commensurable (Sullivan, 2016).

However, valuation through 'pricing' might not capture the 'intrinsic value' of nature under the economic construct. Something has intrinsic value "if it is valuable in and for itself—if its value is not derived from its utility, but is independent of any use or function it may have in relation to something or someone else. …an intrinsically valuable entity is said to be an 'end-in-itself,' not just a 'means' to another's ends" (Callicott, 1989, p. 131). More concretely, the numerous services that nature provides (Table 2.1) are intangible, and therefore, it is not easy to assess the value in monetary terms only. The instrumental valuation of nature, in fact, excluded aesthetic, spiritual, political and ethical dimensions (Beder, 2011).

Second, the valuation methods also are faulty, as they cannot reflect the value fully in market prices (Cummings et al., 1986; Gowdy, 1997; Hanley & Spash, 1993; Mitchell & Carson, 1989). Third, the technological optimism about solving nature's problems is potentially precarious, given that if a problem of natural resource degradation is assumed resolvable through technology, it is also recognised that it is unnecessary to take action to preserve natural resources and change the political-economic conditions that have created these problems (Clark & York, 2012). It is unlikely that technology alone can remedy the problems, and sometimes, it makes matters worse. Technologies designed to solve a particular problem often generate even more severe problem of another kind (Hornborg, 2010). Fourth, it dismisses the idea that aggregating costs and benefits cloud distributional and equity issues of who gets the benefits and who suffers the losses (Beder, 2011). Finally, the model of perfect competition is itself erroneous, which does not hold in the real world (Bowels & Gintis, 1990, 1993; Gaffard, 2008; Hayek, 1948; Kirzner, 1997; Mises, 1949; Stamate & Musetescu, 2013). Scrutinising the natural resources governance mechanism based on such a model is, therefore, a flawed attempt.

Table 2.1 Five types of service flow to the economy by natural resource system

Five types of service flow	Description
Material inputs	Fossil fuels, wood products, minerals, water, fish
Life support services	Breathable atmosphere, clean water, liveable climatic regime by recycling of nutrients, decomposition of organic materials, generation and renewal of soil fertility, pollination of crops and natural vegetation and biological control of agricultural and other pests
Amenity services	Opportunities for recreation, wildlife observation, pleasure of scenic beauty
Waste receptor service	Disperses, transforms and stores the residuals generated as by-products of economic activity
Repository of genetic information	Helps to determine the stability and resilience of the system in the face of anthropogenic and other shocks

Source Prepared by the authors based on Freeman III et al. (2014)

New institutional economics (NIE) has emerged by challenging the assumption of zero transaction cost of the model of neo-classical economics. It contends that market operations involve transaction costs, which institutions can reduce to optimally allocate resources among self-interested individuals. Institutions create incentives by which a greater level of cooperation can be achieved by curbing uncertainty in human exchanges. The NIE is highly relevant to natural resource governance as it explicitly discusses institutional arrangements for managing natural resources. It attempts to capture the intrinsic interrelationship between nature and institutional arrangement to reveal how people organise themselves to utilise resources when sustaining their way of life (Vatn, 2005). In this context, institutional arrangements specify who control or have access to resources, what can be harvested from the resources, who participate in decision-making, how conflicts are resolved and how the resources are managed (Dietz et al., 2002). However, the concerns of NIE regarding natural resource governance primarily revolve around the concepts of formal institutions, property rights and transaction costs.

Formal institutions are the formal rules such as constitutions, laws and property rights that are explicit and generally enforceable through bureaucratic organisational structures (North, 1991, 2008; Po et al., 2019). The NIE has extensively discussed the issue of property rights, but it has a historical legacy. John Locke first appeared with his book *Two Treatises of Government* where the supporter of absolute monarchy came up with the contention of private property. Locke's view was that as people had a social contract, they could enjoy security, while the government would ensure the protection of property. Therefore, "government has no other end but the preservation of property" (Locke, [1689] 2003, p. 329). Following Locke's natural right to property, Adam Smith later accorded in his language what he called 'natural liberty' that pervades the whole political economy of *The Wealth of Nations* (Smith, 1776). Smith's connotation of private property is a sovereign protected right enjoyed by individuals in society. It is an incentive for capital accumulation, which encourages

Table 2.2 Typology of goods

		Rival in consumption	
		Yes	No
Excludable from consumption	Yes	**Private goods** • Ice-cream • Clothing	**Club goods** • Fire service • Cable TV
	No	**Common goods** • Fish in the ocean • Timber in forest	**Public goods** • Air • National defence

Source Mankiw (2007)

the further division of labour and the growth of the economy. The property, however, has the heterogeneity attribute, as has been found later. As such, addressing the divergence in terms of excludability (the property of a good whereby a person can be prevented from using it) and rivalry (the property of a good whereby one person's use diminishes another's use) (Mankiw, 2007), the properties or goods are classified into four types (Table 2.2).

Among the four types, common goods and public goods are mostly relevant to natural resources. As defining rights in the consumption of these two types of goods is costly due to non-exclusionary attribute, it often leads to overconsumption of resources.

Based on the attributes of goods or property, Alchian, as one of the founders of NIE, first addressed the issue of property rights by deriving the divergence between private ownership and public ownership (Alchian, 1965). The most recognised interpretation of property rights, however, is that they are a part of broader institutional framework that governs human actions in resource use and that they gradually change over time (Mahoney, 2004; North, 1990). In other words, a property rights regime is a set of entitlements that define the owner's rights and duties in the use of resources as well as the rules that govern how those rights and responsibilities are exercised (Bromley, 1991a). It is not a mere concept of 'ownership' of resources, should not be conflated with possession and is granted by legitimate legal authority (Hodgson, 2014; Meinzen-Dick et al., 2001). It is a formal institutional arrangement that guides the agents to behave appropriately in the economy by claiming their own rights and performing responsibilities (Table 2.3).

It can be noted, however, that in the case of natural resources, the term 'open access property' or 'common property' is more pronounced instead of public property. Moreover, there is a long-standing tradition of failing to recognise the critical distinction between the common property (*res communes*) and open access property (*res nullies*) (Bromley, 1991b) (see Ostrom, 2000a; Ostrom & Hess, 2010; Remer, 2012; Stavins, 2011 for details). The common property regime is an institutional arrangement under which resources are allocated to a group of interdependent users (Baland & Platteau, 1996; Bromley, 1991b; Feeny et al., 1998). On the contrary, open access resources are not owned by anyone because they are in a state of non-property, where no one owns or regulates the resources (Burke, 2001). Therefore,

Table 2.3 Property rights regimes and associated rights and duties

Regime type	Owner	Owner rights	Owner duties
Private property	Individual	Socially acceptable uses and control of access	Avoidance of socially unacceptable uses
Common property	Collective	Exclusion of non-owners	Maintenance constrain rates of use
State (public) property	Citizen	Determine rules	Maintain social objectives
Open access property	None	Capture	None

Source Adapted from Hanna et al. (1996)

neither the exclusion is feasible nor the consumption is rival, and the outcome is *Pareto inefficient* in this case too (Baland & Platteau, 1996; Libecap, 1994). There is thus a clear divergence in terms of articulating the property rights and the level of extraction from resources between these two property regimes. Nevertheless, the nature of property rights over the natural resources is usually regarded as a common property regime.

Importantly, the theory of the 'tragedy of the commons' emerged regarding the debate over the degradation of common resources (Hardin, 1968). Hardin argued that people would rationally do the tragedy of the commons because every individual would want to maximise his or her own gain from the common property. The policy prescription for tackling such exploitation (tragedy) is to place the resources under government control or private authority to internalise the externality. The theory has been later criticised from several grounds (e.g., Angus, 2008; Baland & Platteau, 1996; Dietz et al., 2002; Feeny et al., 1996; Ostrom, 1990 [2015], 1998; Walker et al., 2000). Critics from the NIE School argued that the model is applicable only to open access resources (Vink, 1986). The argument goes that if a defined set of resource users exists despite the commons dilemma, then the real issue is a lack of or a fragile resource management regime. Alternatively, the tragedy occurs due to institutional failure rather than inherent flaws in common property rights regimes. Evidence also shows that though many countries have adopted government control policy, the outcome has proved to be disastrous instead of being effective (Grafton, 2000). The privatisation policy is also not practical, particularly in terms of natural resource management (Baland & Platteau, 1996; Dietz et al., 2002; Ostrom, 2003; Tietenberg, 2002). The problem is more severe in those contexts where the transaction costs and information asymmetry are higher.

Transaction costs would seem to be partially responsible for the market failure pointed out by Pigou ([1920] 2013) and so his assertion goes for taxation for the negative externalities. Later, Ronald Coase, on the contrary, emphasised setting up a price mechanism among users of common resources by assuming zero transaction costs by considering the externality issue as a complex one (Coase, 1937, 1960). He argued that within the framework of the market mechanism, the *Pareto efficiency* could be achieved if the property rights of the individuals among parties can be defined, where the parties are small in number and there is zero transaction cost. Alternatively, clearly defined property rights would address environmental problems by internalising externalities and relying on private owners' incentives to conserve resources for the future (Bauler, 2012). Thus, Coase emphasised the privatisation of common resources, which was also advocated by Demsetz, who stated that private ownership could only internalise the external cost associated with the open access property (Demsetz, 1967).

Transaction costs are, however, inevitable in the real-world exchange process. It implies the total costs of making a transaction (Williamson, 1979, 1986) which can be broadly divided into three categories (a) search and information costs, (b) bargaining and decision costs and (c) policing and enforcement costs (Dahlman, 1979). Thus, each type of transaction produces the coordination cost of monitoring, controlling and managing (Young, 2013).

A complex and insightful understanding of transaction costs can be drawn by focusing on the power relationship that remains in the transaction process. In the process of transaction, "price and other terms of exchange often include a payment in excess of at least one agent's next best alternatives" (Bowels & Gintis, 1993, p. 87). The excess payments ascend from enforcement rent. Enforcement rent arises because the market is in a 'contested nature of exchange' not in 'voluntary exchange'. There remains an unequal share of information, which causes the principal-agent problem to emerge. In the same purpose of exchange between two individuals, the agent has relatively more information than the principal does. The agent, therefore, has the power to exert control over the principal.

The implications of transaction costs are essential in the case of natural resources as the failure of contracts due to the presence of such costs implies that the property rights regime is altered and market failure would lead to rent-seeking and monopoly control over natural resources. In developing countries, institutional vulnerability is the consequence of toddling contract enforcement. The jurisdiction process of these countries extends the transaction costs and leads to a failure of the contract. Moreover, the transaction cost is particularly related to the issue of unequal exchange in the realm of natural resource governance. Above all, what is important is that appropriate institutional arrangement is the key to reducing such costs (North, 1991); therefore, institutions matter the most in managing resources.

In the most recent context, institutional economics further developed property rights regimes. It postulates that clearly defined property rights among the community, which owns the resources, can manage the common resources in the best way through collective action (Ostrom, 1990 [2015], 2009, 2000a; Ostrom et al., 2002). The bundle of rights and rights holders can be categorised based on different property rights arrangements (Fig. 2.1). The notion is that if the rights are well defined, the community would preserve the property eventually.

Ostrom's arguments, however, have some critical junctions. She formulated her collective action theory based on some case studies of certain groups. Those groups were small, stable and relatively homogenous, and therefore, the transaction costs were considerably low. However, there can be many common resources where the groups can be heterogeneous, large and unstable. The attributes of the groups can

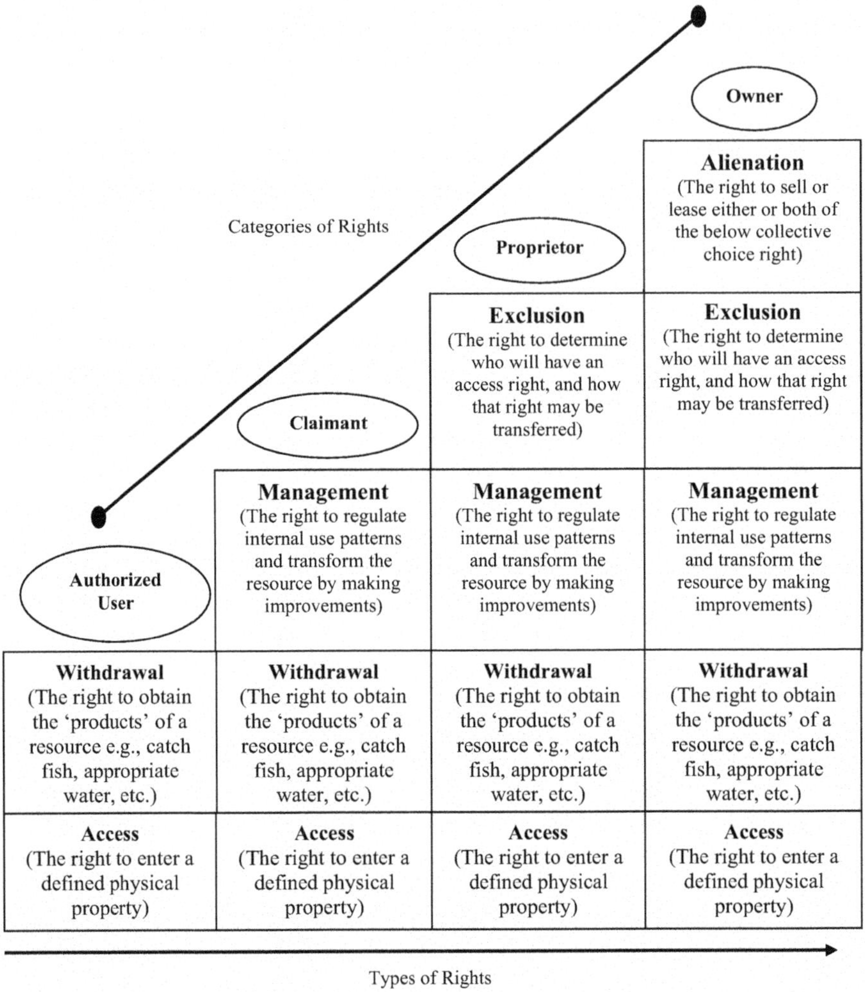

Fig. 2.1 Categories and types of rights. *Source* Prepared by the authors based on Schlager and Ostrom (1992)

make cooperation difficult (Adhikari, 2002). Ostrom also admitted that research on common resources is in its early stages as she focused on limited physical resources such as fisheries and water (Morrison, 2019). Another recent development is the polycentric governance for commons, which is regarded as effective for mitigating institutional failure and resource losses (Carlisle & Gruby, 2017). Nevertheless, the transaction costs of coordination in this governance system can be quite high, and holding decision-makers accountable for their actions can be difficult (Lieberman, 2011; Wyborn, 2015). Moreover, issues such as inequality and equitable distribution remain inconclusive.

Informal institutions, apart from formal institutions, typically comprise habitual, verbalised or customary rules and codes of conduct, which are often implicit and function as de facto rulemaking and enforcement bodies (North, 1991; Po et al., 2019; Rahman et al., 2017). Informal institutions thus evolve internally from society and act in the interest of the community (Yami et al., 2009). In particular, 'social capital' and 'collective action' are the two concepts relevant to informal institutions that the NIE mainly focuses on governing commons. Social capital comprises of the social resources (e.g., networks, membership of groups, the relationship of trust and reciprocity) on which people often draw in pursuit of livelihood (Coleman, 1988; Putnam, 1993; Rakodi, 2002). The idea of social capital can be related to Ostrom's idea of collective action (Ostrom, 2000b, 2009) as such:

> As social capital lowers the costs of working together, it facilitates cooperation. People have the confidence to invest in collective activities, knowing that others will also do so. They are also less likely to engage in unfettered private actions with negative outcomes, such as resource degradation (Pretty & Smith, 2004, p. 633).

The institutional arrangement can also be determined as communal ownership or state ownership or co-management based on the strength of social capital of a particular community (Adhikari, 2001).

The NIE, however, places more of an emphasis on formal institutions to determine the sustainable natural resource governance mechanism, and there is debate surrounding the role of informal institutions in this regard. Some studies argue that informal institutions serve as mechanisms to achieve a sustainable outcome in common resource governance by regulating access to and control over resources, sharing benefits equally and mobilising social capital (Adhikari, 2010; Pretty & Smith, 2004; Tefera et al., 2005; Watson, 2001).

On the contrary, others argue that informal institutions are not always effective to ensure sustainable resource governance (Banana et al., 2007; Campbell et al., 2001). Moreover, there is concern regarding the co-existence of formal and informal institutions. Arguments prevail that if the multi-level interactions between these two types of institutions—existing in the natural resource governance regime (Cash et al., 2006; Poteete, 2012)—are absent or distorted, then disparity, incoherence and even conflict can arise, potentially posing a threat to sustainable natural resource governance (Acheson, 2006; Rahman et al., 2012; Rastogi et al., 2014). Importantly, it is often difficult to discern the role of informal institutions in governing natural resources in comparison with formal institutions (Greif & Kingston, 2011). For instance, there is a relative scarcity of knowledge and policy research on how informal institutions (e.g., local or traditional knowledge) can influence formal natural resource governance systems when formal institutions seek to collaborate with communities to govern natural resources (Po et al., 2019). Furthermore, institutional economics theorises social norms and values in monetary terms. Such entitlement largely ignores the human sociality aspect. The relations between nature and human beings should be conceptualised using a reciprocal framework rather than a monetary framework. Cooperation among resource stakeholders must be maintained in accordance with their values and their relationship with that resource.

Overall, the NIE has brought about some crucial issues in the context of natural resource governance. There is an agreement that 'institutions matter' for sustainable resource management, but the institutions that would do the best job are not identified (Acheson, 2006). In addition, it predominantly emphasises formal institutions such as stable property rights. It argues that some formal rules and obligations can turn the non-cooperative game into a cooperative game and improve natural resource governance performance. This again diverts the management system towards a bureaucratic procedure where collective ownership can only be successful under some rules imposed by an external authority.

On the contrary, NIE considers the informal institutions with less importance and in monetary terms. Moreover, it has not considered the political transaction costs that can also enhance institutional change. Determination of property rights is not sufficient to understand the dynamics of resource usage; instead, it needs to be analysed in the context of a particular form of the power structure and political domain. The powerful groups can arrange the institutional structure to be advantageous for them to make a profit. It is particularly inherent to the relationship of economic agents in the case of developing countries. Overall, environmental actions depend not only on understanding conservation ecology but also on social, economic and cultural processes that influence such behaviour (Scarlett et al., 2013). The main loophole of the NIE, however, is that it considers human beings as 'rational agents' as its unit of analysis.

The alternative resources governance framework developed in this chapter articulates the unit of analysis of human beings as 'reciprocal social beings' (Gintis, 2000) and argues that the social norms and values (Bicchieri, 2010; Binmore, 2010) can affect nature's conservation strategies by shaping the behaviour of human beings.

2.3 Political Economy Approach: Power, Political Settlement and Distribution

Political economy deals with how political decisions, institutional structures and forms of governance influence economic decisions taken by government and citizens (Adam & Dercon, 2009). The approach has a profound significance to explore the accumulation, price mechanism and distribution processes in the realm of natural resources governance. In particular, the political economy of natural resources is concerned with the relationship between nature and the capitalist mode of production (Campling & Baglioni, 2020). The alternative framework developed in this chapter has primarily concentrated on Marxian political economy to examine the issues of power, politics, rent-seeking behaviour, rent distribution and unequal exchange in the process of conservation and sustainable usage of natural resources. In addition, it integrates the political ecology questions and ecological economics' arguments to develop a comprehensive understanding of the factors.

Marx did not explicitly discuss the natural resource governance issue, but he constructed a theory of social development, which is a powerful tool to explore the contested interplay among polity, economy and society (Borak, 2002) even in the realm of natural resources. Natural resource governance is a critical issue, and therefore, the mechanism of allocation, distribution and regulation of these resources is not a simple process that can be easily handled by reforming the market mechanism only. It is necessary to scrutinise, which agents of the market are in a powerful position. Marxian political economy can throw light in this regard.

It is often claimed that Marx was only concerned with 'human nature' and left 'nature' alone. This debate has emerged along with the rise of 'ecological economics' which is quite the opposite (though sometimes seems to be overlapping) to 'environmental economics'. The underlying argument is that Marx could not make any fundamental contributions to the development of ecological thought. On the contrary, others have acknowledged that he had a thoughtful insight into the environmental problem (Burkett, 2006; Foster, 2002). Marx criticised the capitalist economic system, and in doing so, raised questions about the relations between society and nature. The capitalist mode of production has an antagonistic relation to the environment and the concomitant result of it is the current crisis of natural resources degradation. Moreover, Marx's theory of 'metabolic rift' and discussions on 'commodity fetishism' and 'primitive accumulation' illustrated that he was concerned about the interrelationship between nature and society.

The *theory of metabolic rift* developed by Marx can be identified as a critique of environmental degradation. Marx noted that natural systems have their own metabolism, which functions in relation to human society and allows for its regeneration. Humans actively interact with and transform nature. Each mode of production organises human actions and the exchange of matter and energy between society and nature, creating a distinct 'social metabolic order' that influences the reproduction of society and ecosystems (Foster, 2000; Marx, 1976). The rise of capitalism, however, has altered this social metabolic order by eliminating non-capitalist modes of production as it establishes an integrated socio-economic system comprised of interconnected subsystems held together through opposing forces and long-term historical processes (Clark & York, 2012). While economically unified, this system creates social and ecological divisions and contradictions that facilitate the unequal accumulation of capital (Baran, 1957; Baran & Sweezy, 1966). Everything is evaluated under this system based on the maximisation of profit, and accordingly, many social, ethical and spiritual relations are valued in the same way, which is fallacious. It simplifies natural conditions by imposing a division of nature to increase economic efficiency that "tends to destabilize ecological balances in hazardous ways" (Burkett, 1999, p. 87). Capitalism, in fact, fails to regulate its social metabolic relation with nature in an environmentally sustainable manner as a persistent effort to renew the process of capital accumulation. It increases the scale of degradation by imposing capitalist needs on nature, regardless of the consequences for natural systems (Campling & Baglioni, 2020; Clark & York, 2012). Public wealth, which includes the conditions of nature, is thus diminished in creating private riches (Foster et al., 2010). As long as private entities own the resources, the scarcity principles are operated and the monopoly over resources jacks up the prices (the exchange value) (Foster & Clark, 2009).

Such commercialisation of natural resources can also be related to the Marxian concept of *commodity fetishism* (Marx, 1867). Marx used to refer to the process of objectifying human beings and human relations in a modern capitalistic society through the production of commodities (Kosoy & Corbera, 2010). He argued that the fetishistic nature of commodities emerges with the exchange value when they are manufactured for the market. People in capitalist societies treat commodities as if their value is inhered in the commodities themselves, rather than in the amount of real labour invested in their production (Felluga, 2002). Thus, the capitalist society is obsessed with commodities, and accordingly, social relations exist between commodities instead of between people. In the same way, natural resources are treated through monetary values under the market economy, which leads to commodity fetishism by obfuscating multiple values of such resources (Martinez-Alier et al., 1998; Vatn, 2000). Moreover, commodity fetishism often makes it difficult for ethical consumers to be effective in both evaluating objects on offer and influencing the world around them, including environmental protection (Carrier, 2010). Overall, commodity fetishism is a powerful concept for studying the commodification of natural resources and the environment.

Finally, Marx used the concept of *primitive accumulation* to uncover the origins of the capitalist mode of production through historical analysis. Primitive accumulation was thought to be taking place in the 'underdeveloped' world during the twentieth century, where the transition to capitalism was accomplished through the blatant use of force and coercion by the state on behalf of emerging capitalists by tearing apart and uprooting communities, common properties and non-market social institutions (Bhattacharya & Seda-Irizarry, 2014). In the same manner, the concept can be related to the expropriation of natural resources through the coercive power of the state in transitional economies.

In recent decades, however, Marxian literature discussed the issue of *unequal exchange* (both economic and ecological), and in the context of this book, 'unequal ecological exchange' (Bunker, 1985, 2007; Odum, 1988; Odum & Arding, 1991) is relevant (see Foster & Holleman, 2014, p. 205 for details). In the capitalist society, as Marx argued, nature's parts are being extracted, processed and exchanged for money as marketed commodities, and as a result, the resources are being often removed from local habitats to distant places. People either use it for immediate usage for subsistence or exchange its products and services as commodities for long-term personal gain (Merchant, 1987). The unequal ecological exchange at its root discussed the asymmetric relationship between developed and developing countries in terms of resource consumption, degradation and human well-being (Givens et al., 2019; Jorgenson, 2009). The developed countries have been able to extract the resources from the developing countries, and their consumption level is above the global sustainable threshold. On the other hand, the developing countries are experiencing environmental degradation with a lower level consumption, which directly affects the well-being and quality of life of the domestic people.

Ecological economics, in particular, is also concerned with the co-evolution of human economy and nature under the energy transfer procedure based on the entropy law and the second law of thermodynamics (as energy is transformed or transferred, more and more of it is wasted [Lucas, 2015]). Accordingly, this is relevant to the concept of 'metabolic rift'. It contends that there exists mutualistic symbiosis between ecosystem and economy (Martinez-Alier, 1987), which is distorted by the expansion of capitalist accumulation. The industrial expansion is, therefore, the main cause of releasing entropy into the environment (Georgescu-Roegen, 1971, 1986). Therefore, it called for moving towards a steady-state economy (Daly, 1977, 1980; Costanza, 1980, Costanza et al., 1991; Martinez-Alier, 1987; Ropke, 2004; Smith, 2010) through degrowth and decelerating growth processes by developed and developing countries, respectively (O'Neil, 2012).

The particular contributions of ecological economics, however, are to the concepts of natural capital and sustainability. It proclaimed that human-made capital is the subsystem of natural capital other than being a factor of production. The intuition is that the natural environment plays a dual role as a source of productive factors and a waste assimilator. Accordingly, it argues that natural capital cannot be substituted with any capital, and thus, they have argued for 'strong sustainability', while vehemently opposing the 'weak sustainability' perspective of neo-classical economics (Neumayer, 2003). Thus, there is an implication of Marxism in ecological economics, and the contributions can be summarised into four fundamental issues: (a) the relations between nature and economic value, (b) the significance of entropy law for economic system, (c) the treatment of nature as capital and (d) the concept of sustainable development (Burkett, 2006).

Thus, Marx's engagement with the ecological problems can be identified from several grounds, but the approach could not articulate any solution regarding governance strategy. In particular, ecological economics examined the interrelations between economy and ecology only but could not internalise the relations between humans and nature (social and behavioural aspect) (Siebenhuner, 2000). Moreover, the steady state of the economy does not prevent the exhaustion of resources (Georgescu-Roegen, 1975), and it purposely leaves out the human economy (Sagoff, 2012). Therefore, it can be identified only as a neo-classical critique.

Based on the critical scrutiny of both market-centric and political economy approaches, it can be argued that much of the present problem lies in the under-theorisation of the political-economic factors in the resource governance framework in a transitional economy in particular and of sustainability in general. The alternative framework developed in this chapter, accordingly, has tried to provide the basis for a more comprehensive theory of unequal exchange based upon human–nature relations as well as relations among different powerful groups both at the international and national levels. The argument is that the inherent reasons for natural resource depletion have not been explicitly discussed which may emerge from specific social and political contexts of a particular country. The asymmetric power relation between social groups is important to consider in this context, especially to examine its implications in understanding the institutional vulnerability and management of resource rent.

Rent is an income that is greater than the minimum, which an individual or a firm would have accepted, if given alternative opportunities, and it can be both growth-promoting and growth-retarding (Khan, 2000). Natural resource rent is considered efficient and desirable as it helps reduce the degradation of resources by hindering overconsumption. The critical point is that such types of rent are only efficient under some specific and well-defined conditions, and if they do not exist, they may turn into inefficient rent. The existences of clear definition and enforcement of property rights, political stability and equal power structure are necessary to have efficient resource rent. The transitional economy largely lacks the conditions that cause the resource rent to be dissipated and captured by the rent-seekers.

Rent-seeking induces unequal distribution of wealth and income in the economy. The consequence of extensive rent-seeking in illegal forms (bribing, coercion, etc.) in developing economies is that resources are diverted towards expropriations and accumulations of powerful groups. Their power, however, is not derived from their wealth, but rather acquired by political means, and in this case, the existing nature of political settlement—a specific distribution of power between relevant organisations in a country—matters. The rent-seekers influence government decisions that are favourable to their interest by exercising their political power. The idea of political settlement that this book considers is adopted from Khan as follows:

> A political settlement is a combination of power and institutions that is mutually compatible and also sustainable in terms of economic and political viability (Khan, 2010, p. 4).

It is an explicit or implicit agreement among powerful groups or contending social groups or elites about the rules of the political game (di John & Putzel, 2009; Kelsall, 2018; Laws, 2012). It, thus, implies the distribution of power between different groups and classes (clientelist). This results in the coercive power of the state. It signifies why informal organisations based on patron-client networks have been influential here, and they set critical limits on the operation of evolving formal institutions (Khan, 2010). Marxian argument entails that by using coercion (state power), the state derives some capital from people and provides it to the hand of the private sector so that they can invest and initiate industrialisation (Khan, 2000). Such use of coercion by the state can be attributed to the idea of 'monopoly on violence' (Weber, 1919 [2015] as cited in Waters & Waters, 2015).

Weber (1919 [2015]) describes the state as any organisation that succeeds in holding the exclusive right to use, threaten or authorise physical force against the citizens. In developing countries, the state is involved in redistributing incomes as well as creating new property rights, and in the process, the power holders use the state apparatus to secure their interests. Such means of asset transfer or seizure can be described as rent transfer and rent-seeking. The intermediate class is the one which is primarily involved in this process of accumulation (can be linked to the primitive accumulation) by exercising their organisational and political power, and this can be attributed to the extraction of natural resources as well. As long as natural resource rent is greater than the cost of extracting, the powerful entities try to utilise the resources as much as possible without being concerned about the extent of available natural resources. These political economy factors help determine which individuals or groups are likely to win in distributive contests over the natural resources, as the book argues.

Moreover, this book recognises that nature does not exist in isolation from human society or the economic systems that operate within it. The environment both defines and is shaped by the activities of human beings (Lovett & Ockwell, 2010). The book considers the political ecological propositions here to argue that human relations to the environment and human social relations are not separate phenomena but inextricably intertwined (Braun, 2015; Hornborg et al., 2012; Low & Gleeson, 1998; Paulson & Gezon, 2004; Peet & Watts, 1996; Robbins, 2012; Wolf, 1972). Political ecology is a critical approach that emphasises nature–society interactions, recognising the significant role of political economy in shaping human behaviour, ecological conditions and dynamic interactions between the two (Rose & Carr, 2018). In another way, it scrutinises the relationships between social, political and economic factors as well as environmental changes, by combining ecological and social sciences with political economy (Robbins, 2012). It also foregrounds the role of power disparities and international political economy in determining the nature–society relationship (Rose & Carr, 2018). In line with that, the neo-liberalisation of conservation has recently emerged within political ecology also (Fairhead et al., 2012) that examines how the conservation of nature under neoliberalism has become a source of capital accumulation (Brockington & Duffy, 2010). It is necessary to consider questions on political ecology, as nature's governance system needs to conceptualise by defining the interrelations among economy, society and nature. The underlying implication is that recognising the role of human beings as social beings is a prerequisite for ensuring sustainability or sustainable use of natural resources.

2.4 The Complementarity of Human and Nature Well-Being: A New Approach

The critical reviews of the two approaches in the preceding sections allow a classification of theories based on the divergence between market and non-market comprehension of resource governance mechanisms (Table 2.4). Overall, they cover the issues of valuation, institutional arrangement and the emerging notion of sustainability.

Specifically, the new approach vehemently challenges the neo-classical tradition and criticises the NIE but considers the significance of institutional vulnerability issue. The neo-classical rationalistic approach has the flaw to address the issue adequately and is unlikely to lead to an environmentally attuned economy. The new institutional economics mainly emphasised a formal institutional set-up to ensure the sustainable utilisation of natural resources. It, however, could not sufficiently address the issues concerning political transaction costs and informal institutions. Both branches evade social norms and reciprocal behaviour as a collective organisation and identify human beings as rational beings who always behave according to their self-interests.

Table 2.4 Classification of theories (under two approaches) regarding natural resources

Approach		Arguments regarding natural resources	Underlying gaps
Market centrism	Neo-classical economics	*Problem:* common resource problem; negative externality (market failure) *Solution:* private or public provision	Higher price; loss of consumer surplus; rent dissipation; depletion; unequal extraction of resources
	New institutional economics	*Problem:* articulating property rights is costly; transaction cost is important *Solution:* private; collective ownership	Threat of breaking the contract; avoids political transaction cost
Political economy	Marxism	Indirectly addresses the antagonistic relation between nature and capitalism under concepts such as power, primitive accumulation, alienation and commodity fetishism; unequal ecological exchange in global context	Unequal exchange in a particular country context; no focus on human sociality aspect
	Ecological economics	Identification of uniqueness of natural capital (not substitutable); strong sustainability	A critic of neo-classical economics only; focus on human and ecology relations only ignoring the societal factors
	Political ecology	Considers nature, society and economy on the whole	Do not explicitly discuss human sociality as regards to governing natural resources

Source Prepared by the authors

Moreover, all these features are still waiting for an intensive analysis in the context of a transitional economy. Accordingly, this book considers the political economy approach to examine the interrelationships among power, politics, rent-seeking behaviour and rent distribution regarding the conservation and management of natural resources in such economies. The discussion signifies that only efficient management is not enough for natural resources, rather equitable distribution is also important to make the management process sustainable. The Marxian political economy, however, does not explicitly define the environmental degradation aspect though there is a focus on it while criticising capitalism. In addition, there is a gap in considering the 'human sociality' aspect under this approach. Finally, ecological economics and political ecology are the emerging fields within the political economy that are yet to contribute to developing the solution for ending natural resource degradation.

Against this backdrop, the chapter attempts to move towards a novel approach that sustains the complementary relations between humans and nature and, hence, ensures the sustainable governance of natural resources. The new approach tries to explain the usage, management and degradation pattern of natural resources in a transitional economy and identifies the important institutions and political factors that mediate these processes. In particular, it uses the political economy lens to explain the degradation of resources under market centrism and attempts to define the sustainable usage and conservation of the resources by taking the human–nature relationship as the centrepiece. Overall, it is a hybrid framework that integrates power structure and the political arrangement, social organisations, institutions and human sociality into the analysis of the resource governance mechanism.

The framework addresses a structure of production and exchange relationships, with unequal power relations of different agents to focus on three key issues (a) conservation, (b) sustainable management of the natural resource and (c) equitable sharing. In order to address the production and exchange relationships among different classes in the context of natural resource governance of a transitional developing economy, the book has tried to identify and describe the social relations of production—a key term in Marx's theory of history (Chitty, 2000)—in this particular context. Here, social relations of production refer to (a) the relationship to the means or forces of production and (b) the mode of value appropriation. These two elements essentially serve as the material basis of class and define class interests. It is, therefore, essential to find out the characteristics of the 'agents', the 'means' by which they exercise and fulfil their choices, in which way surplus value is being appropriated and finally what the 'outcomes' are (Fig. 2.2).

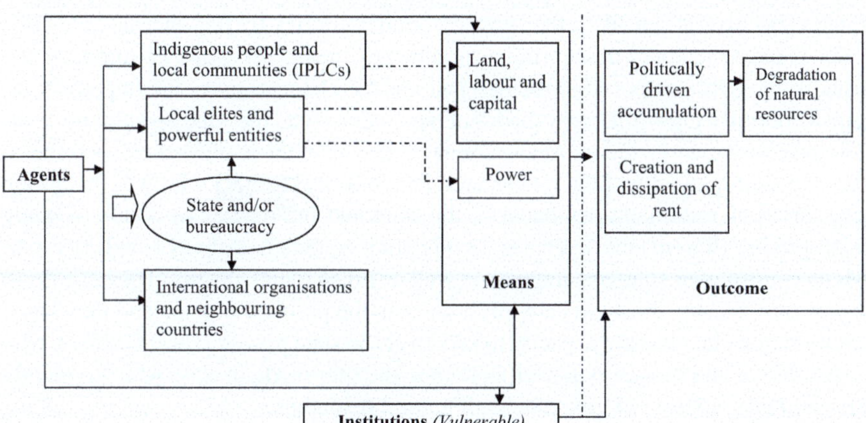

Fig. 2.2 Social relations of production and elements of the model. *Source* Prepared by the authors

2.4.1 Agents

Several agents have direct or indirect relations with or influence over natural resources (Fig. 2.2). The agents for the production and exchange relationship in the case of private goods are relatively easy to identify, whereas in the case of open access or common resources, this is more difficult. Because this book is primarily concerned with three cases of natural resources in the context of Bangladesh, the agents involved in these three relevant cases have been considered. The agents can be at both national and international levels, and for a particular case, one agent could be more important compared to the others and those 'other' agents can again be important for multiple cases simultaneously. For example, international agencies' role is more relevant to climate change and transboundary water governance issues. On the other hand, the role of Indigenous Peoples and Local Communities (IPLCs) is more inherent to the case of biodiversity and water resources governance in the national and local contexts. Such differential roles would be examined in the next three chapters.

The identified agents play diverse roles in different contexts. Therefore, it does not require describing who the key agents are in managing a particular type of resource. The *state* plays a central role–economic and non-economic—as an antecedent agent by influencing the process of accumulation of resources and accordingly, excessive extraction of them. It is engaged in the process of structural reproduction of a system, which is needed for some particular agents to continue appropriating surplus through the enforcement of power and the legitimisation of inequalities. Then, the *intermediate class* is appropriating surplus by exploiting resources by exercising state power. Post-colonial countries comprise a broad range of groups between the two polar classes (proletariat and bourgeoisie) who can be termed as the 'intermediate class' (Gramsci, 1979). In the absence of a well-developed industrial capitalist class, the intermediate class exercises state power to their advantage (Kalecki, 1972). This class used the political space to promote its own interests (clientelist politics). Bangladesh also is run like other post-colonial countries by the intermediate class which is interested only in securing wealth by any means necessary (Titumir, 2015). Broadly, they are composed of the rich and middle-class peasants, urban petty bourgeoisie, political class, different shades of bureaucracy and the educated middle class, who have a greater degree of organisational ability than those of workers, poor peasants, the unemployed and the uneducated (Titumir, 2015, 2021). In fact, this group only serves as intermediaries between the state and the masses by virtue of being employed in different state agencies. In this process, the class always tends to establish its control over state apparatuses as well as other social classes. They can do so as they have the capability to gain importance in the political domain and they know that no single class has the capacity to take over the state power. *Powerful elites,* who have political and monetary power, belong to the intermediate class who are mainly relevant in the context of governance of natural resources.

Thereafter, *Indigenous Peoples and Local Communities (IPLCs)*—defined by IPBES (n.d) as "the original inhabitants of a given region, in contrast to groups that have settled, occupied or colonised that area recently"—are the ones greatly affected by the policy measures taken for the natural resource governance. The debate is that whether the local people are the protectors or destroyers of natural resources and the environment. Therefore, they have also been taken into consideration as potential agents. Finally, in this age of globalisation, the decision-making process of the states is also influenced by international stakeholders (e.g., different international organisations and powerful states), which is also evident in the realm of resource governance. Taken together, it can be argued that the agents of natural resources are diversified, and their roles are complex in nature.

2.4.2 Means

The agents fulfil their choices by having control over several means. Economic resources (land, labour, capital) are the key components through which the agents can exercise their choices. As discussed already, the intermediate class always tries to accumulate rent and, thereby, have been capturing a significant amount of capital in their hand. *Capital* as a factor of production has often been regarded in economic models, but in addition, *power* is also an important variable, which can influence the behaviour and choices of the agents. The intermediate class, having an unequal share of economic and political power, has more access to the state as well as can define the rules of the power game in its favour. The powerful agents have exercised oligopolistic market power to determine the price of natural resource goods according to their interests. On the other hand, the local people have utilised their *labour* and *land* to engage in the production of natural resources. The ownership of land, however, can be found to be in the hands of state authorities. The control over the means in this case, therefore, is essentially in the hands of the powerful groups (national and international) and the state.

2.4.3 Outcomes

Natural resources directly provide human beings and other creatures with various amenities that are difficult to measure as their value is intrinsic in nature. Under the process of market economy, the resources are turned into marketable goods, and subsequently, it fails to appreciate the positive prices for the economic functions performed by nature. In particular, the existing nature of the transitional economy results in weak and broken functioning of the environment. The power dimension and institutional vulnerability induce politically driven accumulation and rent-seeking behaviour. Moreover, the rent is being diverted towards unproductive sectors, and the powerful agents are less concerned about the degradation or loss of natural resources. They are only bothered about capturing more wealth and money, and in this process, they take a larger share of benefit from natural resources by exploiting others. The outcomes are the inequitable distribution and gradual degradation of natural resources.

2.4.4 Institutions

There are extractive (both economic and political) institutions (Acemoglu & Robinson, 2012) that result in inefficient and wasteful usage, management and extraction of natural resources, as well as inequitable benefit sharing of the resources in a transitional economy. In another way, the situation of natural resource exploitation has been aggravated as the fragility of institutions escalates. The key attributes in this regard can be identified as the absence of law and order, insecure property rights, regulations that prevent markets from functioning efficiently and the creation of a non-level playing field. The powerful elite groups design the institutional arrangements that are conducive to securing their own benefits in these states. Overall, the institutional vulnerability is a key factor behind the continuous degradation of natural resources in such economies.

2.4.5 Power and Political Settlement

As is evident from the definition of political settlement, as defined earlier, power remains at its core. When power is uneven, whichever group is in power seeks to convert as many natural assets as possible into irreversible specific capital that benefits itself (Collier, 2010). The structure of power-sharing arrangements and the nature of political settlement determine who would control the resource governance system in the particular context of a transitional economy. In transitional economies, the intermediate class, who are only interested in securing wealth by any means possible, holds organisational power. The never-ending desire for primitive accumulation that entails clinging to power has led to a coterie of the ruling elite to justify this process by inventing this sugar-coated unfounded conjecture (Titumir, 2015). Therefore, the ruling elite in developing economies seek to establish a state based on power concentration. Thus, in such economies, the particular form of materialist incentives of primitive accumulation of resources with monopoly power and coercion has resulted in a system of clientelist networks (Titumir, 2021). Overall, the outcome is the escalation of the state's coercive power with the intermediate class enjoying the benefits. The factors, indeed, have caused the non-functioning institutions and the greater scope for primitive accumulation of resources.

The underlying theme is that the nature of political settlement shapes institutional and policy performance. As institutions and policies describe rules that determine resource allocation (Khan, 2017), it is essential to scrutinise the influence of political settlement on the institutions and policies regarding natural resources. In understanding the failure of resource governance in a developing country, the book adopts these variables to understand why some institutions (e.g., property rights) are not well defined and fragile in that particular context. Overall, what can be argued, firstly, is that the existing type of political settlement requires that power should be concentrated in the hands of a particular party. Second, the intermediate classes, which are backed by the party, are constantly looking for ways to secure wealth. Third, these two intertwined phenomena result in alienation between the ruling party and the public, as well as an excessive reliance on the coercive apparatus of the state (Titumir, 2015). All these result in the excessive extraction of natural resources, which are a major source for securing wealth with the least costs. Moreover, the imbalance of power among different states also causes degradation and unequal distribution of resources, where the international political-economic factors are relevant.

2.4.6 *Human Sociality*

Human groups maintain a high level of sociality despite a low level of relatedness among group members, which signifies that pro-social behaviour exists in human beings (Gintis, 2000). Taking essence from here, the book considers the variable 'human sociality' where:

> "Human sociality refers to human beings, as a collective organisation, and is part of the larger ecosystem, which possesses distinct knowledge and practices that systematically and sustainably contributes to the conservation and regeneration of resources. It stresses that societies in harmony with nature contribute to the resource conservation through revitalisation where informal institutions can play a crucial role" (Titumir et al., 2020, p. 75).[1]

The argument is that incorporating a micro-level perspective of human behaviour into integrated environmental models can yield a better understanding and management of environmental degradation processes (Jager et al., 2000). Social norms are the end results of a largely unrecognised process of cultural evolution, and once a norm is present, its continuous grasp of human beings will be dependent on a web of expectations, and as long as the expectations are self-fulfilling, humans will continue to conform to the norm (Binmore, 2010). A norm can thus be thought of as an equilibrium in the sense that each player maximises her expected utility if she accepts the other players' actions as given and their beliefs as correct (Bicchieri et al., 2004). Thus, Bicchieri asserts "social norms....transforms mixed-motive games into coordinated ones" (Bicchieri, 2005, p. 3).

The interpretation of social norms can be made more elaborate in this context by relating to the idea of convention or social convention theory (Hume, 1739, 1777). Hume suggested that a convention corresponds to a pattern of mutually beneficial behaviour that a group of agents follows when they know that such a pattern is mutually beneficial and that they expect each other to follow this instead of another. Thus, the social convention is defined as a customary, arbitrary and self-enforcing rule of behaviour that is generally followed and expected to be followed in a group

---

[1] The definition is slightly modified here from the authors' own previous work.

or in a society at large (Tummolini, 2013). When a social convention is established, people act in a quasi-agreed-upon manner, even if they did not explicitly agree to do so, and this can be recognised as a *tacit* agreement (*ibid*). Experiments also showed that people generally adhere to cultural rules in their societies even when there is no chance of a deviation being punished (Gintis et al., 2015).

In line with the human reciprocity towards social norms and the informal institutional evolution of cooperation by social beings, the book argues that the human as a social being can preserve natural resources sustainably. Humans are a cooperative species, and the cooperative tendencies appear to be evolutionarily ancient, psychologically distinct and cross-culturally universal (Curry et al., 2019; Skoggard, 2016). Self-interest-seeking behaviour and the non-cooperation occur when a majority of people behave unfairly, and in this case, one may feel that one's efforts to be fair are useless as one's fair action will bring no change, and probably, most people do not even expect to be treated fairly (Bicchieri, 2010). In fact, the non-cooperative behaviour is a product of the culture of an advanced complex society (Gintis et al., 2015). Therefore, what is required, as this book argues, is to establish a convention to revitalise the human-nature relations that exist historically, so that the human beings themselves can be the preserver of nature and not the destructor.

Proposition 1: Exchange and Natural Resources

When non-marketable goods are transformed into marketable goods, it creates rent and deadweight loss. This valuation through pricing mechanisms results in higher prices, loss of consumer surplus, loss in growth and loss in the endowment of the resources (i.e., the producer surplus).

Panel A explains the reality of the natural resources markets (Fig. 2.3). The *x*-axis in the horizontal line indicates the price, whereas the *y*-axis in the vertical line indicates the quantity. Initially, at a very low level of price or sometimes no price, the demand for natural resources is infinite, measured through a perfectly elastic demand curve. Simultaneously, the earth herself is endowed with a fixed amount of resource base (there are some resources, however, that have the regenerative capacity, but are depleted when the rate of harvest is greater than the regeneration rate). Consequently, the supply curve measures as perfectly inelastic.

Panel B of the same figure attempts to depict the price effect on natural resources if they are metamorphosed into marketable goods. When non-market goods transform into marketable goods, the positive demand curve will change and consumers get to interact with different prices. The changing demand curve produces some alternative distribution and sharing. First, the new equilibrium point sets at *E*, spiralling up the prices from *P1* to *P2*. Consequently, the consumer surplus is reduced from *PEHP1* to *PEP2*. Thus, the loss of consumer surplus as rent is *P2EHP1*. Another important change in distribution occurs if non-marketable goods transform into marketable goods. The change generates rent and deadweight loss (shaded area) due to higher demand and lack of supply, causing consumption inefficiency. The rent is a mischievous distribution; hence, the study is concerned with who gets this natural resource rent, how it is distributed and how government loses the share of the rent.

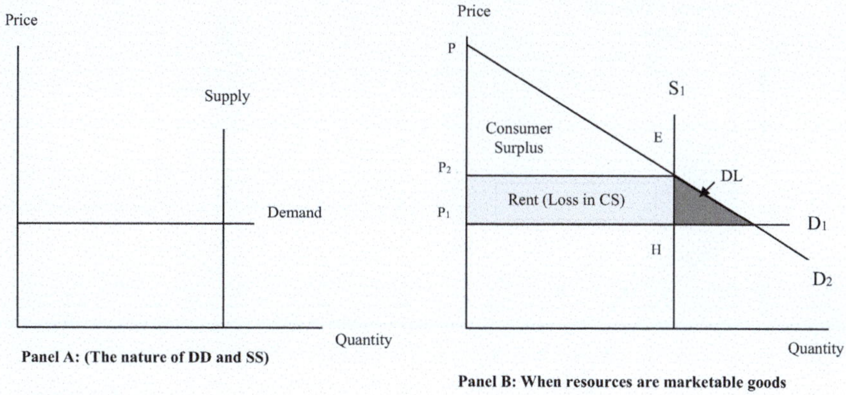

Fig. 2.3 Price effect if natural resources are turned into marketable goods. *Source* Prepared by the authors

Proposition 2: Institutions, Political Settlement and Primitive Accumulation

The agents assert controls over the potential rent under the extractive institutional arrangements that are historically prevailing, not only by the dominant goals of production but also by the prevailing social relations and the scale of production, as well as relations of distribution and property regimes.

Specifically, the likelihood of unstable property rights for natural resources is very high in a developing economy. A strong institutional arrangement can check the stability. Under the neo-classical rationalistic view, rationality drives the individual to break the contract under a fragile institutional arrangement to secure their own interest which can be explained under the game theoretic approach depicting the prisoner's dilemma and Nash equilibrium (Fig. 2.4) and which ultimately results in the destruction of natural resources. More concretely, the figure helps to visualise the way in which different agents behave in a strategic manner under a vulnerable institutional arrangement.

Here, the vertical axis measures the resource extraction by agent *A*, and the horizontal axis measures the resource extraction by agent *B*. Under the mutual contract, both of the agents extract resources on the A_1B_1 line. The next scenario is that as a rational being, agent B can maximise his extraction if he thinks that by holding agent *A* constant, he (*B*) can extract more resources as there is no one to monitor it. Accordingly, agent B will go for more extraction at B_1*, and keeping the A_1 fixed is the symptom of cheating, pointing to B_1*/A_1 in the graph. Similarly, another agent of this model A will do the same at A_1*, keeping the B_1 fixed is another symptom of cheating, pointing to A_1*/B_1. Therefore, under a weak institutional arrangement, the contract does not hold. In the graph, the new resource extraction line is now A_1*/B_1 and B_1*/A_1, galloping up from A_1B_1. Despite the Nash equilibrium, it is also not stable. Until the complete extraction of natural resources happens, the shift and alteration of a non-cooperative game will persist.

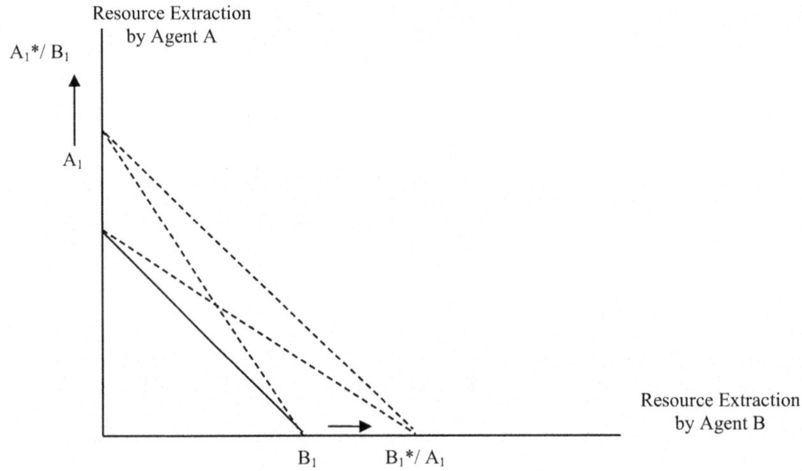

Fig. 2.4 Institutional vulnerability and destruction of resources. *Source* Prepared by the authors

The key underlying essence of this proposition is that the faulty persistence and the adverse development of the institutions of a developing country are the reasons for massive resource destruction.

Proposition 3: Power, Market Structure and Benefit Sharing

The asymmetry of power alters the conditions of functioning of the pure market to imperfect market, as negative externalities (cost shifting) are not taken into account and positive externalities are insufficiently utilised. The powerful agents dictate the price and the sharing of benefits by undermining community, common property and collective action, resulting in institutional fragility and ignoring social interdependence.

Political institutions in developing countries are shaped by the political actors and the business community (Fig. 2.5). By using power, these agents accumulate the resources. The extractive institutions exclude the households and the primary producer by enclosure and draw a line of disparity to enjoy equitable benefits from natural resource rent and the state loses a significant share of the revenue. For communal ownership, the powerful agents impose heavy enforcement rent to access the resource base. The agents are able to marginalise the public or people of the community successfully. The household and the primary producer, therefore, face an interlocked market system with a heavy transaction cost to enter the system. The ultimate result is the inequitable distribution of resources or benefits. The principle of equity, however, is important in the distribution of resources to ensure sustainable management of it. Resources are to be utilised for maximum welfare ensuring equitable benefit sharing. This proposition signifies that pricing cannot address the distributional problem; rather, it can be addressed by the fragile institution, which is a function of power.

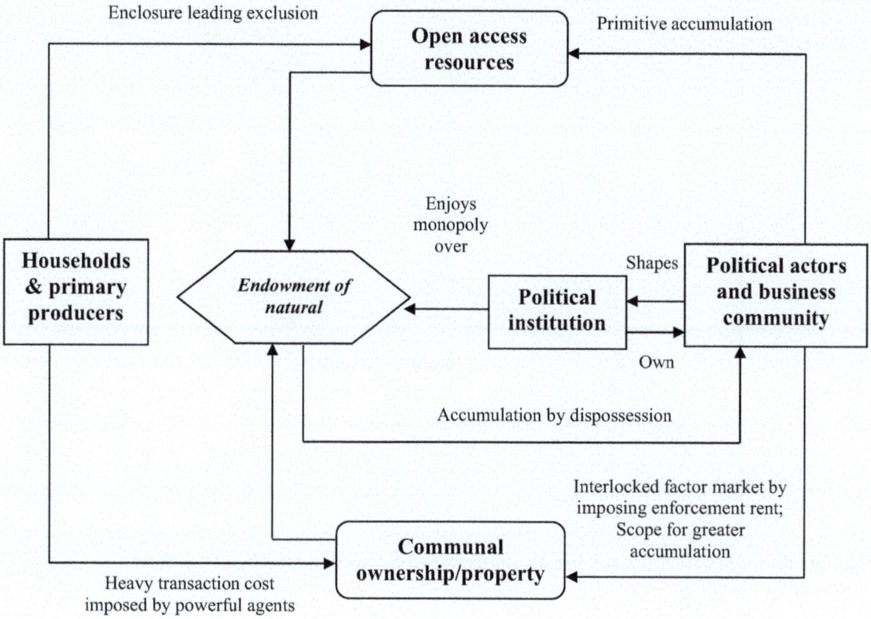

Fig. 2.5 Economics of power in natural resource governance in a transitional economy. *Source* Prepared by the authors

This proposition leads the book to investigate the political basis of the state, the process of rent-seeking and rent dissipation. Another associated theme that merits enquiry is the politics of international organisations and neighbouring countries in terms of sharing and management of natural resources at both domestic and global levels. The book, therefore, examines under this proposition that weak institutions and the power of political negotiations widen the extent of unequal sharing not only within the state's borders but also beyond that.

Proposition 4: Well-Being, Sustainability and Human Sociality

The existence of mutual interrelationships between human sociality and nature is the key to the optimal usage and conservation of natural resources.

The book considers humans as social beings possessing socially constructed norms and values. The key understanding is that these social conventions can solve the natural resource problem. The norms and values create a collective organisation that can preserve the natural resources sustainably and equitably. The authority to impose credible threat and the sanction by the stakeholders of the resources on the resource distribution can immensely solve the natural resource problem and distribution of benefits.

The underlying reasoning can be understood through the explanation of the rational choice view versus choice under social cooperation (Fig. 2.6). The vertical axis measures the individuals' preferences and the horizontal axis welfare as well as

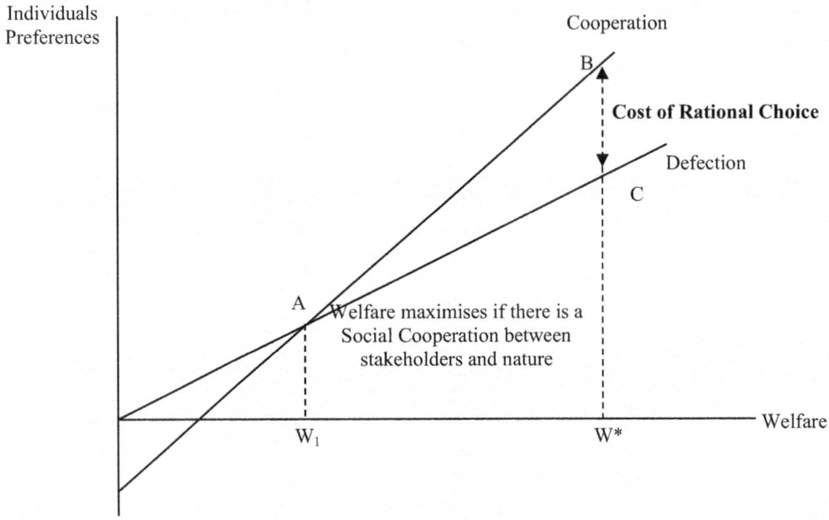

Fig. 2.6 Rational choice versus social cooperation. *Source* Prepared by the authors

the cooperation and defection that depend on two different functions of the individual preferences. At point *A*, the expected cooperation and the expected defection intersect, showing a lower level of welfare and a low level of individuals' consensus. On the contrary, the welfare can be maximised when the individuals have institutional cooperation rather than rational engagement. The book argues that individuals as social beings belong to the web of an ecological and social environment. The nature carries the function of social being as the process of reservation and reiteration. As a social being in a particular ecosystem, the individual can enlarge their welfare at W^*, which is greater than point *A*. As a social organisation, the individuals have more consensuses and hardly any cost of defection. Therefore, the total welfare generates the area of AW_1W^*B.

The essence of this proposition is that social norms can organise the equitable benefit sharing of the resources. The traditional evolution of norms and knowledge can efficiently manage the resource rent and the sustainability of resources. The power of sanction imposed by the stakeholders can informally organise the rights of resource consumption and accretion. Social norms are so powerful that social beings are ready to sacrifice to prolong the relationship between nature and human beings. The recognition of social norms and social ownership, thus, can sustainably manage natural resources and the equitable sharing of natural resource rents. Overall, the inherent unity of humanity and nature needs to be realised both consciously and socially and internalised into the resource governance framework.

2.5 Concluding Remarks

This chapter attempts to contest a number of theoretical positions that explain natural resources governance mechanisms, specifically considering the context of a developing economy. The core thesis of the study is that political economy factors are necessary to consider for analysing the causes of degradation or unsustainable use of natural resources, while a human sociality-centred framework needs to be built up to ensure the sustainable management of those resources. Accordingly, the dependent variable is the 'state of natural resources', and the independent variables are 'institutions, power, political settlement and human sociality'.

The key argument that can be drawn from this chapter is that the market-centric solution is not efficient; rather, the configuration of power in society is a critical variable determining the usage, allocation and management of natural resources. In a transition economy, in particular, the powerful agents most often capture the resource and extract from it without considering the loss of that resource. The specific type of political settlement and the existing nature of the social relationship is important in this regard. Moreover, there is an intrinsic and supportive interrelation between human beings and nature; however, under a market-based economy, humans are being turned into destructive opponents of nature by exploiting natural resources to gain personal or group interests. The value of natural resources is expressed in terms of market value, and the human sociality is found to be eroding, which needs to be revitalised.

The next three chapters provide empirical analyse of different cases of the natural resource governance system and challenges considering the context of Bangladesh under the new conceptual framework built in this chapter. The analysis will demonstrate that the current accumulation process has resulted in a non-sustainable natural resource base due to market failures caused by power imbalances, institutional vulnerability, and non-recognition of interrelations between human beings and nature.

Acknowledgements The 'propositions' that are built in this chapter can be found available in one of the authors' previous work (Titumir et al., 2019). The published article was prepared by taking relevant materials from the draft of this book. Thus, the propositions are mainly developed for this book, which is also recognised in the reference list of the published article.

References

Acemoglu, D., & Robinson, J. A. (2012). *The origins of power, prosperity and poverty: Why nations fail*. Crown Publishers.

Acheson, J. M. (2006). Institutional failure in resource management. *Annual Review of Anthropology, 35*(1), 117–134. https://doi.org/10.1146/annurev.anthro.35.081705.123238

Adam, C., & Dercon, S. (2009). The political economy of development: An assessment. *Oxford Review of Economic Policy, 25*(2), 173–189. https://doi.org/10.1093/oxrep/grp020

Adhikari, B. (2001). *Literature review on the economics of common property resources: Review of common pool resource management in Tanzania*. Report Prepared for NRSP Project. UK Department for International Development (DFID).

Adhikari, B. (2002). *Property rights and natural resource: Socio-economic heterogeneity and distributional implications of common property resource management in Nepal*. Discussion Paper. Beijer Institute Advanced Workshop on Property Rights Structures and Environmental Resource Management, 28–30 May, South Africa.

Adhikari, B. (2010). How do institutions affect the management of environmental resources? In J. C. Lovett & D. G. Ockwell (Eds.), *A handbook of environmental management* (pp. 119–167). Edward Elgar Publishing Limited.

Adler, M. D. & Posner, E. A. (2006). *New foundations of cost-benefit analysis*. Harvard University Press.

Alchian, A. A. (1965). Some economics of property rights. *Il Politico, 30*(4), 816–829.

Angus, I. (2008). The myth of the tragedy of the commons. *Climate & Capitalism*. Retrieved February 15, 2021, from https://climateandcapitalism.com/2008/08/25/debunking-the-tragedy-of-the-commons/

Arrow, K. J., Cropper, M. L., Eads, G. C., Hahn, R. W., Lave, L. B., Noll, R. G., Portney, P. R., Russell, M., Schmalensee, R., Smith, V. K., & Stavins, R. N. (1996). Is there a role for benefit-cost analysis in environmental, health, and safety regulation? *Science, 272*(5259), 221–222.https://doi.org/10.1126/science.272.5259.221

Baland, J. M., & Platteau, J. P. (1996). *Halting degradation of natural resources: Is there a role for rural communities?* Food and Agriculture Organisation of the United Nations.

Banana, A. Y., Vogt, N. D., Bahati, J., & Gombya-Ssembajjwe, W. (2007). Decentralised governance and ecological health: Why local institutions fail to moderate deforestation in MPIGI district of Uganda. *Scientific Research and Essays, 2*(10), 434–445.

Baran, P. A. (1957). *The political economy of growth*. Monthly Review Press.

Baran, P. A., & Sweezy, P. (1966). *Monopoly capital*. Monthly Review Press.

Bauler, T. (2012). Property rights, mapping environmental justice. *EJOLT Global Research Project*. Retrieved June 20, 2022 from http://www.ejolt.org/2012/12/property-rights/

Beder, S. (2011). Environmental economics and ecological economics: The contribution of interdisciplinarity to understanding, influence and effectiveness. *Environmental Conservation, 38*(2), 140–150. https://doi.org/10.1017/S037689291100021X

Bhattacharya, R., & Seda-Irizarry, I. J. (2014). *Primitive accumulation*. Working Paper No. 3. John Jay College of Criminal Justice, New York.

Bicchieri, C. (2005). *The grammar of society: The nature and dynamics of social norms*. Cambridge University Press. https://doi.org/10.1017/CBO9780511616037

Bicchieri, C. (2010). Norms, preferences, and conditional behavior. *Politics, Philosophy & Economics, 9*(3), 297–313. https://doi.org/10.1177/1470594X10369276

Bicchieri, C., Duffy, J., & Tolle, G. (2004). Trust among strangers. *Philosophy of Science, 71*(3), 1–34. https://doi.org/10.1086/381411

Binmore, K. (2010). Social norms or social preferences? *Mind & Society, 9*(2), 139–157. https://doi.org/10.1007/s11299-010-0073-2

Borak, N. (2002). Book review on Marxian political economy: Theory, history and contemporary relevance by Bob Milward. *Europe-Asia Studies, 54*(4), 658–659.

Bowels, S., & Gintis, H. (1990). Contested exchange: New micro-foundations of the political economy of capitalism. *Politics and Society, 18*(2), 165–222. https://doi.org/10.1177/003232929001800202

Bowels, S., & Gintis, H. (1993). The revenge of homo economicus: Contested exchange and the revival of political economy. *The Journal of Economic Perspectives, 7*(1), 83–102. https://doi.org/10.1257/jep.7.1.83

Braun, B. (2015). From critique to experiment? Rethinking political ecology for the Anthropocene. In T. Perreault, G. Bridge, & J. McCarthy (Eds.), *The Routledge handbook of political ecology* (pp. 102–114). Routledge.

Bresnihan, P. (2017). *Valuing nature-perspectives and issues*. The NESC Research Series Paper No. 11, National Economic and Social Council, Dublin, Ireland.

Brockington, D., & Duffy, R. (2010). Capitalism and conservation: The production and reproduction of biodiversity conservation. *Antipode, 42*(3), 469–484. https://doi.org/10.1111/j.1467-8330.2010.00760.x

Bromley, D. W. (1991a). *Environment and economy: Property rights and public policy*. Oxford University Press.

Bromley, D. W. (1991b). Testing for commons versus private property: Comment. *Journal of Environmental Economics and Management, 21*(1), 92–96. https://doi.org/10.1016/0095-0696(91)90007-6

Brown, R. (1984). *The nature of social laws*. Cambridge University Press.

Bunker, S. G. (1985). *Underdeveloping the Amazon: Extraction, unequal exchange, and the failure of the modern state*. The University of Chicago Press.

Bunker, S. G. (2007). Natural values and the physical inevitability of uneven development under capitalism. In A. Hornborg, J. R. McNeill, & J. Martinez-Alier (Eds.), *Rethinking environmental history: World-system history and global environmental change* (pp. 239–258). AltaMira Press.

Burke, B. E. (2001). Hardin revisited: A critical look at perception and the logic of the commons. *Human Ecology Review, 29*(4), 449–476. https://doi.org/10.1023/A:1013145905257

Burkett, P. (1999). *Marx and nature: A red and green perspective*. St. Martin's Press.

Burkett, P. (2006). *Marxism and ecological economics: Toward a red and green political economy*. Brill.

Callicott, B. (1989). *In defense of the land ethic: Essays in environmental philosophy*. State University Press of New York Press.

Campbell, B. M., Mandondo, A., Nemarundwe, N., & Sithole, B. (2001). Challenges to proponents of common property resource systems: Despairing voices from the social forests of Zimbabwe. *World Development, 29*(4), 589–600. https://doi.org/10.1016/S0305-750X(00)00114-5

Campling, L., & Baglioni, E. (2020). The political economy of natural resources. In A. Kobayashi (Ed.), *International encyclopedia of human geography* (2nd ed., pp. 259–266). Elsevier. https://doi.org/10.1016/B978-0-08-102295-5.10752-8

Carlisle, K., & Gruby, R. L. (2017). Polycentric system of governance: A theoretical model for the commons. *Policy Studies Journal, 47*(4), 927–952. https://doi.org/10.1111/psj.12212

Carrier, J. G. (2010). Protecting the environment the natural way: Ethical consumption and commodity fetishism. *Antipode, 42*(3), 672–689. https://doi.org/10.1111/j.1467-8330.2010.00768.x

Cash, D. W., Adger, W. N., Berkes, F., Garden, P., Lebel, L., Olsson, P., Pritchard, L., & Young, O. (2006). Scale and cross-scale dynamics: Governance and information in a multilevel world. *Ecology and Society, 11*(2), 8. https://doi.org/10.5751/ES-01759-110208

Chitty, A. (2000). Recognition and social relations of production. In T. Burns et al. (Eds.), *The Hegel-Marx connection* (pp. 167–197). Palgrave Macmillan. https://doi.org/10.1057/9780230595934_8

Clark, B., & York, R. (2012). Techno-fix: Ecological rifts and capital shifts. In A. Hornborg, B. Clark & K. Hermele (Eds.), *Ecology and power: Struggles over land and material resources in the past, present and future*. Routledge.

Coase, R. H. (1937). On the nature of the firm. *Economica, 4*(16), 386–405. https://doi.org/10.1111/j.1468-0335.1937.tb00002.x

Coase, R. H. (1960). The problem of social cost. *Journal of Law and Economics, 3*, 1–44. https://doi.org/10.1086/466560

Coleman, J. S. (1988). Social capital in the creation of human capital. *American Journal of Sociology, 94*, 95–120.

Collier, P. (2010). The political economy of natural resources. *Social Research, 77*(4), 1105–1132.

Costanza, R. (1980). Embodied energy and economic valuation. *Science, 210*(4475), 1219–1224. https://doi.org/10.1126/science.210.4475.1219

Costanza, R., Daly, H. E., & Bartholomew, J. (1991). Goals, agenda, and policy recommendations for ecological economics. In R. Costanza (Ed.), *Ecological economics: The science and management of sustainability*. Columbia University Press.

Cummings, R., Brookshire, D., & Schultze, W. (1986). *Valuing environmental goods: An assessment of the contingent valuation method*. Roman and Allanheld.

Curry, O. S., Hare, D., Hepburn, C., Johnson, D. D. P., Buhrmester, M. D., Whitehouse, H., & Macdonald, D. W. (2019). Cooperative conservation: Seven ways to save the world. *Conservation Science and Practice, 2*(1), e123. https://doi.org/10.1111/csp2.123

Cutting, R. H., & Cahoon, L. B. (2005). Thinking outside the box: Property rights as a key to environmental protection. *Pace Environmental Law Review, 22*(1), 55–90.

Dahlman, C. J. (1979). The problem of externality. *Journal of Laws and Economics, 22*(1), 141–162. https://doi.org/10.1086/466936

Daily, G. C., Soderqvist, T., Aniyar, S., Arrow, K., Dasgupta, P., Ehrlich, P. R., Folke, C., Jansson, A., Jansson, B., Kautsky, N., Levin, S., Lubchenco, J., Mäler, K., Simpson, D., Starrett, D., Tilman, D., & Walker, B. (2000). Ecology. The value of nature and the nature of value. *Science, 289*(5478), 395–396. https://doi.org/10.1126/science.289.5478.395

Daly, H. E. (1977). *Steady-state economics: The economics of biophysical equilibrium and moral growth*. W. H. Freeman.

Daly, H. E. (Ed.). (1980). *Economics, ecology and ethics: Essays toward a steady-state economy*. W. H. Freeman.

Demsetz, H. (1967). Toward a theory of property rights. *The American Economic Review, 57*(2), 347–359.

di John, J., & Putzel, J. (2009). *Political settlements*. Issues Paper, Governance and Social Development Resource Centre (GSDRC), International Development Department, University of Birmingham.

Dietz, T., Dolsak, N., Ostrom, E., & Stern, P. C. (2002). Introduction. In E. Ostrom, T. Dietz, N. Dolsak, P. C. Stern, S. Stonich & E. U. Weber (Eds.), *The drama of the commons* (pp. 3–35). National Research Council, Committee on the Human Dimensions of Global Change.

Dixon, J. A., & Sherman, P. B. (1990). *Economics of protected areas—A new look at benefits and costs*. Inland Press.

Fairhead, J., Leach, M., & Scoones, I. (2012). Green grabbing: A new appropriation of nature? *The Journal of Peasant Studies, 39*(2), 237–261. https://doi.org/10.1080/03066150.2012.671770

Feeny, D., Hanna, S., & McEvoy, A. F. (1996). Questioning the assumptions of 'tragedy of the commons' model of fisheries. *Land Economics, 72*(2), 187–205. https://doi.org/10.2307/3146965

Feeny, D., Berkers, F., McCay, B. J., & Acheson, J. M. (1998). The tragedy of the commons: Twenty-two years later. In J. A. Baden & D. S. Noonan (Eds.), *Managing the commons*. Indiana University Press.

Felluga, D. (2002). *Modules on Marx: On fetishism*. Introductory guide to critical theory. Purdue U. Retrieved July 12, 2021, from http://www.purdue.edu/guidetotheory/marxism/modules/marxfetishism.html

Field, B. C. (1997). *Environmental economics: An introduction* (2nd ed.). McGraw Hill International Editions.

Foster, J. B. (2000). *Marx's ecology: Materialism & nature*. Monthly Review Press.

Foster, J. B. (2002). Marx's ecology in historical perspective. *International Socialism, 97*, 71–86.

Foster, J. B. ([1997] 2003). *Valuing nature? Ethics, economics and the environment*. Routledge.

Foster, J. B., & Clark, B. (2009). The paradox of wealth: Capitalism and ecological destruction. *Monthly Review, 61*(6).

Foster, J. B., & Holleman, H. (2014). The theory of unequal ecological exchange: A Marx-Odum dialectic. *The Journal of Peasant Studies, 41*(2), 199–233. https://doi.org/10.1080/03066150.2014.889687

Foster, J. B., Clark, B., & York, R. (2010). *The ecological rift*. Monthly Review Press.

Freeman, A. M., III., Herriges, J. A., & Kling, C. L. (2014). *The measurement of environmental and resource values: Theory and methods*. RFF Press.

Fulton, E. A., Smith, A. D. M., Smith, D. C., & van Putten, I. E. (2011). Human behaviour: The key source of uncertainty in fisheries management. *Fish and Fisheries, 12*(1), 2–17. https://doi.org/10.1111/j.1467-2979.2010.00371.x

Gaffard, J. L. (2008). Innovation, competition, and growth: Schumpeterian ideas within a Hicksian framework. *Journal of Evolutionary Economics, 18*(3), 295–311. https://doi.org/10.1007/s00191-008-0102-z

García-Llorente, M., Harrison, P. A., Berry, P., Palomo, I., Gómez-Baggethun, E., Iniesta-Arandia, I., Montes, C., del Amo, D. G., & Martín-López, B. (2018). What can conservation strategies learn from the ecosystem services approach? Insights from ecosystem assessments in two Spanish protected areas. *Biodiversity and Conservation, 27*, 1575–1597.https://doi.org/10.1007/s10531-016-1152-4

Georgescu-Roegen, N. (1971). *The entropy law and the economic process*. Harvard University Press.

Georgescu-Roegen, N. (1975). Energy and economic myths. *Southern Economic Journal, 41*(3), 347–381. https://doi.org/10.2307/1056148

Georgescu-Roegen, N. (1986). The entropy law and the economic process in retrospect. *Eastern Economic Journal, 12*(1), 3–25.

Gintis, H. (2000). Strong reciprocity and human sociality. *Journal of Theoretical Biology, 206*(2), 169–179. https://doi.org/10.1006/jtbi.2000.2111

Gintis, H., van Schaik, C., & Boehm, C. (2015). Zoon politikon: The evolutionary origins of human political systems. *Current Anthropology, 56*(3). https://doi.org/10.1086/681217

Givens, J. E., Huang, X., & Jorgenson, A. K. (2019). Ecologically unequal exchange: A theory of global environmental injustice. *Sociology Compass, 13*(5), e12693. https://doi.org/10.1111/soc4.12693

Gowdy, J. M. (1997). The value of biodiversity: Markets, society and ecosystems. *Land Economics, 73*(1), 25–41. https://doi.org/10.2307/3147075

Grafton, R. Q. (2000). Governance of the commons: A role for the state? *Land Economics, 76*(4), 504–517. https://doi.org/10.2307/3146949

Gramsci, A. (1979). *Selections from the political writings 1921–1926* (Translated and edited by Quintin Hoare). Lawrence and Wishart.

Greif, A., & Kingston, C. (2011). Institutions: Rules or equilibria? In N. Schofield & G. Caballero (Eds.), *Political economy of institutions, democracy and voting* (pp. 13–43). Springer-Verlag.

Gunningham, N., & Holley, C. (2010). *Bringing the 'R' word back: Regulation, environment protection and NRM*. ASSA.

Hanley, N., & Spash, C. L. (1993). *Cost-benefit analysis and the environment*. Edward Elgar.

Hanna, S. S., Folke, C., & Maler, K. (Eds.). (1996). *Rights to nature: Ecological, economic, cultural and political principles of institutions for the environment*. Island Press.

Hardin, G. (1968). The tragedy of the commons. *Science, 162*(3859), 1243–1248. https://doi.org/10.1126/science.162.3859.1243

Hodgson, G. M. (2014). *The economics of property rights is about neither property nor rights*. Hertfordshire Business School, University of Hertfordshire.

Holley, C., Gunningham, N., & Shearing, C. (2012). *The new environmental governance*. Earthscan.

Hornborg, A. (2010). Toward a truly global environmental history: A review article. *Review (Fernand Braudel Center), 33*(4), 295–323.

Hornborg, A., Clark, B., & Hermele, K. (Eds.). (2012). *Ecology and power: Struggles over land and material resources in the past, present, and future*. Routledge.

Hume, D. (1739). *A treatise of human nature*. Clarendon Press.

Hume, D. (1777). *Enquiry concerning the principles of morals*. Clarendon Press.

IPBES. (n.d). *Glossary: Indigenous peoples and local communities*. Intergovernmental Science-Policy Platform on Biodiversity and Ecosystem Services. Retrieved January 15, 2020 from https://ipbes.net/glossary/indigenous-peoples-local-communities

Jager, W., Janssen, M. A., De Vries, H. J. M., De Greef, J., & Vlek, C. A. J. (2000). Behavior in commons dilemmas: Homo economicus and homo psychologicus in an ecological–economic model. *Ecological Economics, 35*(3), 357–379. https://doi.org/10.1016/S0921-8009(00)00220-2

Jansen, M. A., & Jager, W. (2000). The human actor in ecological-economic models. *Ecological Economics, 35*(3), 307–310. https://doi.org/10.1016/S0921-8009(00)00215-9

Jorgenson, A. K. (2009). The sociology of unequal exchange in ecological context: A panel study of lower-income countries, 1975–2000. *Sociological Forum, 24*(1), 22–46. http://www.jstor.org/stable/40210335

Kalecki, M. (1972). *Selected essays on the economic growth of the socialist and the mixed economy*. Unwin.

Kelsall, T. (2018). Towards a universal political settlement concept: A response to Mushtaq Khan. *African Affairs, 117*(469), 656–669. https://doi.org/10.1093/afraf/ady018

Khan, M. H. (2000). *Rents, rent-seeking and economic development: Theory and evidence in Asia*. Cambridge University Press.

Khan, M. H. (2010). *Political settlements and the governance of growth-enhancing institutions*. School of Oriental and African Studies, University of London.

Khan, M. H. (2017). Political settlements and the analysis of institutions. *African Affairs, 117*(469), 636–655. https://doi.org/10.1093/afraf/adx044

Kirzner, I. M. (1997). Entrepreneurial discovery and the competitive market process: An Austrian approach. *Journal of Economic Literature, 35*(1), 60–85.

Kosoy, N., & Corbera, E. (2010). Payments for ecosystem services as commodity fetishism. *Ecological Economics, 69*, 1228–1236. https://doi.org/10.1016/j.ecolecon.2009.11.002

Kula, E. (1994). *Economics of natural resources, the environment and Policies* (2nd ed.). Chapman & Hall.

Laws, E. (2012). *Political settlements, elite pacts, and governments of national unity: A conceptual study*. Background Paper 10. Developmental Leadership Program.

Li, M. (2008). Climate change, limits to growth, and the imperative for socialism. *Monthly Review, 60*(3), 51–67.

Libecap, G. D. (1994). The conditions for successful collective actions. *Journal of Theoretical Politics, 6*(4), 563–592.

Lieberman, E. S. (2011). The perils of polycentric governance of infectious disease in South Africa. *Social Science & Medicine, 73*(5), 676–684. https://doi.org/10.1016/j.socscimed.2011.06.012

Locke, J. ([1689] 2003). *Two treatises of government* (Edited by Laslett, P.). Cambridge University Press.

Lovett, J. C., & Ockwell, D. G. (Eds.). (2010). *A handbook of environmental management*. Edward Elgar Publishing Limited.

Low, N., & Gleeson, B. (1998). *Justice, society and nature: An exploration of political ecology*. Routledge.

Lucas, J. (2015). What is the second law of thermodynamics? Retrieved March 15, 2020, from https://www.livescience.com/50941-second-law-thermodynamics.html

Mahoney, J. T. (2004). Property rights theory. In Economic Foundations of Strategy (pp. 109–137). SAGE Publications.

Mankiw, N. G. (2007). *Principles of microeconomics*. Akash Press.

Martinez-Alier, J. (1987). *Ecological economics: Energy, environment and society*. Blackwell Publishers.

Martinez-Alier, J., Munda, G., & O'Neill, J. (1998). Weak comparability of values as a foundation for ecological economics. *Ecological Economics, 26*(3), 277–286. https://doi.org/10.1016/S0921-8009(97)00120-1

Marx, K. (1867). *Capital* (Vol. I). International Publishers.

Marx, K. (1976). *Capital* (Vol. 1). Vintage.

Meinzen-Dick, R. S., Knox, A., & Gregorio, M. D. (Eds.). (2001). *Collective action, property rights and devolution of natural resource management: Exchange of knowledge and four implications for policy*. ZentralstellefürErnährung und Landwirtschaft.

Merchant, C. (1987). The theoretical structure of ecological revolutions. *Environmental Review, 11*(4), 265–274. https://doi.org/10.2307/3984135
Milner-Gulland, E. J. (2012). Interactions between human behaviour and ecological systems. *Philosophical Transactions of the Royal Society of London, Series B, 367*, 270–278. https://doi.org/10.1098/rstb.2011.0175
Mises, L. V. (1949). *Human action: A treatise on economics*. Ludwig Von Mises Institute.
Mitchell, R., & Carson, R. (1989). *Using surveys to value public goods: The contingent valuation method*. Resources for the Future.
Morrison, H. (2019). *Open access versus the commons, or steps towards developing commons to sustain open access*. Sustaining the Knowledge Commons. Retrieved May 12, 2021, from https://sustainingknowledgecommons.org/2019/04/23/open-access-versus-the-commons-or-steps-towards-developing-commons-to-sustain-open-access/
Nadeau, R., & Cleveland, C. J. (2007). Environmental and ecological economics. In C. J. Cleveland (Ed.), *Encyclopedia of earth*. Environmental Information Coalition, National Council for Science and the Environment.
Neumayer, E. (2003). Weak versus strong sustainability: Exploring the limits of two opposing paradigms. Edward Elgar.
North, D. C. (1990). Institutions, institutional change and economic performance. Cambridge University Press.
North, D. C. (1991). Institutions. *The Journal of Economic Perspectives, 5*(1), 97–112. https://doi.org/10.1257/jep.5.1.97
North, D. C. (2008). Institutions and the performance of economies over time. In C. Menard & M. M. Shirley (Eds.), *Handbook of new institutional economics* (pp. 21–30). Springer-Verlag.
Odum, H. T. (1988). Self-organization, transformity, and information. *Science, 242*(4882), 1132–1139. https://doi.org/10.1126/science.242.4882.1132
Odum, H. T., & Arding, J. E. (1991). Energy analysis of shrimp mariculture in Ecuador. *Working Paper*. Coastal Resources Center, University of Rhode Island, Narragansett.
O'Neil, D. W. (2012). Measuring progress in the degrowth transition to a steady state economy. *Ecological Economics, 84*, 221–231. https://doi.org/10.1016/j.ecolecon.2011.05.020
Ostrom, E. (1990 [2015]). *Governing the commons: The evolution of institutions for collective action*. Cambridge University Press.
Ostrom, E. (1998). Reflections on the commons. In J. A. Baden & D. S. Noonan (Eds.), *Managing the commons*. Indiana University Press.
Ostrom, E. (2000a). *Private and common property rights*. Center for the Study of Institutions, Population and Environmental Change, Indiana University.
Ostrom, E. (2000b). Collective action and the evolution of social norms. *Journal of Economic Perspectives, 14*(3), 137–158. https://doi.org/10.1257/jep.14.3.137
Ostrom, E. (2003). How types of goods and property rights jointly affect collective action. *Journal of Theoretical Politics, 15*(3), 239–270. https://doi.org/10.1177/0951692803015003002
Ostrom, E. (2009). Collective action theory. In C. Boix & S. C. Stokes (Eds.), *The Oxford handbook of comparative politics*. https://doi.org/10.1093/oxfordhb/9780199566020.003.0008
Ostrom, E., & Hess, C. (2010). Private and common property rights. In B. Bouckaert (Ed.), *Property law and economics (encyclopedia of law and economics)*. Edward Elgar.
Ostrom, E., Dietz, T., Dolsak, N., Stern, P. C., Stonich, S., & Weber, E. U. (Eds.). (2002). *The drama of the commons*. National Academy Press.
Paulson, S., & Gezon, L. L. (Eds.). (2004). *Political ecology across spaces, scales, and social groups*. Rutgers University Press.
Pearce, D. (Ed.). (1991). *Blueprint2: Greening the world economy*. Earthscan.
Pearce, D. (2002). An intellectual history of environmental economics. *Annual Review of Energy and the Environment, 27*, 57–81. https://doi.org/10.1146/annurev.energy.27.122001.083429
Pearce, D., Markandya, A., & Barbier, E. (1989). *Blueprint for green economy*. Earthscan.
Peet, R., & Watts, M. J. (Eds.). (1996). *Liberation ecologies: Environment, development, social movements*. Routledge.

Pigou, A. ([1920] 2013). *The economics of welfare*. Palgrave Macmillan.

Po, J. Y. T., Ville, A. S. S., Rahman, H. M. T., & Hickey, G. M. (2019). On institutional diversity and interplay in natural resource governance. *Society and Natural Resources, 32*(12), 1333–1343. https://doi.org/10.1080/08941920.2019.1667463

Poteete, A. (2012). Levels, scales, linkages, and other 'multiples' affecting natural resources. *International Journal of the Commons, 6*(2), 134–150. https://doi.org/10.18352/ijc.318

Pretty, J., & Smith, D. (2004). Social capital in biodiversity conservation and management. *Conservation Biology, 18*(3), 631–638.

Putnam, R. D. (1993). *Making democracy work*. Princeton University Press.

Rademaekers, K., van der Laan, J., Smith, M., van Breugel, C., & Pollitt, H. (2011). *The role of market-based instruments in achieving a resource efficient economy*. European Commission.

Rahman, H. M. T., Hickey, G. M., & Sarker, S. K. (2012). A framework for evaluating collective action and informal institutional dynamics under a resource management policy of decentralization. *Ecological Economics, 83*(C), 32–41. https://doi.org/10.1016/j.ecolecon.2012.08.018

Rahman, H. M. T., Ville, A. S. S., Song, A. M., Po, J. Y. T., Berthet, E., Brammer, J. R., Brunet, N. D., Jayaprakash, L. G., Lowitt, K. N., & Rastogi, A. (2017). A framework for analyzing institutional gaps in natural resource governance. *International Journal of the Commons, 11*(2), 823–853. https://doi.org/10.18352/ijc.758

Rakodi, C. (2002). A livelihoods approach-Conceptual issues and definitions. In C. Rakodi & T. Llyod-Jones (Eds.), *Urban livelihoods: A people-centred approach to reducing poverty*. Earthscan Publications.

Rastogi, A., Hickey, G. M., Badola, R., & Hussain, S. A. (2014). Understanding the local socio-political processes affecting conservation management outcomes in Corbett Tiger Reserve, India. *Environmental Management, 53*(5), 913–929. https://doi.org/10.1007/s00267-014-0248-4

Remer, T. L. (2012). *Do stronger collective property rights improve household welfare? Evidence from a field study in Fiji*. Graduate Program in International Affairs, New School University.

Robbins, P. (2012). *Political ecology: A critical introduction*. Wiley-Blackwell.

Roma, A. (2006). Energy, money, and pollution. *Ecological Economics, 56*(4), 534–545.

Ropke, I. (2004). The early history of modern ecological economics. *Ecological Economics, 50*(3–4), 293–314. https://doi.org/10.1016/j.ecolecon.2004.02.012

Rose, J., & Carr, A. (2018). Political ecologies of leisure: A critical approach to nature-society relations in leisure studies. *Annuals of Leisure Research, 21*(3), 265–283. https://doi.org/10.1080/11745398.2018.1428110

Sagoff, M. (2012). Environmental economics. *Encyclopedia of Applied Ethics*, 97–104.https://doi.org/10.1016/B978-0-12-373932-2.00354-9

Scarlett, L., Boyd, J., & Brittain, A. (2013). *Catalysts for conservation: Exploring behavioral science insights for natural resource investments*. RFF Report. Resources For the Future.

Schlager, E., & Ostrom, E. (1992). Property rights regimes and natural resources: A conceptual analysis. *Land Economics, 68*(3), 249–262. https://doi.org/10.2307/3146375

Schlüter, M., Mcallister, R. R. J., Arlinghaus, R., Bunnefeld, N., Eisenack, K., Hoelker, F., Milner-Gulland, E. J., & Mueller, B. (2012). New horizons for managing the environment: A review of coupled social-ecological systems modeling. *Natural Resource Modeling, 25*(1), 219–272.

Siebenhuner, B. (2000). *Homo sustinens*- Towards a new conception of humans for the science of sustainability. *Ecological Economics, 32*(1), 15–25. https://doi.org/10.1016/S0921-8009(99)00111-1

Skoggard, I. (2016). *Human sociality and cooperative behavior: A cross-cultural Approach*. Retrieved November 12, 2019, from https://hraf.yale.edu/human-sociality-and-cooperative-behavior-a-cross-cultural-approach/

Smith, A. (1776). *An inquiry into the nature and causes of the wealth of nations*. W. Strahan and T. Cadell.

Smith, R. (2010). Beyond growth or beyond capitalism? *Real-World Economics Review*, (53).

Solow, R. M. (1974). Intergenerational equity and exhaustible resources. *The Review of Economic Studies, 41*(5), 29–45. https://doi.org/10.2307/2296370
Stamate, A., & Musetescu, R. (2013). A short critique of perfect competition model from the perspective of Austrian school of economics. *Romanian Economic and Business Review, 6*(4), 112–122.
Stavins, R. (1989). Harnessing market forces to protect the environment. *Environment: Science and Policy for Sustainable Development, 31*(1), 5–35. https://doi.org/10.1080/00139157.1989.9929926
Stavins, R. (2003). Experience with market-based environmental policy instruments. In K. G. Mäler & J. R. Vincent (Eds.), *Handbook of environmental economics* (Vol. 1, pp. 355–435). Elsevier.
Stavins, R. N. (2011). The problem of the commons: Still unsettled after 100 years. *The American Economic Review, 101*(1), 81–108. https://doi.org/10.1257/aer.101.1.81
Stavins, R., & Whitehead, B. (1992). Dealing with pollution: market based incentives for environmental protection. *Environment, 34*(7), 7–11, 29–42.
Sullivan, S. (2013). Banking nature? The spectacular financialisation of environmental conservation. *Antipode, 45*(1), 198–217. https://doi.org/10.1111/j.1467-8330.2012.00989.x
Sullivan, S. (2016). What's ontology got to do with it: Nature, knowledge and 'the green economy.' *Journal of Political Ecology, 24*(1), 217–242. https://doi.org/10.2458/v24i1.20802
Tefera, M., Demel, T., Hultén, H., & Yonas, Y. (2005). The role of communities in closed area management in Ethiopia. *Journal of Mountain Research and Development, 25*(1), 44–50. https://doi.org/10.2307/3674368
Tietenberg, T. H. (1990). Using economic incentives to maintain our environment. *Challenge, 33*(2), 42–46. https://doi.org/10.1080/05775132.1990.11471412
Tietenberg, T. (2002). The tradable permits approach to protecting the commons: What have we learned? In E. Ostrom, T. Dietz, N. Dolsak, P. C. Stern, S. Stonich & E. U. Weber (Eds.), *The drama of the commons* (pp. 197–132). National Research Council, Committee on the Human Dimensions of Global Change.
Tietenberg, T., & Lewis, L. (2012). *Environmental & natural resource economics* (9th ed.). Pearson Education Inc.
Titumir, R. A. M. (2015). 'Economic growth, social development need democratic public order'. *The New Age.* Retrieved June 20, 2022 from http://archive.newagebd.net/184900/economic-growth-social-development-need-democratic-public-order/
Titumir, R. A. M., Paran, M. S., & Pash, M. W. (2019). "The Sundarbans is our mind": An exploration into multiple values of nature in conversation with traditional resource users. In Satoyama Initiative Thematic Review (SITR) (Vol. 5). United Nations University Institute for the Advanced Study of Sustainability (UNU-IAS)
Titumir, R. A. M. (2021). *Numbers and narratives in Bangladesh's economic development.* Palgrave Macmillan. https://doi.org/10.1007/978-981-16-0658-8
Titumir, R. A. M., Afrin, T., & Islam, M. S. (2020). Traditional knowledge, institutions, and human sociality in sustainable use and conservation of biodiversity of the Sundarbans of Bangladesh. In O. Saito, S. M. Subramanian, S. Hashimoto & K. Takeuchi (Eds.), *Mapping socio-ecological production landscapes, stakeholders, policy and action.* Science for Sustainable Societies, Integrated Research System for Sustainability Science (IR3S), University of Tokyo.
Tummolini, L. (2013). Social conventions. In B. Kaldis (Eds.), *Encyclopedia of philosophy and the social sciences* (pp. 900–902). SAGE Publications Inc. https://doi.org/10.4135/9781452276052
Vatn, A. (2000). The environment as a commodity. *Environmental Values, 9*(4), 493–509. https://doi.org/10.3197/096327100129342173
Vatn, A. (2005). *Institutions and the environment.* Edward Elgar.
Vink, N. (1986). *An institutional approach to livestock development in Southern Africa.* Doctoral Dissertation Report. University of Stellenbosch, Stellenbosch.
Von Hayek, F. A. (1948). The meaning of competition. In F. A. Von. Hayek (Ed.), *Individualism and economic order.* The University of Chicago Press.

Walker, J. M., Grander, R., Herr, A., & Ostrom, E. (2000). Collective choice in the commons: Experimental results on proposed allocation rules and votes. *The Economic Journal, 110*, 212–234.

Waters, T., & Waters, D. (2015). Max Weber's sociology in the twenty-first century. In T. Waters & D. Waters (Eds.), *Weber's rationalism and modern society*. Palgrave Macmillan.

Watson, E. (2001). *Inter institutional alliances and conflicts in natural resource management: Preliminary research findings from Borana, Oromia Region, Ethiopia.* Working Article No. 4. Marena Research Project.

Weber, M. (1919 [2015]). *Weber's rationalism and modern society* (Trans. and Eds. By Tony Waters and Dagmar Waters). Palgrave Macmillan. (as "Poltics as Vocation").

Williamson, O. (1979). Transaction-cost economics: The governance of contractual relations. *Journal of Law and Economics, 22*(2), 233–261.

Williamson, O. (1986). Transaction-cost economics: The governance of contractual relations. In J. Barney & W. Ouchi (Eds.), *Organizational economics* (pp. 98–129). Jossey-Bass.

Wolf, E. (1972). Ownership and political ecology. *Anthropological Quarterly, 45*(3), 201–205.

Wyborn, C. (2015). Cross-scale linkages in connectivity conservation: Adaptive governance challenges in spatially distributed networks. *Environmental Policy and Governance, 25*(1), 1–25. https://doi.org/10.1002/eet.1657

Yami, M., Vogl, C., & Hausera, M. (2009). Comparing the effectiveness of informal and formal institutions in sustainable common pool resources management in Sub-Saharan Africa. *Conservation and Society, 7*(3), 153–164. https://doi.org/10.4103/0972-4923.64731

York, R., & Clark, B. (2010). Critical materialism: Science, technology, and environmental sustainability. *Sociological Inquiry, 80*(3), 475–499. https://doi.org/10.1111/j.1475-682X.2010.00343.x

Young, S. (2013). Transaction cost economics. In S.O. Idowu, N. Capaldi, L. Zu & A.D. Gupta (Eds.), *Encyclopedia of corporate social responsibility*. Springer. https://doi.org/10.1007/978-3-642-28036-8_221

Zhang, B. (2013). Market-based solutions: An appropriate approach to resolve environmental problems. *Chinese Journal of Population Resources and Environment, 11*(1), 87–91. https://doi.org/10.1080/10042857.2013.777526

Chapter 3
Biodiversity Resources: Degradation, Restoration and Sustainable Conservation

3.1 Introduction

The contemporary era is characterised by biodiversity crises with diverse species facing imminent risk of extinction across the world. The objective of this chapter is to examine the underlying causes of the rapid and enormous reduction of biodiversity resources by illustrating the case of Bangladesh in general and that of the Sundarbans in particular, in the context of global discussions as the latter is an example of a diverse ecosystem. The chapter, in this regard, identifies the drivers of the degradation of biodiversity resources through the lens of political economy, considering the unique context of developing countries. Moreover, it provides an alternative conceptualisation of the measures for sustainable utilisation, conservation and restoration of those resources based upon the interface between human beings and nature.

The United Nations Convention on Biological Diversity (UN-CBD) defines biodiversity as: "the variability among living organisms from all sources, including, inter alia, terrestrial, marine and other aquatic ecosystems and the ecological complexes of which they are a part; this includes diversity within species, between species and of ecosystems" (United Nations, 1992, p. 3). A precise and operational definition of biodiversity, however, remains obscure, as it is a series of overlapping public goods from the local to the global scale (Helm & Hepburn, 2012). The biodiversity resources are of critical importance to the health of ecosystems and the sustainability of human settlements (Gagné et al., 2019).

Nevertheless, it is well documented that these resources have been disappearing expeditiously in the form of degraded ecosystems and their corresponding services in many parts of the world posing a threat to humanity's future (Higgins et al., 2013; IPBES, 2019). There is clear evidence that economic growth causes the loss of biodiversity, but the majority of international biodiversity and sustainability policies advocate economic growth (Otero et al., 2020). The three most important direct

The chapter, mainly the empirical analysis sections, draws on the authors' previous work (Titumir et al., 2020).

R. A. M. Titumir et al., *Natural Resource Degradation and Human-Nature Wellbeing*,
https://doi.org/10.1007/978-981-19-8661-1_3

drivers of recent global anthropogenic biodiversity loss are identified as land/sea use change, direct resource exploitation and pollution (Jaureguiberry et al., 2022). Resource extraction in a selfish manner, in fact, endangers societies and livelihoods while causing a serious distortion in the ecosystem (Battersby, 2017). The situation has been acutely felt in the face of the COVID-19 pandemic as it has exposed that it is more than an ethical commitment for humanity to conserve biodiversity (WWF, 2020). It is evident that newer types of diseases are emerging and are likely to emerge more in future because of hunting, selling, illegal poaching and eating of wild animals and destroying their habitats through illegal encroachment and deforestation (Grandcolas & Justine, 2020; Vidal, 2020). On the other side, it is also a matter of concern whether the countries are now concentrating on the protection of biodiversity resources, or whether they should address the issue of economic recovery. Overall, the pandemic, its subsequent lockdown and the post-pandemic urgency to return to the new normal all would have critical implications for biodiversity conservation (Bang & Khadakkhar, 2020).

Developing countries around the world, however, lie in the greatest peril due to the consequences of the loss of biodiversity resources and Bangladesh is no exception. The problem had become more acute in the context of the pandemic. The poverty situation has been deteriorating, accompanied by increased food insecurity and loss of livelihood opportunities (Irfanullah, 2020) like many other poor but biodiverse regions of the world (Lu, 2021). These factors might push poor people to rely more on natural resource harvesting, resulting in the overexploitation of biodiversity resources and further degradation of different ecosystems. Against this backdrop, the key requirements arising at both the national and global levels are to identify the underlying causes of degradation as well as to devise appropriate policy measures to minimise degradation and ensure the restoration and sustainable use of these resources.

Various schools of thought have yet to identify answers for sustainable biodiversity resource management that can result in a long-term conservation process, secured livelihood options for stakeholders and a balanced environment (Titumir et al., 2020). The current theoretical approach to understanding the dynamics of biodiversity resource usage and management is largely dominated by a variety of neoclassical traditions, including environmental economics and institutional economics. The argument of this market-centric approach is that biodiversity resources degrade predominantly due to the absence of market and negative externality (Perrings et al., 1992; Sadmo, 2015). It further contends that pricing-based valuation methods can offer useful insights to support policy initiatives by measuring the economic value of the resources and developing an exchange rule connected with the protection of biological resources (e.g., Barbier, 2009; Bräuer, 2003; Costanza et al., 1997; Dasgupta, 2021; Kumar, 2010; McAfee & Shapiro, 2010; Pearce, 2001).

An exchange based on such economic valuation of ecosystem services, however, remains controversial (e.g., Gómez-Baggethun & Ruiz-Pérez, 2011; Kosoy & Corbera, 2010; Neuteleers & Engelen, 2015). The key underlying problem is the undervaluation of biodiversity resources by putting a price tag on them, which makes them lucrative for maximising profit by downplaying their inherent ecological values.

The institutional economics, additionally, emphasised establishing a formal property rights regime or sustaining an appropriate institutional arrangement that can efficiently manage biodiversity resources (Ituarte-Lima et al., 2014; Muradian & Gómez-Baggethun, 2013; Ostrom, 2000; Vatn, 2010). Nevertheless, it could not sufficiently explain the causes behind the instability of institutions, particularly in a transitional economy. Moreover, it places less emphasis on informal institutions in managing such resources. This book acknowledges that institutions (formal and informal) matter in the management of biodiversity resources. Simultaneously, it argues that politico-economic factors should also be considered in this regard, as the factors have significant impacts on the process of usage, management, conservation, protection and distribution of resources. Finally, the most important missing point in the existing theoretical understanding (in both market-centric and political economy approaches) is the lack of recognition of the relationship between nature and human beings. Indeed, the dynamic aspects of human behaviour are frequently overlooked in biodiversity resource management (Bieg et al., 2017) although only recently it drew attention to how human behaviour and ecological systems interact (Nyborg et al., 2016; Schill, 2017; World Bank, 2015). This chapter has attempted to formulate a new theoretical construct in view of these issues and the stated hypotheses under each proposition have been tested considering the case study of the Sundarbans.

The Sundarbans, known as the lungs of Bangladesh, is an important case of an ecologically critical area in terms of biodiversity resource degradation. Several studies have indicated already that the Sundarbans' resources have been steadily declining (e.g., Aziz & Paul, 2015; Giri et al., 2007; Gopal & Chauhan, 2006; Iftekhar & Islam, 2004; Islam, 2014; Rahman et al., 2010a; Sarker et al., 2016; Uddin et al., 2013). Most of the studies, however, have fallen short to establish a sound theoretical foundation for analysing the underlying reasons for forest degradation, let alone proposing an alternative conservation framework (Titumir et al., 2020).

Considering this gap, the chapter demonstrates through theoretical constructs and empirical evidence that fragile institutions, lax regulatory regimes, trends of political settlements and unequal power-sharing arrangements in Bangladesh cause the dissipation of natural resource rent resulting in the degradation of biodiversity in the Sundarbans. The chapter also exhibits that Indigenous Peoples and Local Communities (IPLCs) living adjacent to the forest area have been practising different livelihood strategies based upon traditional knowledge that can significantly contribute to the sustainable management and conservation of biodiversity resources through reciprocal human–nature relationships. Hence, the alternative framework developed in this chapter posits that the sustainability of the Sundarbans fundamentally hinges on human sociality constructed by societal norms and values along with formal institutions.

3.2 State of Biodiversity Resources in Bangladesh

Bangladesh—a developing country in South-Asia—is one of the ten global hotspots for biodiversity having nearly 7000 endemic plant species, 1952 species of invertebrates, 113 mammals, 628 birds, 126 reptiles, 22 amphibians, 708 species of fish and 2493 species of insects (BFD, 2016a). The biodiversity resources of the country can be broadly categorised into four types based on the nature of the ecosystem (Fig. 3.1). The current state of such biodiversity resources can be presented using the Driving Forces-Pressures-State-Impacts-and-Response (DPSIR) framework (Box 3.1).

Box 3.1 DPSIR Framework

The Driving Forces-Pressures-State-Impacts-Response (DPSIR) framework—developed by the Organisation for Economic Cooperation and Development (OECD)—is one type of systems approach that can be used to analyse how society is using natural resources and the various implications of this use (IRP, 2019). The components of the framework are:

- **Driving forces**: The forces that act indirectly on the environment such as social and economic development.
- **Pressures**: Human activities that directly affect the environment.
- **State**: The observable change in the environment including trends.
- **Impacts**: Impacts may be environmental, social and economic, contributing to the vulnerability of people.
- **Responses**: Responses indicate the problem solving attempts by society in case of environmental management.

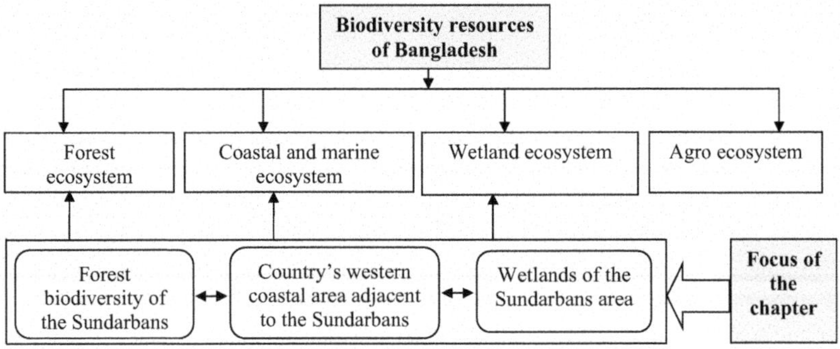

Fig. 3.1 Taxonomy of biodiversity of Bangladesh. *Source* Prepared by the authors

3.2.1 Forest Biodiversity

A government estimate shows that the area under forests in Bangladesh is 2.6 million ha, representing approximately 17.5% of the total geographic area of the country (BFD, 2017) but there are other estimates as well which signify less coverage of the forest area. For instance, according to World Bank statistics, the total forest cover of the country is 1.47 million ha (11% of the total land area) (World Bank, 2018). Despite the existence of diverse statistics on the precise spatial coverage of the forest, it is apparent that the forest cover has been dwindling gradually.

A study considering a period of eight decades has reported that forests in Bangladesh covered an area of 23,140 km^2 in 1930 which dwindled to 14,086 km^2 in 2014 signifying a net loss of 9054 km^2 (39.1%) (Reddy et al., 2016). Accordingly, the percentage of the forest of the total land area of the country signifies a decreasing trend from 15.7% in 1930 to 9.5% in 2014 (Fig. 3.2a). Moreover, the trend of deforestation continues showing an increasing trend in recent times in spite of several conservation policies (Fig. 3.2b). The highest net annual rate of deforestation (calculated as the gross rate of deforestation minus the rate of afforestation) was 0.75% during 2006–2014.

The major forest areas which are under the jurisdiction of the country's Forest Department (FD) are of three types: hill forests, plain land *sal* forests and mangrove forests representing 10.54% of the total land area of the country (Table 3.1) and supporting a diverse species of flora and fauna. A brief description has been provided below on the key attributes and the present state of these three types of forest based on different sources (see BFD, 2016a, 2021; DoE, 2010, 2015; MoEF, 2001 for details).

Tropical evergreen and semi-evergreen hill forests cover an area of 670,000 ha in the eastern parts of the country, including the divisions of Chittagong, Chittagong Hill Tracts, Cox's Bazar and Sylhet, accounting for 4.54% of the country's total landmass and 44% of the national forest land.

Some of the prominent plant species of this forest are garjan (*Dipterocarpus* spp.), chapalish (*Artocarpus chaplasha*), jarul (*Legarstromia speciosa*), gamar (*Gmelina arborea*), koroi (*Albizzia* spp), civit (*Swintonia floribunda*), telsur (*Hopea odorata*), chandul (*Tetramelesnudi flora*), boilam (*Anisoptera scaphula*), jam (*Palaquium polyanthrum* sp.), rata (*Calophyllum polyanthum Mangifera*), tali (*P. polyanthrum*), champa (*Michelia champaca*), shimul (*Salmalia insignis*) and different species of bamboos, canes and ferns. It also supports a large number of wild mammals, including Asian elephant (*Elephas maximus*), Asiatic black bear (*Ursus thibetanus*), hoolock gibbon (*Hoolock hoolock*), samvar (*Cervus unicolor*), Indian leopard (*Panthera pardus*), monkey (*Macaca mulatta*) and wild dog. Among the reptiles, king cobra (*Ophiophagus hanna*), monitor lizard (*Varanus salvator*) and Bengal monitor lizard (*Varanus bengalensis*) are also available in the forest.

The natural features of the forest, however, have been undergoing massive alterations due to excessive human interventions. A majority of the areas have already been converted to plantations under various development projects since 1871, but some remnants of native forests still exist. Fundamental causes of the shrinking of the

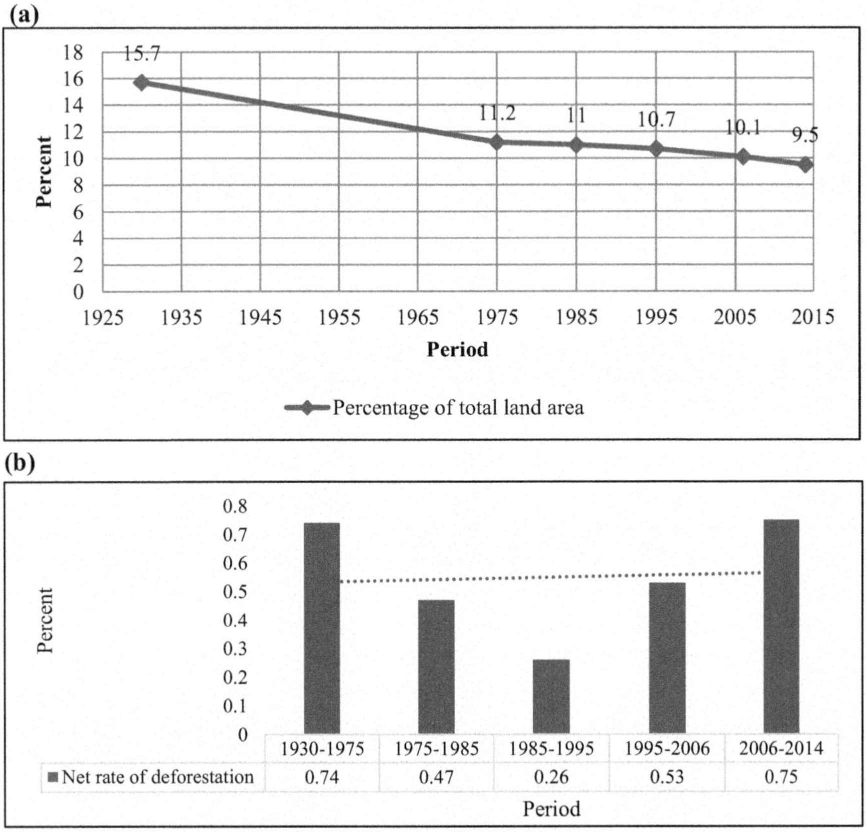

Fig. 3.2 **a** Percentage of forest of total land area of Bangladesh (1930–2014); **b** net rate of deforestation (1930–2014). *Source* Prepared by the authors based on data collected from Reddy et al. (2016)

Table 3.1 Area of forests under the forest department

Forest type	Area (million ha)	Share of total land (%)
Hill forest	0.67	4.65
Natural mangrove forest	0.60	4.07
Mangrove plantation	0.14	0.97
Sal forest	0.12	0.83
Total	1.53	10.54

Source DoE (2015)

hill forest are identified as encroachment by migrants; cutting of trees for the timber industry; and the ineffective and corrupt role of the Forest Department (Salam et al., 1999). Some locations of this forest area have been declared as Protected Areas (PAs) to conserve the resources. These are Pablakhali Wildlife Sanctuary, Kaptai National Park and Sangu Valley Wildlife Sanctuary.

Tropical moist deciduous *sal* forest covers an area of 120,000 ha (0.81% of the country's total land) in the central and northern districts. The single dominating plant species is sal (*Shorea robusta*). Other commercially valuable tree species are koroi (*Albizzia procera*), azuli (*Dillenia pentagyna*), sonalu (*Cassia fistula*), bohera (*Terminalia belerica*), haritaki (*Terminalia chebula*), kanchan (*Bauhinia acuminata*), kaika (*Adina cardifolia*), amlaki (*Phyllanthus emblica*), palash (*Butea frondosa)*, chambal (*A. chaplasha*), etc. Once this forest area used to be very diverse in terms of faunal diversity. In the last century, the forest was home to the Indian rhinoceros, but now only five species of globally threatened wildlife species can be found in this forest. However, it has been identified as home to the largest population of capped langur in Bangladesh.

The National Forest Resources Assessment of 2007 reported that the area of *sal* forests was 34,000 ha in Bangladesh (Altrell et al., 2007). Between 1989 and 2007, 14% of this forest area was reduced (DoE, 2010). On the contrary, a recent report shows that only about 17,500 ha of the original *sal* forest exist (BFD, 2016b). Indeed, this forest has diminished in terms of both tree density and stand quality (Hossain, 2009). Some of the causes behind this change are the overexploitation of timber, fuel wood, bark tannin and animal fodder; illegal encroachment and poaching; expansion of agriculture and urbanisation, etc. (Rahman et al., 2010b). Moreover, the area has also been used for rubber monoculture and the plantation of exotic commercial fuel-wood species. Under such circumstances, the participatory management strategy has been adopted in three divisions of *sal* forests in Dhaka, Tangail and Mymensingh districts in order to conserve it (Alam et al., 2008; Hossain, 2009). Furthermore, both the Bhawal National Park and the Madhupur National Park have been declared PAs.

The mangrove forests in Bangladesh can be categorised into three zones based on their location and origin (natural or man-made): (a) the Sundarbans (b) the *Chakaria* Sundarbans and (c) the coastal mangrove plantation. The Sundarbans is the richest in terms of the country's biodiversity resources, but the area is degrading in terms of both quantity and quality (see "Biodiversity Degradation of the Sundarbans: A Micro-Case Study" section for details). The *Chakaria* Sundarbans is the second-largest mangrove forest zone situated in Cox's Bazar covering an area of 85.10 sq. km (Hossain et al., 2004). Over the last century, however, both locals and outsiders have been haphazardly exploiting the natural resources leading to its complete destruction. It is evident that the area of mangrove forest and agricultural land have lost out to salt cultivation and shrimp culture over time (Table 3.2).

Another portion of the mangrove forest has been initiated through coastal mangrove plantations (Altrell et al., 2007). The plantations have been practised on the newly accreted char lands.

According to the discussion, the forest ecosystem, on the whole, remains in a critical condition although some initiatives have been taken at the government level

Table 3.2 Changes in land use (in hectare) pattern over time in *Chakaria* Sundarbans

Land use	Years			
	1974	1984	1994	2012
Mangrove forest	6127	3048	123	170
Agriculture	7847	6556	2832	1582
Salt pan	432	1484	4363	5078
Shrimp culture	0	3456	5583	4601
Water bodies	3096	2484	2463	3308
Accreted land	1180	1254	1949	2113
Bare land	136	421	1350	1680
Reserve (hill) forest	2650	2677	2625	2646
Settlement	422	510	602	712
Total area	21,890	21,890	21,890	21,890

Source Rahman and Hossain (2015)

(Fig. 3.3). The overall situation can be also understood in terms of a recent government estimate, which shows that over 1.60 lakh individuals and organisations have grabbed forestland of 2.57 lakh acres illegally across the country for various purposes including housing, agriculture and industrialisation (The Daily Star, 2021). The estimate also denotes, in particular, that 88,215 individuals and organisations occupied 1.38 lakh acres of reserved forests. It signifies that the forest areas of the country are under immense pressure because of human intervention.

3.2.2 *Coastal and Marine Biodiversity*

Bangladesh is located at the head of the Bay of Bengal. Bay of Bengal is a northern extended arm of the Indian Ocean occupying an area of about 2.2 million sq km (Banglapedia, 2021). The coastal zone along with the Bay of Bengal covers 32% of the country and the maritime boundary of Bangladesh has been extended in recent times by 118,813 sq. km (DoE, 2015). The area provides diverse types of biodiversity resources, and the identified number of species varies across different studies (Table 3.3). The biological resources of this ecosystem comprise not only fish but also crustaceans, elasmobranches and cetaceans. In particular, it supports more than ten globally threatened migratory shorebirds, including spoon-billed sandpiper, Asian dowitcher, spotted redshank, goliath heron and Indian skimmers (DoE, 2015).

The resources of this ecosystem, though diverse in nature, are available across various locations. The St. Martin's island in particular nurtures and supports diverse species of fauna among which fish constitutes the highest number followed by seaweed (Table 3.4). Other important faunal species are mollusk, coral, algae, crabs

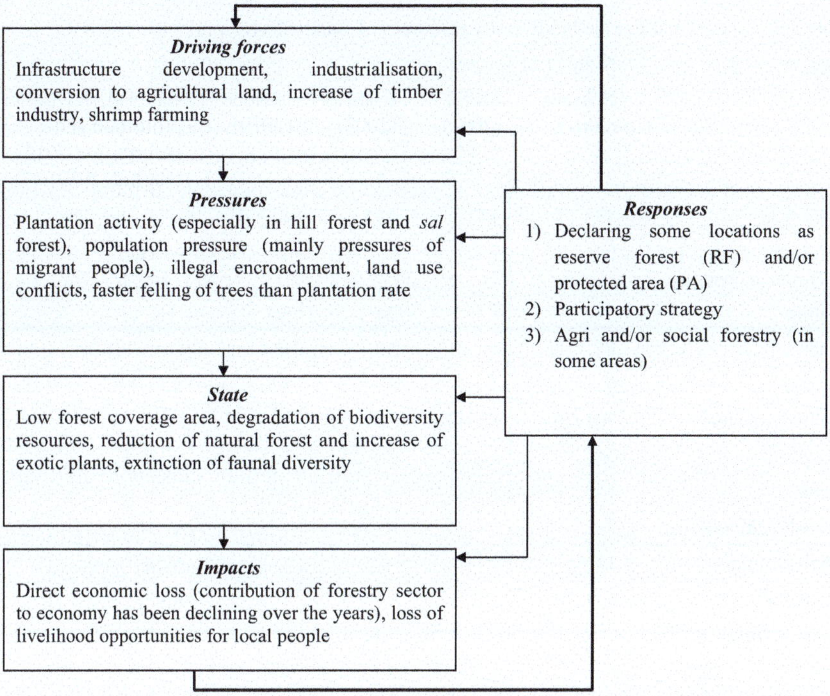

Fig. 3.3 State of forest ecosystem biodiversity of Bangladesh. *Source* Prepared by the authors

and lobsters. About 66 species of corals from 22 generations have been identified at this location.

The coastal ecosystem, however, remains under both natural and human-induced burdens such as socio-economic drivers, institutional drivers and climate change (Behera & Haider, 2012). A significant number of species have already been identified as vulnerable (Table 3.5).

The specific anthropogenic factors that have a negative impact on the marine biodiversity are indiscriminate catching of juveniles, siltation, extensive use of pesticides, industrial and oil pollution, ship breaking industry and discharging of municipal waste (Quader, 2010). Moreover, the rapid growth of the tourism industry, construction of hotels, collection of sand and stone, etc. have a harmful effect on this ecosystem. Finally, the sea level rising has compromised the ecological stability of the coastal zone and the marine ecosystem (Behera & Haider, 2012; Minar et al., 2013).

The government has taken an Integrated Coastal Zone Management (ICZM) policy in order to protect the coastal region. The policy attempts to rationalise and coordinate the environmental and development initiatives taking place in the coastal zone more effectively. Moreover, some locations of the coastal and marine areas have been declared as PAs (e.g., *Nijhumdweep, Himchari, Teknaf, Swatch of No Ground*).

Table 3.3 Coastal and marine biodiversity resources of Bangladesh

Category	Number of species				
	Hossain (2001)	MoEF (2001)	Islam (2003)	Quader (2010)	Bangladesh wildlife protection and security Act 2012 (Rashid, 2019)
Bony fish	475	442	475	442	–
Cartilaginous (soft-boned) fish	50	–	–	–	–
Shrimp	25	–	24	56	–
Crab	15	–	50	16*	–
Lobster	5	3	–	3	22
Mollusk (oyster)	301 (6)	336 (7)	301 (3)	336	–
Algae/seaweed	56	–	20–22	168	–
Coral	13	–	–	66	–
Starfish/echinoderms	3	–	–	–	–
Whale/dolphin	11	–	–	–	6/14
Shark and ray	–	–	–	–	30
Squids (cuttlefish)	–	–	7 (2)	–	–
Reptiles	–	17	–	17	–
Mammals	–	3	–	–	–
Sponges	–	–	–	3	–
Amphibians	–	–	–	22	–
Birds	–	–	–	628	–

*Both marine and freshwater
Source Authors' compilation from different sources

Table 3.4 Faunal species of St. Martin's island

Name of the species	Number of species
Fish	234 (of which 98 are coral associated)
Mollusk	187
Crabs	7
Coral	66 (of which 19 are fossil coral and 36 living coral)
Algae	14
Lobsters	3
Seaweed	200

Source Prepared based on Quader (2010)

Table 3.5 Vulnerable marine biodiverse species

Category of species	Name of vulnerable species	State of vulnerability
Fish	Yellow tailed pangas (*Pangasius pangasius*)	Endangered
	Anguilla bengalensis, *Plotosus canius*, *Carcharhinus limbatus* and *C. melanopterus*	Vulnerable
Reptiles	Hawksbill turtle (*Eretmochelys imbricate*), green turtle (*Chelonia mydas*), leatherback turtle (*Dermochelys coriacea*)	Critically endangered
	loggerhead turtle (*Caretta caretta*) and olive ridley turtle (*Lepidochelys olivacea*)	Vulnerable
Mammals	Irrawaddy dolphin (*Orcaella brevirostris*), Bryde's whale (*Balaenoptera edeni*), sperm whale (*Physeter microcephalus*) and finless porpoise (*Neophocaena phocaenoides*)	Vulnerable

Source Prepared based on Rashid (2019)

The loss of biodiversity in the coastal and marine ecosystem can be depicted comprehensively with a cause-and-effect relationship by a DPSIR framework (Fig. 3.4). Overall, the current state of this ecosystem is deteriorating in terms of both quantity and quality, and the remedies to the problems remain flawed.

3.2.3 Wetlands Biodiversity

The wetlands in Bangladesh comprise of different types of ecosystems, which can be broadly identified as inland freshwater wetlands (e.g., floodplains, *haor*, *baor* and *beel*), tidal salt-water wetlands and man-made wetlands (Haroon & Kibria, 2017). They support diverse species of biodiversity resources. Among all the wetland-based resources, fish stands at the top due to its quality, quantity and socio-economic contribution although the wetlands support diverse faunal species (Fig. 3.5). Additionally, about 300 plant species and some 400 vertebrate species are identified as being dependent on the wetlands (Byomkesh et al., 2009; Haroon & Kibria, 2017). Some plant species are relatively uniform in all types of wetlands such as hizal (*Barringtonia acutangula*), tamal (*Diospyros cordifolia*), barun (*Crataeva nurvala*), madar (*Erythrina variegate*), dumur (*Ficus hispida*), paniphal (*Trapa bispinosa*), etc. Also, several medicinal species exist in the wetlands (Islam, 2010b).

The biodiversity resources of the wetland ecosystem are on the decline primarily due to man-made development interventions. For instance, there is an increasing trend of cereal production particularly rice mono-cropping in the floodplain (Rahman, 1995). Some other anthropogenic pressures are constructing closures and regulators, unplanned making of rural roads, so-called fisheries development projects (e.g., carp stocking program), overfishing and pollution, (Byomkesh et al., 2009; Rahman, 1995). The existing leasing system is also detrimental to the wetlands. Most of the

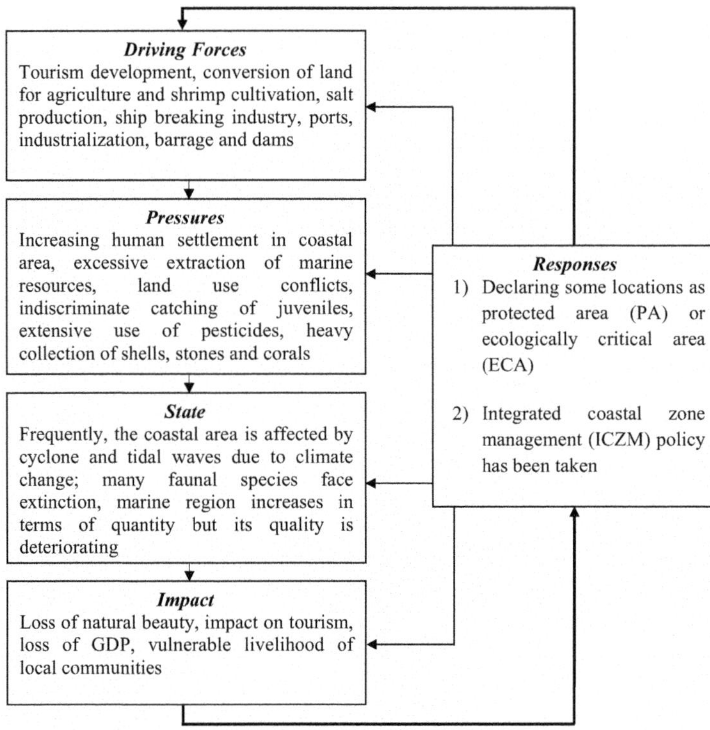

Fig. 3.4 State of coastal and marine ecosystem biodiversity of Bangladesh. *Source* Prepared by the authors

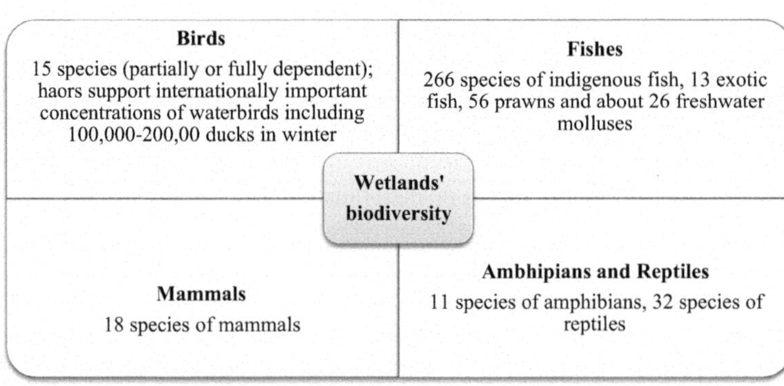

Fig. 3.5 Faunal species available in the wetlands of Bangladesh. *Source* Prepared by the authors based on Rahman (1995), MoEF (2001), DoE (2010)

wetlands are leased out to the local elites and powerful groups who are not concerned about the conservation of the wetlands' resources. Finally, climate change in particular negatively impacts the biodiversity of wetlands by causing the loss or shift of breeding grounds, destruction of habitats, salinisation of aquifers, and deterioration of water quality of small wetlands (e.g., ponds) (Kibria & Haroon, 2017; Rabbani et al., 2013).

Little effort has been given to date to protect the wetlands in Bangladesh. There is no particular policy or management approach for the conservation of this ecosystem. The government, however, has introduced a community-based wetland conservation model in some contexts, which has been successful in protecting some of the threatened species (DoE, 2015). Mangrove wetlands have actually been successful in gaining some attention at the policy level, but other types of wetlands continue to be overlooked.

The overall state of the wetland ecosystem (Fig. 3.6) fairly exhibits that its resources are under a massive threat of destruction, which primarily affects the country's fish production as well as the livelihood of the wetland communities.

3.2.4 Agricultural Biodiversity

Agriculture is a major sector of the Bangladeshi economy. The chosen case study of the Sundarbans explored in this chapter, however, is not closely related to this ecosystem and therefore, the ecosystem receives less attention here.

Bangladesh is rich in biodiversity of crops, fish, farm animals and forest trees used for food and agriculture (BARC, 2016). Bangladesh has 30 agro-ecological zones and 88 subzones based on physiography, soil properties, soil salinity, depth and duration of flooding. Over the centuries, people have cultivated, preserved, and used over 1364 plant species from both local and exotic origins. People over the centuries have been cultivating, preserving and using more than 1364 plant species coming from both endemic and exotic origins. There are also approximately 1000 species of medicinal herbs (DoE, 2010, 2015).

Farming practices in Bangladesh are both complex and varied, owing primarily to cropping seasons. A year has two such seasons: *kharif* (monsoon) and *rabi* (winter). The *kharif* season's major crops are paddy and jute, whereas the *rabi* season's crops are far more diverse, including paddy, vegetables, wheat, pulses, oilseeds, spices, potatoes and others. Rice is the most widely grown crop due to its adaptability to a wide range of environmental conditions. Moreover, several species of jute, sugarcane, cotton, linseed, mustard, cucumber, beans, etc., are cultivated (DoE, 2010, 2015). The biodiversity resources of this agro-ecosystem are of immense importance for the ecological balance of the country. However, the resources have been facing threats due to the adoption of modern agricultural practices, including the use of hybrid seeds, chemical fertilisers and pesticides (BARC, 2016; *Unnayan Onneshan*, 2010a).

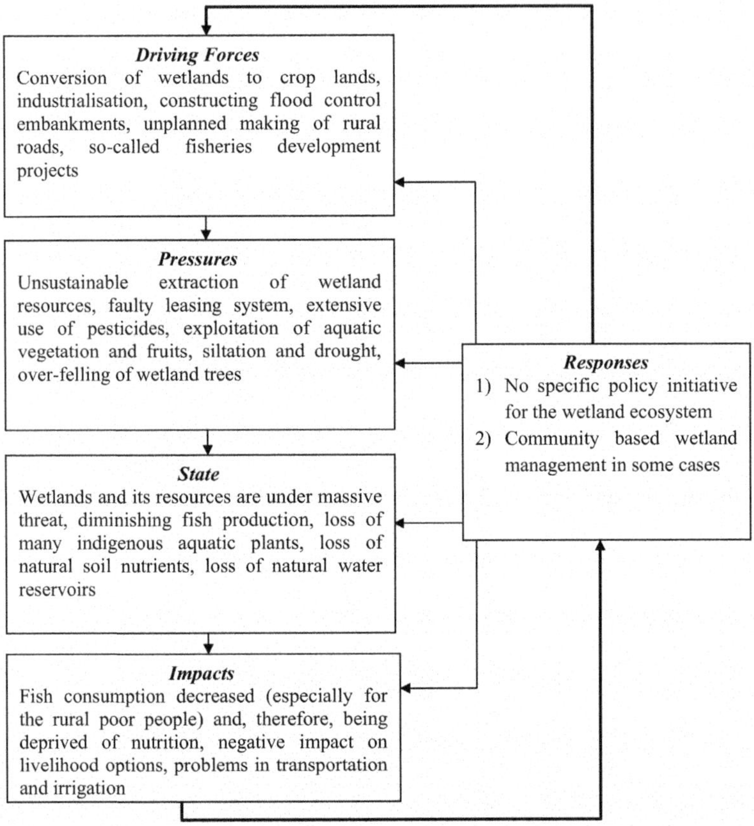

Fig. 3.6 State of wetland ecosystem biodiversity of Bangladesh. *Source* Prepared by the authors

3.3 Biodiversity Degradation of the Sundarbans: A Micro-case Study

The Sundarbans was chosen as a case study for the biodiversity resources for a number of reasons. First, because the scope of biodiversity is colossal in nature, it would be challenging to conduct a fruitful empirical analysis by incorporating data for various types of biodiversity resources. Accordingly, a specific case would be useful in scrutinising the dynamics of biodiversity resource governance. Second, it is necessary to ensure that the selected case study is a representative one. The ecosystem of the Sundarbans in Bangladesh is the single most important one that supports diverse biological resources. Finally, the Sundarbans alone contains forest biodiversity, marine and coastal biodiversity as well as wetland biodiversity, representing nearly all types of biodiversity in the country.

The Sundarbans is the world's largest contiguous single-tract mangrove ecosystem, located at the edge of the Bay of Bengal in the great delta of the

Ganges, Brahmaputra and Meghna (GBM) rivers (Map 3.1) (Titumir et al., 2020). This mangrove ecosystem is located in both India (in the state of West Bengal) and Bangladesh. The Bangladesh part is larger than that of the India, with an area of 6071 km^2 (62% of the total area), accounting for 39.5% of Bangladesh's total forest region (Roy & Alam, 2012). Of that part, 70% is land area and the remaining (30%) is water (Kabir & Hossain, 2008). The wetlands of the Sundarbans are made up of about 200 islands separated by around 400 interconnected tidal rivers, creeks and canals (Rahman et al., 2010a). In 1997, UNESCO designated the Sundarbans as a Natural World Heritage Site, and it was recognised as a Ramsar Site of international importance in 1992 (IUCN Bangladesh, 2014).

The amalgamation of different types of ecosystem such as forest, coastal and wetland makes the Sundarbans a home to several uniquely adapted aquatic and terrestrial flora and fauna. It harbours 334 species of trees, shrubs, herbs and epiphytes (Behera & Haider, 2012). The most important tree species is the sundori (*Heritiera fomes*). Some other notable species are gewa (*Excoecaria agallocha*), baen (*Avicinnia officinalis*), passur (*Xylocarpus mekongensis*), keora (*Sonneratia apetala*), goran (*Ceriops decandra*), ora (*S. caseolaris*) and hental (*Phoenix paludosa*). In addition, non-timber forest products of high value include golpata (*Nypa fruticans*), honey, wax, fish and crabs. It is the country's largest honey-producing habitat with giant honey bees (*Apis dorsata*) (Titumir et al., 2020).

This forest region is also rich in its faunal diversity. There are 448 species of vertebrates, including ten amphibians, 58 reptiles, 339 birds, 41 mammals and 291

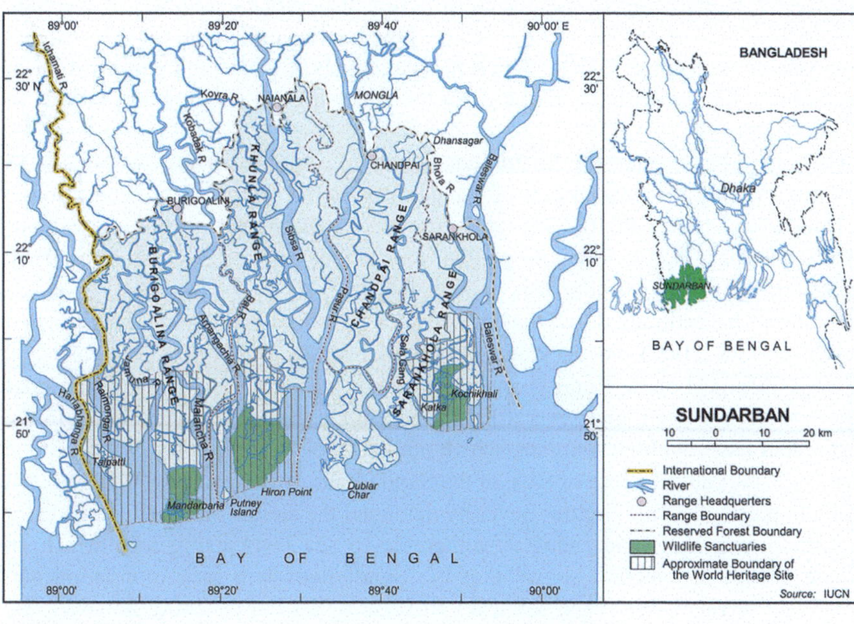

Map 3.1 Location of the Sundarbans. *Source* IUCN as cited in Rahman et al. (2010a)

fish as per the latest government statistics (BFD, 2010; DoE, 2015; Titumir & Afrin, 2017). Among the invertebrates, there are 20 species of prawns, 8 species of lobsters, 7 species of crabs and several species of gastropods (Kabir & Hossain, 2008; Rahman & Asaduzzaman, 2010). Apart from fish, it also serves as a habitat for a wide variety of aquatic wildlife including estuarine crocodiles (*Crocodylus porosus*), turtles (*Lepidochelys olevacea*), dolphins (*Platanista gangetica* and *Peponocephala electra*) and molluscs such as the giant oyster (*Crassostrea gigas*). The Royal Bengal Tiger (*Panthera tigiris*), however, is the most magnificent animal. According to the 2004 Census, there were approximately 440 tigers in the Bangladesh part, but the most recent estimate puts the number at 114 tigers (BFD, 2019). Thousands of spotted deer (*Axis axis*) and barking deer (*Muntiacus muntjak*) also live there.

These *biotic* along with other *abiotic* resources of the Sundarbans contribute directly or indirectly to both the local and national economy (Titumir et al., 2020). Altogether, the resources have been used for numerous purposes (Fig. 3.7) and thus benefitting the life and livelihood of the IPLCs as well as the country's economy.

However, the Sundarbans has been undergoing major ecological and physiographical alterations (Baten & Kumar, 2010) because of diverse 'anthropogenic drivers' and 'natural direct drivers' (see Diaz et al., 2015 for details about drivers) which have had a significant impact on the forest's regenerative capacity and ability to sustain itself (Titumir et al., 2020).

Commercial shrimp farming is one of the greatest anthropogenic pressures on the mangrove ecosystem of the Sundarbans. Overexploitation, illegal encroachment, industrialisation, oil spill and incidents of fire are some of the other such pressures, which occur regularly. Moreover, the construction of significant numbers of upstream dams, polders and embankments has reduced the freshwater flow in the downstream rivers (Baten & Kumar, 2010), which has negatively affected the composition of vegetation in the Sundarbans, particularly by increasing the siltation and salinity of soil and water.

The Sundarbans resides in the immediate path of cyclonic storms. Among the 45 cyclones crossing the coastal belt of Bangladesh in 135 years, 13 trekked through the Sundarbans (BFD, 2010). *Sidr* and *Aila*, among the most devastating cyclones in the recent past, had a devastating effect on the Sundarbans. Cyclone *Sidr* struck the eastern parts of the Sundarbans on November 15, 2007, leaving a trail of severe devastation. One estimate shows that around 31% of the total Sundarbans was affected by *Sidr* (CEGIS, 2007). Then, cyclone *Aila*, which struck on May 25, 2009, also had a disastrous impact on the Sundarbans, primarily on the western part. A large number of trees were uprooted, and several floral and faunal species lost their lives. In the most recent case, cyclone *Amphan,* which passed through Bangladesh and India on May 21, 2020, caused a loss of 83% of the Sundarbans of Bangladesh (Mishra et al., 2021). In particular, this cyclone severely damaged the *sundari* trees of the forest.

Furthermore, the Sundarbans is vulnerable to sea level rising resulting from climate change. Projection shows that the Sundarbans will face inundation and succeeding wetland loss if the sea level will rise by 98 cm by 2100; however, the magnitude of the loss remains unclear (Mukhopadhyay et al., 2018). Climate change also negatively affects the Sundarbans by increasing water and soil salinity. Globally

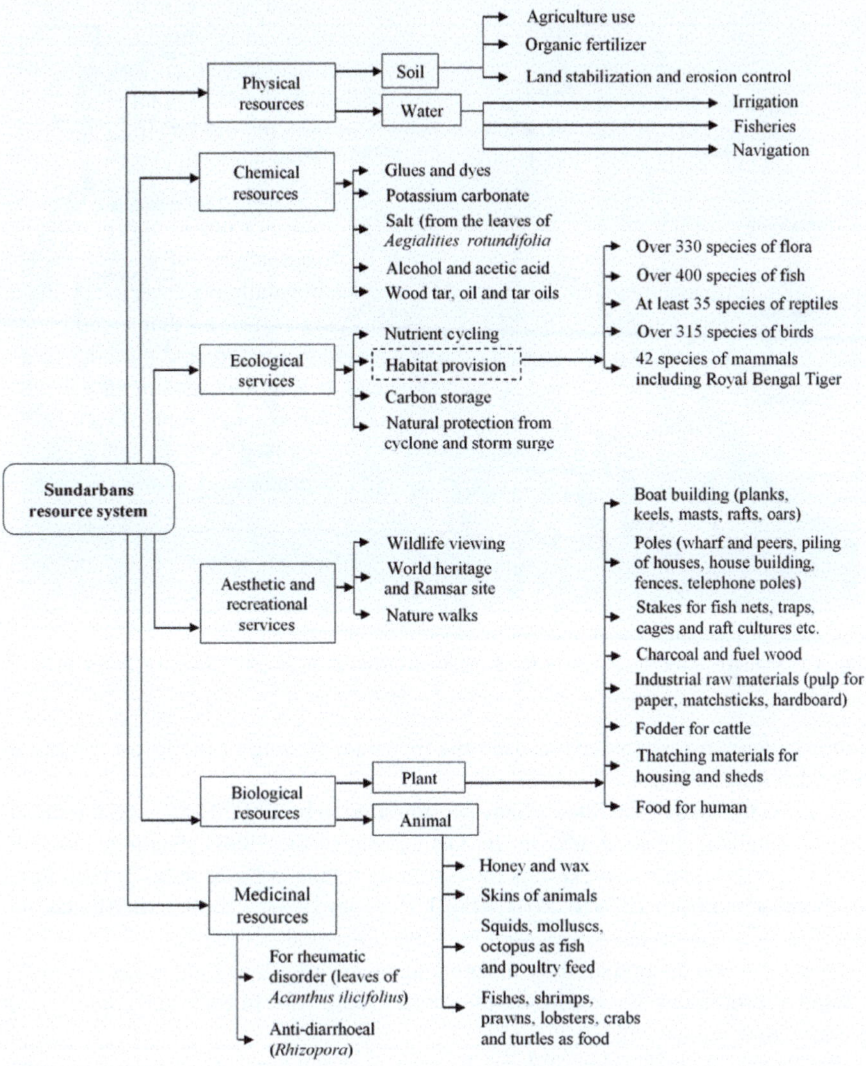

Fig. 3.7 Resource system of the Sundarbans. *Source* Adapted from the data reservoir of *Unnayan Onneshan* as cited in Titumir and Afrin (2017)

endangered *H. fomes* declined (about 70% of trees are affected) because of 'top dying' disease caused by increased salinity (Sarker et al., 2016). The increase in temperature, too, seriously affected the Sundarbans' ecosystem. Generally, the Sundarbans has the ability to recover from the damages incurred by natural disasters. Natural disasters, however, have increased in frequency and intensity in the advent of climate change, leaving little time to recuperate. Hence, the natural recovery capacity of the Sundarbans is eroding gradually (Baten & Kumar, 2010). Overall, man-made as well as

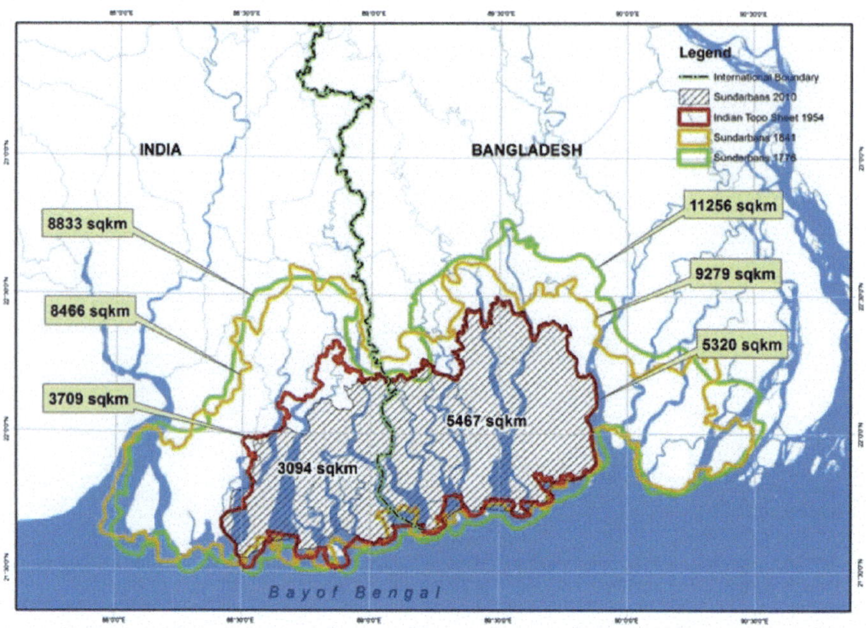

Map 3.2 Mangrove forest change of the Sundarbans from 1776 to 2010. *Source* CEGIS 2016

nature-induced pressures have caused the incessant waning of the forest cover and its biodiversity resources.

In 1776, the size of the Sundarbans was reported to be 17,000 km^2, and it reduced to almost half of the total area in the last decade of the 1990s (Islam & Gnauck, 2009). Another report also exhibits the declining trends in forest areas in both India and Bangladesh from 1776 to 2010 (Map 3.2). From 1970 to 2000, in particular, the forest cover is said to have decreased by 1.2% (Giri et al., 2015).

Forest inventories prepared by different agencies also depict the reduction in the volume of prominent tree species in the forest (Table 3.6). In each case, the trend in tree growth is decreasing.

The decline in the number of floral diversity also puts a negative impact on faunal diversity (Titumir & Afrin, 2017). The Sundarbans are home to up to 20 globally threatened species. *Batagur baska* (turtle), Ganges River dolphin (*Platanista gangeticus*) and Irrawaddy dolphin are the most endangered species. Other threatened wildlife species include python, king cobra, adjutant stork, white-bellied sea eagle, clawless otter, masked fin-foot, ring lizard, river terrapin, fishing cat, spoon-billed sandpiper and eagle (DoE, 2015). The Royal Bengal Tiger, the most important faunal species, is also listed as endangered by the IUCN. Most strikingly, a number of animals have already become extinct from the whole Sundarbans during a 100-year period (Table 3.7).

The Public Participation Geographic Information System (PPGIS)-based research by the *Unnayan Onneshan* also reveals similar vulnerable and depleting scenarios of

Table 3.6 Growing stock of the Sundarbans according to different inventories

Year of publication of inventory results	Inventory done by	*Sundari* (number of trees per hectare)	*Gewa* (number of trees per hectare)	All tree species (number of trees per hectare)
1959	Forest and Forestal Engineering, Canada	211	61	296
1983	Overseas development authority	125	35	180
1996	Forest resource management project, FD, GoB	106	20	144

Source FAO (2011)

Table 3.7 Floral species that became extinct in the Sundarbans

Scientific name	Local name
Rhinoceros sondaicus	Javan rhinoceros
Bubalus bubalis	Water buffalo
Cervus duvauceli	Swamp deer
Bos gaurus	Guar
Axixporcinus	Hog deer
Crocodiles palustric	Marsh crocodile

Source Adapted from Behera and Haider (2012)

forest cover and aquatic resources in the Sundarbans. The decadal changes in forest cover in the Koyra region of the Sundarbans—a part of the Khulna administrative range—demonstrate widespread degradation of trees in the last two decades (Map 3.3). The dark green part of the map refers to the dense forest areas, whereas the light green part refers to the areas covered by medium-dense forests. On the other hand, the areas where the forest is very thin are marked with white part, indicating that the number of trees here is much less or not present at all. These areas have become empty fallow lands.

During that period, the value of different waterbodies declined significantly. The value is defined here as the availability of fish and crabs. The declining trend of values implies that aquatic resources are degrading gradually, resulting in their less availability in the waterbodies of the forest (Map 3.4). The dark red part refers to the most valuable part of the waterways where fishes and crabs are abundant. The medium red part means that crabs are also found here but fewer than the dark red part. Some crabs are also seen in the light red part. The higher the red part, the greater the amount of crab harvesting in those areas.

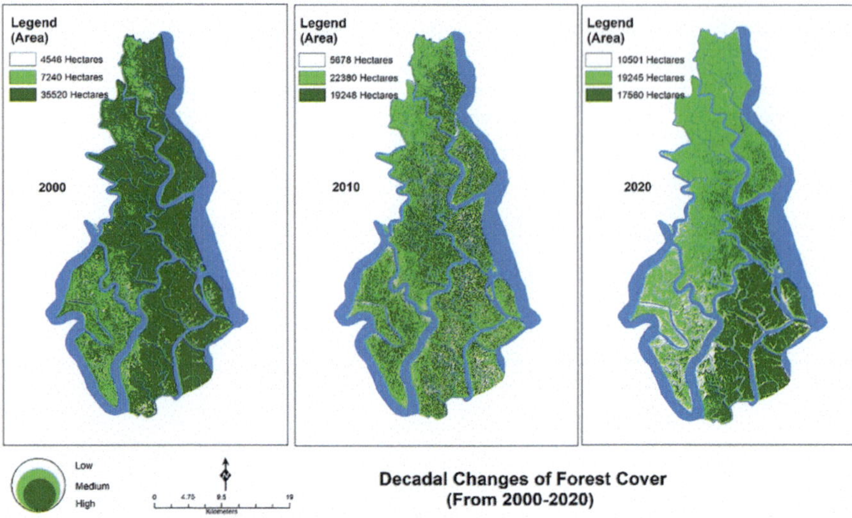

Map 3.3 Decadal changes of forest cover in a part of Khulna administrative range of the Sundarbans (2000–2020). *Source Unnayan Onneshan* (2020)

This finding is also congruent with another recent study that used Landsat satellite data (from January 1988 to June 2018) covering the entire Sundarbans to uncover the health decline in this mangrove forest (Awty-Carroll et al., 2019). The study reported that there is a decline in the health of about 25% of the mangrove trees.

3.4 Biodiversity Under Market: Commodification and Institutions

A market-centric approach assesses the value of biodiversity resources based on the neo-classical economic framework by considering that such resources degrade predominantly due to the absence of market and negative externality (Perrings et al., 1992; Sadmo, 2015). The value is accordingly determined by adding up the market price multiplied by the quantities traded of all the various attributes of biodiversity. Thus, biodiversity is considered just as a commodity, which can be substituted with another. It fails to recognise that if specific and special kinds of species of a particular ecosystem are traded for monetary gains, it is possible that they will not be replaced (Titumir et al., 2020). For instance, according to this framework, utility from standing forest coverage is the same as the one derived from the sale of timber collected from that forest area. The ecosystem resources, therefore, have value to the extent that they are useful for human beings, and as such, they are defined as 'ecosystem services' under the approach. The branch argues that such can offer useful insights to support policy initiatives by measuring the economic value of the

(a) Fish

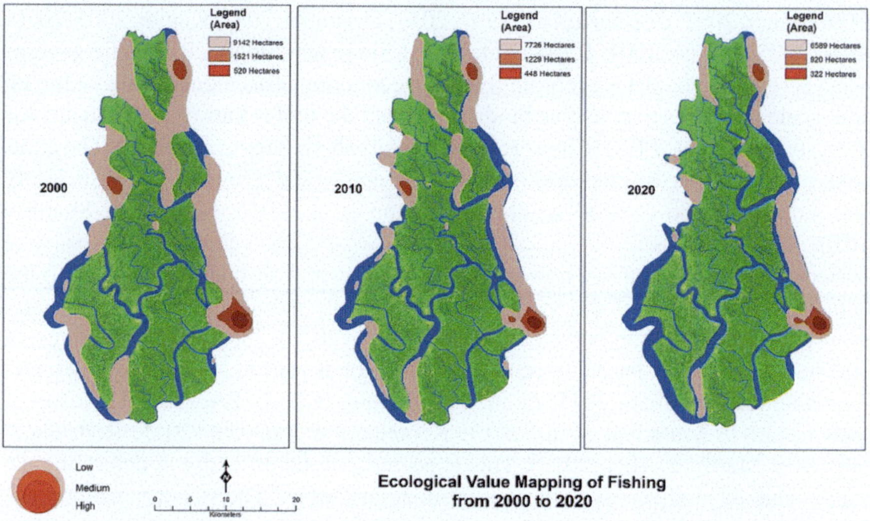

(b) Crabs

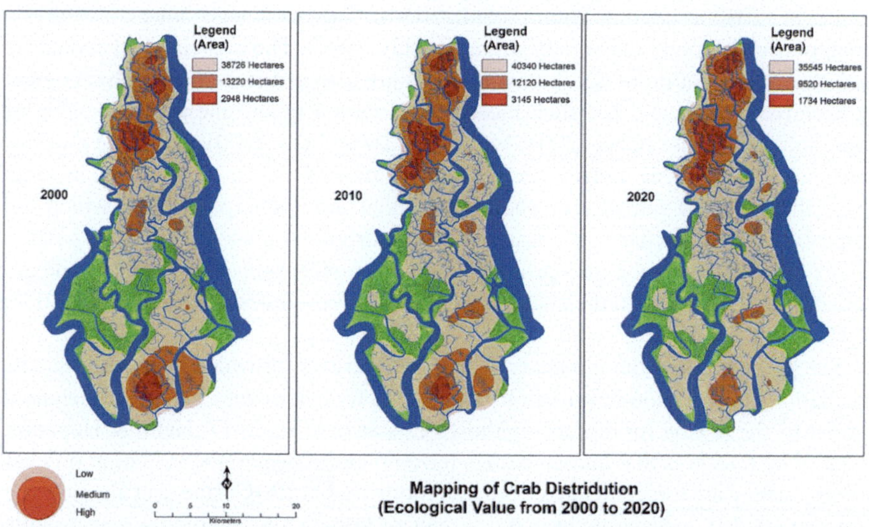

Map 3.4 Ecological value mapping of **a** fish and **b** crabs in a part of Khulna administrative range of the Sundarbans (2000–2020). *Source Unnayan Onneshan* (2020)

resources, which is a fundamental step in conserving such resources (Barbier, 2007, 2009; Bräuer, 2003; Costanza et al., 1997, 2017; Dasgupta, 2021; Kumar, 2005, 2010; MEA, 2005; Pearce, 2001; Rasul et al., 2011; Singh and Singh, 2017). The governments across the countries accordingly adopt economic instruments while taking up large-scale measures to conserve biodiversity and the tool is known as Payments for Ecosystem Services (PES) (Chen et al., 2020; Purushothaman et al., 2013). The argument, overall, is that the market allocates scarce ecological resources more efficiently than 'command-and-control' regulations (Hahn et al., 2015; McAfee & Shapiro, 2010).

Neo-classical economics, however, fails to provide a sustainable solution to the unique characteristics of the symbiotic relationship among humans, biodiversity resources and ecosystem services (Titumir et al., 2020). Indeed, the exchange of biodiversity resources based on economic valuation is proved to be flawed (Gómez-Baggethun & Muradian, 2015; Gomez-Baggethun & Ruiz-Perez, 2011; Kosoy & Corbera, 2010; Muradian et al., 2013; Neuteleers & Engelen, 2015; Rodríguez-Labajos & Martínez-Alier, 2013). This is because of the difficulty to determine the actual value of biodiversity resources in monetary terms. The standard economics, in fact, "reduces biodiversity into a number of quantifiable parts, subjecting to the utilitarian usage and reducing social–natural relations to market transactions" (Turnhout et al., 2013 as cited in Titumir et al., 2020). Thus, exchange value constitutes a small portion of total biodiversity value (Gowdy, 1997). The neo-classical economic tools for the valuation of ecosystem services are also criticised from an ecological perspective by arguing that such means of valuation ignore the intrinsic value of some specific entities (Chee, 2004; McCauley, 2006; Vatn, 2000) and, in the process, can also erode intrinsic motivations for conservation (Bowles, 2008). Even, in some cases, the monetary valuation can be controversial and counterproductive, which can undervalue the objectives of conservation (Rodríguez-Labajos & Martínez-Alier, 2013). Overall, such measures provide "a narrow conception of biodiversity and are potentially detrimental to the conservation of these resources" (Titumir et al., 2020, p. 75).

The proponents of the monetary valuation, however, provide counter arguments by claiming that economic valuations are essentially incomplete, but incompleteness cannot be the reason for discarding those (Costanza et al., 2017; Helm & Hepburn, 2012). The reason is that human beings have no alternative to the valuation process as long as they are forced to make choices among trade-offs (Costanza et al., 1997; de Groot et al., 2012). Nevertheless, it does not address the distributional aspect while promoting access and benefit sharing of such resources, which is important not only for extending equity but also for conserving the resources. It promotes unequal access to resources favouring those with the ability to pay (Corbera et al., 2007a; Gómez-Baggethun & Muradian, 2015; Martínez-Alier, 2002). Broadly, the debate largely disregarded the socio-political and institutional dimensions of ecosystem services (Purushothaman et al., 2013).

There is also an emphasis on the process of financialisation under a market-based approach and the CBD indicates (Article 20.4) that the obligation of biodiversity-related international financial resource flows lies with the developed countries. The post-2020 Global Biodiversity Framework (GBF), under the CBD, reiterated the issue of finance flow from the developed to developing countries to raise it at least $30 billion per year by 2030 (SCBD, 2022). This process aims at conserving the biodiversity resources through investment. The argument is that private sector financing is important for biodiversity conservation as there is not enough public sector money to conserve biodiversity, especially in developing countries (Rubino, 2000). The financialisation of ecosystem services ultimately leads to its commodification and marketisation (Khor, 2011; McCauley, 2006) resulting in excessive extraction of resources. There is little evidence that the developed countries have maintained the commitment according to the CBD principles regarding the financial resource flow to developing countries for the conservation of biodiversity. Rather, they have come up with the so-called private sector-based 'innovative financial mechanism', full of social risks, lacking economic sustainability and ecological feasibility. Moreover, it is also evident that the conservation of resources, for example, forests, does not always depend on investment (Lovera et al., 2015) and which is obvious, as forest conservation will simply happen for free, provided that anthropogenic pressure is not put on the forest (Lovera, 2020).

Overall, using market-based instruments for valuing biodiversity contributes to the undesirable commodification of nature (McAfee, 1999, 2012), which ultimately results in the excessive extraction of these resources. More specifically, biodiversity becomes a market commodity subject to free trade that induces different (powerful) agents to accumulate the resources as much as possible for earning profit under the neo-liberal political regime. Thus, it raises an ethical argument about transforming human–nature relations by 'commodity fetishism' (Kosoy & Corbera, 2010) and crowding out moral obligations as motives for nature protection (Luck et al., 2012; Rode et al., 2014).

The understanding of neo-classical economics later has been complemented by the New Institutional Economics (NIE), which argues that the creation of a formal property rights regime can help efficiently manage natural resources whereas the absence of property rights results in resource degradation (Ituarte-Lima et al., 2014; Ostrom, 2000; Vatn, 2010). The argument is that ensuring equitable access to and sustainable management of such resources requires clearly defined and equitably enforced property rights (appropriate and stable institutional arrangements).

The NIE, however, ignores the political transaction cost and the political nature of rights. It is not sufficient to have clearly defined property rights, whether through official documentation or collective action. People cannot exercise their rights if they are unaware of their existence due to a lack of knowledge, communication or access to information. Moreover, the NIE also fails to identify the nature of conflicting interests existing among the agents over a biodiversity-rich area. Conflict exists due to the overlapping property rights regime between opposing interest groups at the local level as well as at the national or international level. These issues are particularly important in the context of a transitional economy. Accordingly, it is necessary to examine in which power structure and political domain the property rights exist. The political economy factors can influence whether property rights over a biodiverse area are effective or vulnerable. The argument is that institutional arrangements (e.g., property rights) are usually more vulnerable in transitional economies due to some political economic factors, causing excessive accumulation by various agents through non-cooperative behaviour and it induces the degradation of biodiversity resources.

Moreover, the informal institutional arrangements (e.g., values, norms, customs, etc.) are necessary to consider in developing the framework for sustainable biodiversity resource management. When formal institutions fail to enforce collectively desirable outcomes, informal institutions, such as social norms, can play an important role (Nyborg et al., 2016). Moreover, norms can sometimes promote cooperation and increase stock biomass in common pool resource systems (Tilman et al., 2017) though they cautioned about the impossibility of long-term sustainable use of those resources. In line with this, the role of IPLCs and their traditional knowledge have assumed importance in protecting biodiversity-rich regions in recent times (e.g., Hill et al., 2020; Williams et al., 2020). Nevertheless, there exist debates about whether local people are the protectors (e.g., Ancrenaz et al., 2007; Kohlar & Brodizio, 2019) or destroyers (e.g., Basu & Cetzal-lx, 2018; Walters, 2004) of the ecosystem.

The issue of co-management is relevant here as it seeks to integrate local people's participation and knowledge with the system of governance. Co-management, however, can suffer from inappropriate conflict-resolution mechanisms, fragile institutional arrangements, inefficient accountability mechanisms, insufficient local engagement and poor collaboration between the regulating authority and local communities (Bhattacharya et al., 2010; de Freitas et al., 2020; Terborgh & Peres, 2017). The transaction cost is also important in the case of common resource management, which is not incorporated into the economic analysis of participatory resource management (Adhikari & Lovett, 2005). Particularly, transaction costs largely exist in a transitional economy where the organisation of politics matters significantly. The underlying cause is that people in advanced countries bear the cost of bargaining and negotiating under a state of law whereas the people in a transitional economy become victims of predation, appropriation and racketeering in this process (Vahabi, 2008). Moreover, there exist equity and distributional problems, especially in the distribution of costs and benefits of ecosystem benefits, which are also beyond the analysis of mainstream economic thoughts.

Overall, three major issues concerning biodiversity resources and institutional arrangements must be addressed right away. First, unstable property rights create favourable conditions for the reclamation and conversion of biodiversity-rich area to alternative uses without considering the underlying ecological value. Second, institutional instability or mismatch hinders the rights of the poor to stake claim to the biodiversity resources. Third, indigenous and customary institutions are frequently disregarded in the conservation process of those resources; however, such informal institutions need to be taken into account to guarantee the conservation and sustainable use of biodiversity resources.

3.5 Political Economy of Biodiversity: Accumulation and Distribution

In addition to the market-centric strand, there have been attempts of late to investigate biodiversity through the lens of political economy, but they are limited in number (e.g., Corbera et al., 2007b). The political economy analyses contend that conflicting resource management and degradation are exacerbated by the existence of overlapping property rights regimes. It offers insights into the political components that exist in resource management regimes and emphasises the hierarchical relationship that exists in society in fathoming the nature of benefit sharing among multiple entities. The argument is that institutional arrangements (property rights) are vulnerable to some political-economic factors arising from accumulation by different agents in the presence of a non-cooperative solution (Titumir et al., 2020). The discussion further stresses upon the roles of formal political institutions as well as narratives regarding changes in ecosystem services (Robbins, 2012).

Indeed, political institutions determine whether the state's ability to coerce will be used to maximise net benefits to the population at large or to benefit specific groups while shifting costs to the rest of society (Deacon & Mueller, 2003). The groups in charge of the political process might consider their own benefits and costs only that arise from a change in property rights, ignoring the implications for society as a whole. Thus, the country's political institutions shape the nature of property rights by determining transaction costs (Deacon & Mueller, 2003). Such opportunities to acquire private profit associated with conservation territories can be related to Marx's (1976) concept of 'primitive accumulation' and Harvey's (2003, 2005) reformulation as 'accumulation by dispossession'.

The key argument here is that primitive accumulation is characterised by the enclosure of a common and thus the enclosure of conservation territories can be viewed as a source of accumulation (Newmann, 2017). Primitive accumulation, however, can be regarded as a decades-long process of deepening commoditisation that may or may not include enclosures (Newmann, 2017). When the power is centred in the hands of particular groups, the scope is created for primitive accumulation to take place.

Second, when the political system in a country is unstable or non-representative, people cannot be assured of the future return of a particular type of resource. They always, therefore, remain afraid of tenure insecurity and have very little inclination to conserve the resource base. Third, power relations also matter in making the access and benefit sharing of biodiversity resources equitable. The expectations, however, have not been fulfilled under market centrism because equity and legitimacy are critical dimensions in the design and implementation of the market for ecosystem services if social development goals other than economic gains are to be attained (Corbera et al., 2007b). Finally, political economy is essential because international coordination is often required in governing commons—for example, ecosystems do not respect national borders always and many of the world's biodiverse ecosystems are located in poor countries (Helm & Hepburn, 2012). Overall, such literature

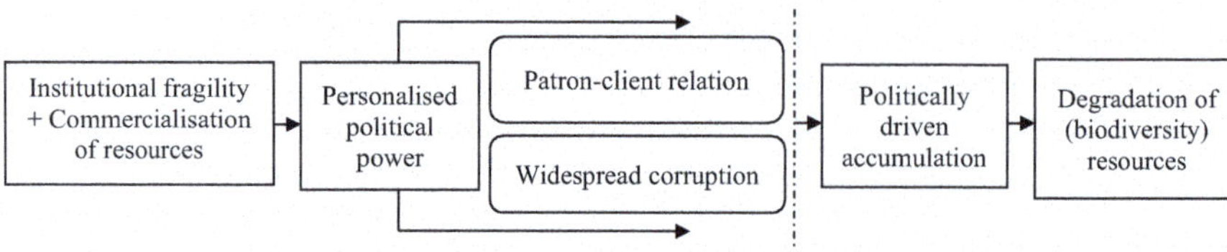

Fig. 3.8 Political economy factors inducing biodiversity resources degradation. *Source* Prepared by the authors

provides a prism through which to describe the bio-environmental relationship in the presence of distribution of the power to production activities and its relevance to ecological analysis (Greenberg & Park, 1994; Titumir et al., 2020).

These political economy factors are vital in understanding the accumulation and distribution processes of biodiversity resources, particularly in a transitional economy. They help understand the way in which power and politics shape the decision-making process in terms of management, use and conservation of these resources. Indeed, the factors behind the institutional vulnerability in developing countries are the existing nature of the power structure and the political settlement. Primitive accumulation is practised in transitional economies either through the use of force, coercion, abuse of state power, or accumulation through encroachment. Moreover, in such a context, land grabbing can also be analysed as the process of primitive accumulation and accumulation by dispossession (Adnan, 2013). In particular, land alienation along with grabbing is produced by a variety of actors, processes and mechanisms, as part of the complex interactions of neo-liberal globalisation, state policies and interventions, and involving power struggles, threats, violence and resistance (Adnan, 2016). The approach exhibits, overall, that the degradation of biodiversity resources is "not only about the non-existence of the market but also about unequal power-sharing by the stakeholders over the management of resources" (Fig. 3.8) (Titumir et al., 2020, p. 74).

The existence of power-based vertical relations in such societies and the upward enforcement of rules enable the powerful group to capture resources with impunity (Adhikari & Goldey, 2010). The process prioritises the rule of individuals over the rule of law, resulting in institutional fragility, increased rent dissipation, rent-seeking and property rights seizure. Insecure property rights cause rent dissipation by providing inducements to usurp, defend and lobby for more secure rights.

Based on the above discussion, a brief overview of the relevant concepts in the context of this study in terms of examining land alienation and land grabbing process in the biodiversity-rich area has been presented in brief (Table 3.8) along with some other concepts such as de jure and de facto rights. These interconnected and partially overlapping concepts are critical for comprehending the complex interactions and transformations involving land use, claims and rights in the biodiversity-rich region of the Sundarbans, which will be examined in the empirical sections.

Overall, the political economy premise aids in identifying the major factors behind the degradation of biodiversity resources from a critical and broader perspective. It,

Table 3.8 Important concepts under political economy approach regarding biodiversity

Concept	Elaboration
Primitive accumulation	PA precedes capitalist accumulation; an accumulation which is not a result of the capitalist mode of production but its point of departure (Marx, 1976, 873–895); entails not only the *quantitative* transfer of resources but also their *qualitative transformation* in terms of property rights
Accumulation by dispossession	The contemporary processes feeding the expansion of capitalist production under neo-liberal globalisation. A distinctive characteristic of this phase has been the unprecedented manipulation of finance capital and the credit system as critical mechanisms of dispossession (Harvey, 2005)
De jure rights	*De jure* rights are based on formal laws concerning landed property and are enforced by the state
De facto rights	*De facto* rights are informal rights, i.e. without titles, settlement documents or other kinds of formal land records
Land grabbing	The process in which an area has already been taken over by others forcibly
Land alienation	The mechanism in which a group is prevented from gaining *access* to land to which it is *entitled* (e.g., in accordance with land reform measures)

Note For more clear understandings about the concepts can see: Kelly (2011), Benjaminsen & Bryceson (2012), Adnan (2013, 2016)
Source Prepared by the authors based on Adnan (2013, 2016)

however, does not suggest any measures to solve the problem. Moreover, it could not internalise the 'human–nature' relations into the resource governance framework.

3.6 An Alternative Framework

In the backdrop of shortcomings in the existing theoretical understandings (Table 3.9), specifically of the market-centric framework, it is necessary to devise an alternative understanding to determine the drivers of biodiversity resources degradation, particularly in a transitional economy and on the policy measures for ensuring the sustainable utilisation and conservation of those resources. At its core, the alternative framework focuses on the political economy factors to identify the causes of degradation while it argues for revitalising the complementary interrelations between human beings and nature for the sustainable utilisation, conservation and distribution of biodiversity resources.

Table 3.9 Underlying gaps in approaches regarding biodiversity

Branches' name	Arguments for biodiversity	Underlying gaps
Neo-classical economics	(a) Economic valuation of biodiversity is important as it, for example, raises awareness about conservation by expressing the importance in monetary terms (b) Biodiversity as a 'natural capital' is decreasing (c) A 'market' for ecosystem services can determine gain and loss and which improves management mechanism (d) Conservation tools, for example, are taxes, subsidies, quotas, regulatory protections	(a) Ignores the weak governance and the political transaction cost in developing countries (b) Political feasibility of market-based instruments is questionable (c) Treating biodiversity resources as a marketable commodity is fallacious (d) Ignores the inherent values of biodiversity resources and identifies them as natural capital which is substitutable (e) Optimisation method allows partial representation of the complexity
Institutional economics	(a) Emphasises institutional stability to ensure sustainable management and equitable distribution of biodiversity resources (b) Policy tools are (re) allocation of property rights, land use change (e.g., declaration as protected regions)	(a) Could not incorporate distributional aspects (b) Gap in explaining political economy factors (power and political settlement for example) behind unstable nature of institutions (c) Human sociality aspect is missing; less emphasis on informal institutional set-up
Political economy	(a) Address the equity and distributional aspect	(a) Could not explicitly internalise the 'human–nature' mutuality into the sustainable utilisation of resource framework (b) Does not provide any measure but a broad understanding of the contributing elements of the degradation of biodiversity resources

Source Prepared by the authors

3.6.1 *Proposition 1: Pricing and Rent*

Biodiversity resources, if assumed to be a commodity by transforming use value into the exchange value, result in economic rent, loss of consumer surplus and destruction of resources.

The proposition challenges the neo-classical thesis of the commodification of biodiversity resources. The argument is that treating biodiversity resources as 'marketable goods' results in the loss of consumer surplus through monopoly pricing. Certainly, the commodification transforms the use value into the exchange value, compelling consumers to buy the biodiversity resources at a higher price and the increased price reduces consumer surplus. In this process, the producers get the scope to earn monopoly profit and therefore, have the opportunity to accumulate excessive rent by extracting the biodiversity resources without any concern for its conservation. This, in turn, puts pressure on biodiversity.

Generally, the demand is perfectly elastic for natural resources (see Chap. 2 for details). When the use value of resources, however, turns into exchange value, the demand curve becomes downward sloping, signifying that demand responds to the price (Fig. 3.9). In this process, the producers produce or extract resources up to the Q_0 point where marginal revenue (MR) equals marginal cost (MC), the intersection point of E^*. Consequently, the MC curve represents the resource extraction cost. It then uses the demand curve to find the price P_0 that will induce consumers to buy that quantity. The agent(s) of production, in this case, is maximising larger profit. In other words, the powerful agents create an oligopolistic market structure and by charging a spiralling price, accumulate a huge amount of resource rent. The total economic profit or rent is the area of $A_0B_0C_0P_0$, accumulated by the power owner. It can be noted that economic profits can be received as long as the price is greater than the average total cost (ATC). Since the agents operate where the price is greater than the marginal cost, consumers have to pay higher prices to access biodiversity resources and therefore, their surplus decreases. In addition, it can be noted that the socially efficient point is D_0. However, the producers in this case produce less and charge a higher price. Hence, the deadweight loss (the triangle $E^*C_0D_0$) is the total welfare loss of society.

The hypothesis, accordingly, is that because of the commodification of biodiversity resources of the Sundarbans, the powerful groups find the incentive to extract the resources to earn higher profits. They are not aware of the conservation of resources though. The loss of the resources ultimately results in the loss of revenue collected by the government in the long run. Overall, the traditional communities are losing their

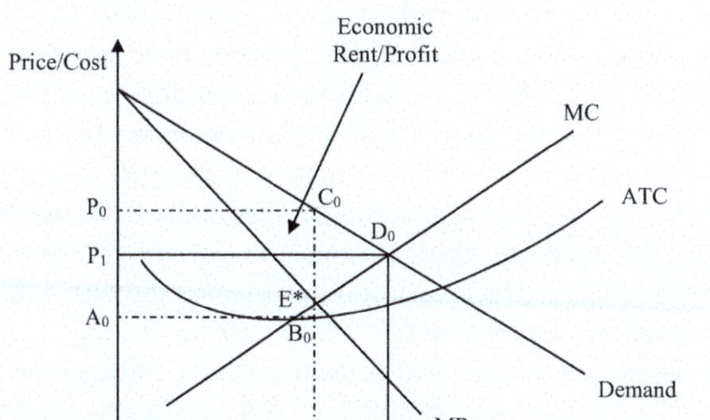

Fig. 3.9 Economic rent or profit and loss of consumer surplus. *Source* Prepared by the authors

consumer surplus, the government its share of revenue and the forest its endowment of resources, whereas the powerful groups are enjoying the lion's share of the rent.[1]

3.6.2 Proposition 2: Rent, Institutions and Regulation

The extraction of resources and the collection of revenue are the functions of the inefficient regulatory regime and the instability of property rights.

The second proposition postulates that when the market is in practice, weak institutional arrangements and the absence of efficient regulatory regimes induce both the rent dissipation process (missing revenue) and the extraction of biodiversity resources in developing countries. Weak institutional arrangements induce overexploitation of natural resources as they create chances for the accumulation of negative or extractive rents (see Khan, 2010, p. 11 for details). Alternatively, when there exist unstable property rights or weak institutions, the users have the tendency to overconsumption of the biodiversity resources. Overall, the absence of property rights creates damaging effects of rent dissipation resulting in allocative inefficiency, adversely impacting the sustainability of natural resources (excessive extraction instead of conservation) and collection of revenue.

Further explanation on the loss of revenue (rent) due to additional exploitation under the absence of effective regulatory regimes together with weak property rights has been provided (Fig. 3.10). In this case, the demand and supply of biodiversity resources are depicted as downward and upward sloping, respectively, considering the case that biodiversity resources are marketable goods that respond to price. The vertical axis shows the rent (revenue), earned through pricing and the horizontal axis, shows the quantity of biodiversity resource that is being extracted. When the agents of production or extractors extract at point Q_0, they can earn revenue (resource rent) at a maximum of R_0. Then, if the production point moves to Q_1 (extraction increases due to unstable property rights), the revenue or rent falls to R_1. Further extraction up to point Q_2 results in the complete missing of revenues. The total loss of rent (revenue) is $R_0D_1D_0R_2$ due to the additional exploitation of OQ_2–OQ_0 amount of resources. In this case, the share of loss of revenue can be explicated into two segments (a) extra payment or unofficial rent of $R_0D_1D_2R_1$ and (b) rent through intermediation $R_1D_2D_0R_2$. The interpretation is that in the first case of increasing extraction under unstable property rights (from Q_0 to Q_1), the producers or capturing agents find out ways of extraction in an unofficial manner through collusion among themselves. The extraction becomes more extensive (from Q_1 to Q_2) when the practice of extraction is supported by the management officials, through the process of intermediation.

Accordingly, the stability of rights is a pre-condition for sustainable resource governance, and it calls for the deconstruction of institutions for putting in order

[1] One of the authors used this proposition developed for this book in a previous publication to present a case on multiple values of nature in the context of the Sundarbans. Please see Titumir et al. (2019).

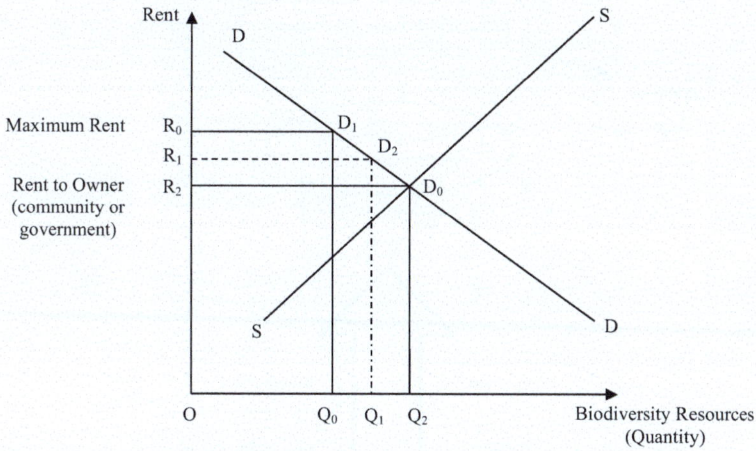

Fig. 3.10 Loss of revenue and additional exploitation. *Source* Prepared by the authors

of avoidance, evasion and missing rents, which could unleash adequate resources for sustainable conservation. The absence of regulatory regimes also allows rent fixation through collusions, captures and intermediation. Overall, necessity lies in the understanding, recognition, enforcement and maintenance of property rights and ensuring effective regulatory regimes.

The underlying hypothesis is that resources in the Sundarbans are being captured through collusion and intermediation by different agents due to the existence of unstable property rights and faulty institutional arrangements. Consequently, the dissipation of rent is the outcome of such faulty institutional arrangements that gives the scope to the agents to exercise power in order to extract the resources.

3.6.3 Proposition 3: Power, Political Settlement and Primitive Accumulation

The composition of power is highly associated with vertical and horizontal collusions that often results in the primitive accumulation of biodiversity resources.

The third proposition is that power relations or political institutions affect the process of property rights being unstable, which in turn affects the biodiversity resource use pattern in a negative way.

The nature of the power structure and political settlement of the transitional economy provide options for vertical and horizontal collusions of biodiversity resource appropriating groups of politicians, bureaucrats, the business community, industrialists, locally powerful agents and even international agents (Fig. 3.11). These factional groups influence the policymaking process, find the scope for primitive

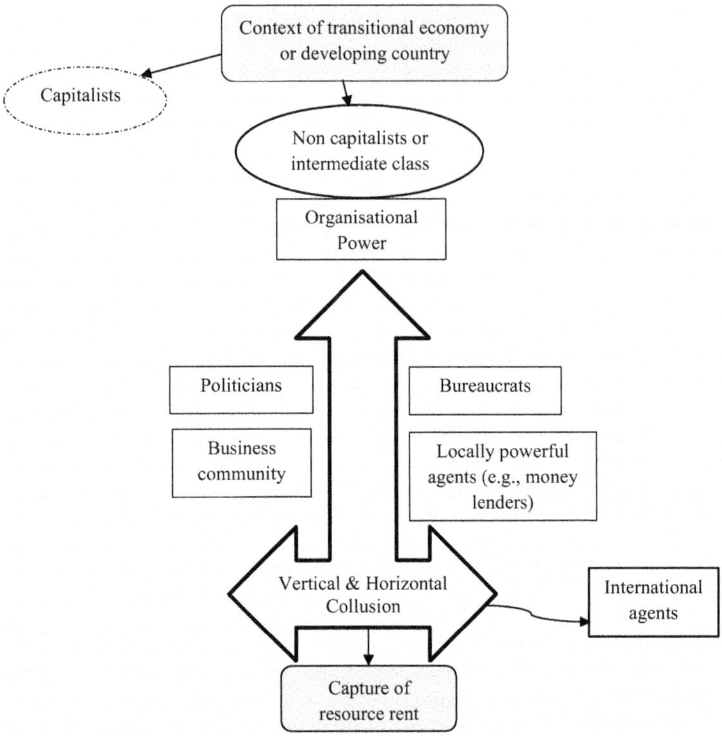

Fig. 3.11 Natural resource rent capture through horizontal and vertical collusion. *Source* Prepared by the authors

accumulation and capture the lion's share of rent earned from the utilisation of biodiversity resources. Some of them are involved in the rent-seeking behaviour.

In the context of a developing country, however, capitalists may have control over significant resources, but they are so few in number that they cannot control the political process. On the contrary, the intermediate class consisting of several groups plays a key organisational role within these factions. In particular, their organisational power involves the construction of cross-class alliances. In some cases, these groups also collude with international agents to maximise profit by both at the expense of loss of resources. In this process, thus, powerful agents (local and international) are allowed to extract or destroy resources even if the process is sometimes illegal. On the other hand, the rights of the local communities are hindered. Overall, the faction, by exercising its organisational power, influences the distribution process by undermining the equitable benefit-sharing mechanism.

The hypothesis is that there is always a competition among the powerful agents to capture the resources of the Sundarbans. They create collusions at different levels to meet this goal. This leads to the primitive accumulation and in turn, excessive extraction of resources.

3.6.4 Proposition 4: Collaboration and Well-Being

The well-being of the biodiversity resources depends on human sociality constructed by traditional norms, values and informal institutions.

The last proposition signifies that socially constructed norms and values, the informal institutions, are important for the conservation and sustainable management of biodiversity resources. The previous (three) propositions articulated the factors which induce the process of biodiversity resource degradation, and this proposition argues that recognising and revitalising the intrinsic relations between nature and human sociality are the key to sustaining biodiversity resources.

This can be done if informal institutions of a particular socio-economic context are recognised and supported by strong formal institutions under the governance framework of biodiversity resources (Table 3.10). Recognising human beings as social beings implies that they behave in a reciprocal manner and therefore, when their traditional knowledge of resource management would be recognised at a formal level through the effective enforcement of their property rights, they would try to conserve the resources in return.

The hypothesis is that the IPLCs of the Sundarbans, by utilising their traditional knowledge, can sustainably manage the resources of the region in the best possible manner and accordingly, their knowledge and experiences need to be institutionalised at the formal level.

3.7 Missing Institutions: Property Rights Instability and Marginalisation of Local People

Since the onset of formulating and defining property rights, the nature of the property rights of the Sundarbans has been remaining ambiguous. During the Mughal period, it was considered as open access forest for harvesting and converting for agriculture. Later, the British colonisers reigning over the Indian subcontinent realised the value of this mangrove forest and designated it as Reserve Forest (RF) in 1878. Thus, the government retained control of the forest, which was also the case during the Pakistan era. Following Bangladesh's independence in 1971, the Forest Act of 1927 declared the Bangladeshi portion of the forest to be RF once more. The Forest Policy of 1994, however, acknowledged community engagement in the management process and accordingly, recognised the rights of the local communities. As a result, the property rights structure of the forest can no longer be characterised in terms of a particular type of property rights (common or public). Rather, they are distributed between the state and the local people. A diagrammatic representation based on Schlager and Ostrom's (1992) typology of the bundle of property rights can explain the overall structure of property rights of the Sundarbans (Fig. 3.12).

Since 1994, the Forest Department (FD) has been responsible as the owner, proprietor, authorised claimant and authorised user on behalf of the state for ensuring the

Table 3.10 Conservation of resources through collaboration and cooperation under strong institutional (formal and informal) set-up

		Informal institutions (traditional knowledge or TK of the IPLCs)	
		Weak set-up (criteria: *TK is not recognised *TK is being regarded as old and unscientific methods *IPLCs are considered as destroyers)	Strong set-up (criteria: *TK is recognised *recognising that IPLCs are the protector)
Formal institutions (nature of property rights)	Weak set-up	**(Case-1: Extreme biodiversity resources degradation)** • Poor management • High extraction rate both by local people and outsiders • Resource depletion	**(Case-2: Resources degradation continues)** • The conservation system is ambiguous • The local powerful groups reap the benefits of TK • Rent seeking and rent dissipation • Impose heavy transaction cost to get access to resources • The IPLCs remain vulnerable and thus put pressure on resources
	Strong set-up	**(Case-3: Resources degradation continues)** • Rights of IPLCs are recognised but are unstable • Cannot utilise their knowledge properly • Sometimes have to follow directions of the concerned authority in case of collecting and using resources	**(Case-4: Sustainable and best utilisation of the biodiversity resources)** • Human obligation to protect the resources • Maximum benefit sharing • Sustainability

Source Prepared by the authors

efficient use of the Sundarbans' resources. On the contrary, the resource users have the right to access and use resources after receiving permission from the FD. Since 1994, however, the local people have had management rights along with access and withdrawal rights.

There are, however, insecure and ill-defined property rights that allow politically and economically powerful groups to encroach on the Sundarbans forest in illegal ways. Exploiters extract resources in an undesirable phenomenon because of legislative flaws and enforcement problems. *De facto* rights are responsible for resource degradation in most cases.

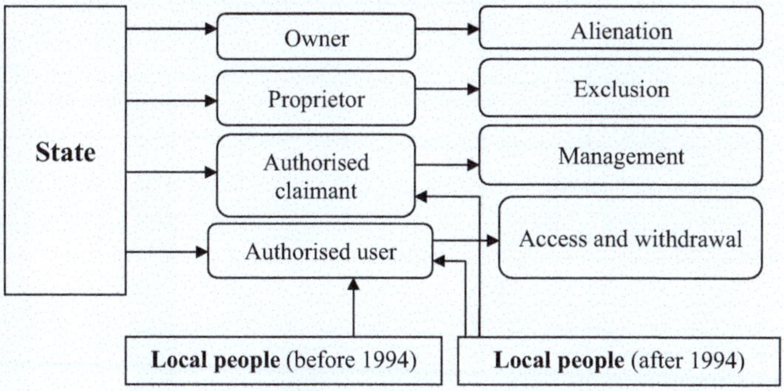

Fig. 3.12 Property rights structure of the Sundarbans. *Source* Prepared by the authors

The Sundarbans spans three districts: Khulna, Satkhira and Bagerhat. The density of population settlement in these three regions has been increasing over the years, and this trend is expected to continue as per the projection (Fig. 3.13). As a result, sheer dependence on the Sundarbans' biodiversity resources is also rising precipitously. The increased habitation is fundamentally the outcome of a fragile property rights regime over this socio-ecological production landscape. A large number of migrated people find it possible to intrude into the forest and therefore, intend to live in the adjacent regions of the Sundarbans.

They are not indigenous locals and thus do not adhere to local customary practices which are beneficial for the conservation of forest resources. As a result, they always intend to extract as many resources as possible accelerating the degradation process. Furthermore, politically and economically powerful groups are constantly encroaching on forest region by forming coalitions at various levels.

Then, from 1992 to 2005, there was an increasing trend of shrimp cultivation (Bagda—*Penaeus monodon*) in areas near the Sundarbans (Fig. 3.14). Shrimp farming contributes to the destruction and loss of mangrove habitats in a variety of ways. A shrimp-cultivating pond, for example, is only useful for three to six years

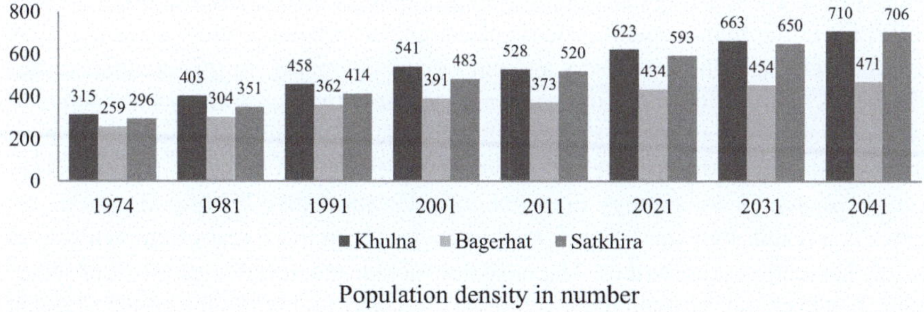

Fig. 3.13 Density of population in the districts encompassing the Sundarbans (in number). *Source* Authors' calculation based on population census of (2001) and (2011) by BBS

Upazila Name	1992	2001	2005
Shyamnagar	3750	22401	23612
Assasuni	4298	18243	21651
Paikgachha	14195	19733	21006
Kaliganj	2250	14488	16117
Rampal	15658	15792	15960
Dacope	7507	15206	15456
Morrelganj	3277	15137	15147
Mongla	9283	12567	12858
Debhata	5793	10442	11436
Koyra	196	7501	8788
Dumuria	2812	7092	7494
Bagerhat Sadar	3836	4835	4840
Batiaghata	2570	4143	4394
Satkhira	1166	3192	3807
Tala	740	3270	3740
Kachua	352	1054	1066
Sharankhola	0	147	172
Fakirhat	2330	89	99
Total (hectares)	**80,015**	**175,331**	**187,644**

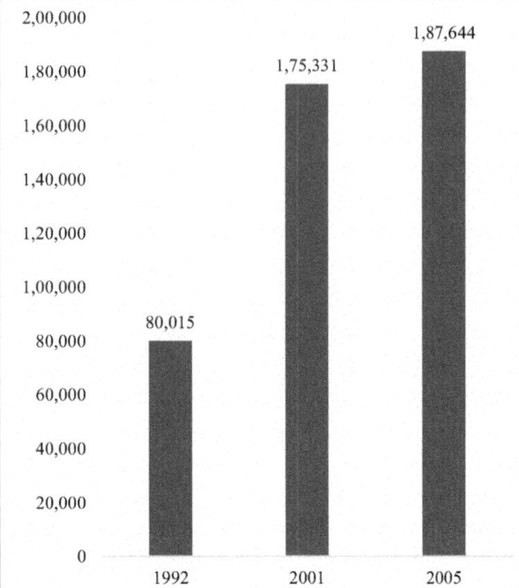

Fig. 3.14 Bagda shrimp cultivated areas adjacent to the Sundarbans (in hectares). *Source* Hussain (2014)

after construction. As a result, the cultivators must move along the coast, destroying mangroves to make space for more ponds. In addition, it affects the soil composition of that location by increasing the salinity-level. The increase in the number of farms is predominantly the result of a quasi-legal intervention. The farms are built by influential local stakeholders, particularly the rich fishermen (who are not part of the IPLCs), being allied with political and administrative structures at the local and national levels. They are violating the structure of property rights and finding scope to convert the mangrove-rich area into a commercial one to make profits without considering the ecological importance of it.

The unstable nature of property rights is also evident through the marginalisation of the IPLCs. The existing management structure of the Sundarbans largely excludes the IPLCs in the management process although the existing policy states that they have the right as authorised users and authorised claimants. In this context, exclusion means that the communities are unable to apply their customary knowledge to resource management. Their exclusion induced them to undermine the conservation process of the forest due to a lack of sufficient representation of their interests. Moreover, the management strategy does not consider alternative livelihood options for them. As a result, they are being forced to encroach on the forest often resulting in the overharvesting of resources. The most destructive and unstable nature of property rights, however, can be depicted through the process of allowing the non-traditional outsiders as resource users marginalising the local people through unfair means and corrupt activities. In this case, marginalisation signifies that the IPLCs have to overcome many a hurdle before they can enter the forest and access the biodiversity resources.

Moreover, the combined impact of climate change and the COVID-19 pandemic has aggravated the marginalisation of IPLCs in recent times. Most of them have lost their income without any alternative income-generating opportunities, leading to an increase in poverty. The absence of adequate, universal social security programmes exacerbated the distress in their lives and livelihoods. In the context of the hardships, approximately 95% of the IPLCs in the Sundarbans incurred a substantial income loss. The pandemic, on the whole, worsened the poverty rate increasing to 40% (Titumir, 2021).

The above discussion clarifies that the existing formal institutional arrangement is not stable in the Sundarbans. The rights of the IPLCs have been violated and external stakeholders have been included in the property rights structure. As a result, local people remain vulnerable in terms of their income status, which has been further aggravated in the face of the pandemic. Indeed, the interactions among multiple factors, inclusive of neo-liberalisation, state interventions and power relations have propelled the land alienation or denial of the right to access of the local people of that region. On the contrary, the outside stakeholders, who have no legal rights over the forest resources, have been getting access to the resource area through unfair means. Such fragile institutional arrangements ultimately induce the process of natural resource degradation in the Sundarbans.

3.8 Power, Politics and Degeneration of Biodiversity Resources

The critical evaluation of the programmes and policies for the forestry sector under different regulatory regimes indicates that the policies have been devised in such a way that some groups managed to reap the benefits.

Mughal Period (1526–1760) No comprehensive policy regarding forests, including the Sundarbans, was formulated during the Mughal period. The state recognised forest clearance as a way to generate substantial revenue by allowing agriculture (Roy et al., 2011). The Sundarbans was first surveyed by Mughal Emperor Akbar's revenue minister Todar Mall, in 1582, in order to settle the revenue (CEGIS, 2016). Thus, the clearance of the forest was given state recognition during that period.

British Colonial Period (1760–1947) The British government's hunger for revenue led to the clearing of the Sundarbans in both West Bengal and Bangladesh (Chacraverti, 2014). However, there is evidence too that the national policies gave importance to the conservation of forests for the first time during British rule. Several policies, acts and initiatives were adopted for the forestry sector of the whole continent including that of Bangladesh over several years (Table 3.11).

Table 3.11 Policy instruments for forest resources during British period

Year	Instruments/initiatives	Description
1855	Charter of Indian Forests	The first policy instrument recognises the importance of Reserve Forests (RF)
1864	Establishment of Forest Department	Actual beginning of forest conservation in Bengal
1865	Indian Forest Act	First Act that classified forests and legalised the reserved and protected forests
1871	Declaration of 6882 sq. miles of Chittagong Hill region as gazetted government forest area	First attempt by the state towards conservation of forests in the territory of Bangladesh
1878	Declaration of the Sundarbans as RF	Giving property rights to the state
1879	The Elephant Preservation Act	Wildlife conservation
1894	National Forest Policy	First forest policy, established the preference to agriculture over forestry, focused on earning revenue, restricted the rights of users
1927	Forest Act	Main purpose was to collect revenue

Note See Mustafa (2002), Kabir and Hossain (2008), Choudhury et al. (2009) for details
Source Authors' compilation from different sources

After 1813, *zamindars*, helped by the British government, started bringing the indigenous Munda tribe from Bihar in India to the Sundarbans to clear the forest for cultivation in order to convert it into a revenue-paying asset, which reduced its forest area considerably (Chacraverti 2014; Roy et al., 2011). Hence, the drive was to get the land settled as far as possible. Later, the British government recognised the importance of RF in 1855 and the Sundarbans was declared a reserve forest in 1878. The forest policy of 1894 and the Forest Act of 1927, however, indicated that the main purpose of the state was to collect revenue by exploiting forest resources.

Pakistan Period (1947–1971) In 1955, a reformed forest policy was adopted but it was also a continuation and outcome of colonial rule that exhibited similar characteristics (Khan, 2001; Rasul, 2005). It discouraged the involvement of local people in the management. Nevertheless, the policy for the first time gave importance to the beneficial aspects of forestry over commercial motives, including the protection and conservation of wildlife (Alam, 2009). The forest policy of 1962 again put the focus back on commercial motives of forest management and overlooked the rights of local communities. The policy was not conducive to the sustainable management of the Sundarbans at all (Hakim, 2007). In this manner, the emphasis on conservation in the forest policy of 1955 was nipped in the bud in 1962.

Post-Independence Period (1971-Onwards) After the liberation of Bangladesh in 1971, it was not possible to preserve wildlife and manage forest resources under the previous legislation. The government, therefore, adopted some new initiatives and policies (Table 3.12).

Table 3.12 Policy instruments for the forestry sector and the Sundarbans in the post-Independence period

Year	Instruments/initiatives	Description
1973	Bangladesh wildlife (preservation) order	Classified National Parks, wildlife sanctuaries, game reserves and private game reserves for protecting wildlife; designated the Sundarbans as 'RF' (SRF) under the Forest Act 1927
1979	Forest policy	Focused mainly on the reformation of forest department; aimed at scientific management and qualitative improvement of the forest; treated local people as threat for the forests
1994	Forest policy (revised)	Seeks participation of local people in forest protection (for the first time); promoted social forestry
1995	Environmental conservation act	10-km wide band surrounding the northern and eastern boundaries of the SRF, with an approximate area of 175,000 ha, was declared Ecologically Critical Area (ECA)
1996	Government notification	Bangladesh wildlife preservation order 1973 was modified; consider some parts of Sundarbans as protected areas
1997	Government notification	32,400 ha of the Sundarbans have been declared as three wildlife sanctuaries
1999	Government notification	The Sundarbans was declared an ECA (amended in 2010)

Note See Iftekhar and Islam (2004), Choudhury et al. (2009), BFD (2010), Roy and Alam (2012) for details
Source Authors' compilation from different sources

The first National Forest Policy of 1979 treated the local people as destroyers of the forests and failed to address the issue of their livelihood security. The reformed forest policy of 1994, however, is considered a landmark in conservation policy. It emphasises the local people's participation in the conservation process, to a limited extent though. The government also considered the conservation policy for the Sundarbans very quickly in the Post-Independence period. The Preservation Order of 1973 declared the Sundarbans as RF under the Forest Act 1927. The community participation and rights of the local people were recognised based on the forest policy of 1994. Nonetheless, the forest department, being the managing authority, uses the same 'command-and-control' policy as it did during the colonial era (Sen, 2010). In particular, none of the resource management plans was prepared in consultation with local people. The ultimate outcome is the continuous degradation of resources. In this way, the policy-level initiatives under different regimes signify that in most cases, the policies induced the resource extraction process.

Power dynamics in the Sundarbans region can also be illustrated through the examination of the industrialisation process and development projects undertaken

Types of Factory	Numbers
Cement factory	6
LPG	7
Gas cylinder	1
Oil refinery	3
Ship building	2
Saw mill	15
Betel nut processing	8
Rice mill	73
Fish farm and hatchery	19
Saline water refinery	7
Brick kilns	3
Others	46

Fig. 3.15 Factories near the Sundarbans. *Source* The Daily Star (2018)

during the past several decades. A number of industries had been established in the adjacent areas of the Sundarbans. In this case, the process of land grabbing is actually legal in its form. The powerful agents have been successfully pursuing the government to approve their projects by forming coalitions at different levels. In defiance of its own policy of not permitting the establishment of any factory in the Ecologically Critical Area (ECA), the government has allowed the setting up of 190 industrial and commercial units in the ECA of the Sundarbans in recent years, posing a major threat to the biodiversity resources (Fig. 3.15).

The majority of these agents and interest groups who are grabbing the lands adjacent to the forest are businessmen and industrial entities with strong political linkage. The competition for the land near the Sundarbans involves not only national and local agents but also international agents. The 'Rampal Power Plan Project', a coal-based power plant, is the most recent and contentious project, having triple risks in the three domains of environmental, economic and technical feasibility (Table 3.13). The project is under the process of implementation by the government with strong support from India. It would result in unequal benefit sharing between the two countries and pose threats to the integrity of the Sundarbans, particularly in the Bangladesh part. The UNESCO has already declared the Sundarbans as a 'World Heritage in Danger' site being discontent with measures taken to protect the Sundarbans in the purview of the implementation of the project (The Daily Star, 2019).

Thus, the expropriated areas through the land grabbing process deployed in capitalist production (industrial development) indicate the realised outcome of primitive accumulation. While the state plays a direct role in making the process possible and legitimate (*de jure* land rights are given), it is partly induced by neo-liberal policy regimes promoted by international stakeholders.

There is also the absence of equal power-sharing arrangements among the local-level stakeholders. These stakeholders can be divided into three groups broadly (a) Indigenous Peoples and Local Communities (IPLCs) as primary stakeholders, (b)

Table 3.13 Triple jeopardises of 'Rampal Power Plant Project'

Theme	Indicators	Remarks
Environmental loss	(a) Loss of fishing and agricultural lands	The ecosystems of the Sundarbans are under threat, especially the wetland ecosystem
	(b) Risk of coal transportation through the rivers	
	(c) Pollution of water through cleaning and discharging of coal (d) Largest source of air pollution burning 5 million tonnes of coal per year	
Economic loss	(a) Oversubsidised and risky	Unequal benefit sharing among Bangladesh and India; Marginalisation of local people and loss of their livelihoods
	(b) 70% of the capital cost of the project is proposed to cover from debt	
	(c) Loan repayment responsibility is not being shared between India and Bangladesh	
	(d) Displacement of 400 households and marginalisation of local people (e) Increase of health hazards of the local people, especially of children due to air pollution	
Technical faults	(a) Credibility gap in Environmental Impact Assessment (EIA)	The Sundarbans as a whole is under threat due to the location of the project at the closest proximity; adverse impacts of carbon emission on the mangrove ecosystem
	(b) Geographically vulnerable location	
	(c) Site clearance approval by the Department of Environment (DoE) without consulting the Forest Department (FD) violates law	
	(d) Gibberish argument about technology use to make the emission level zero	

Source Prepared by the authors based on Bank Tract (2015), Basu (2016), Islam and Al-Amin (2019), Sharda and Buckley (2016), TIB (2015)

the intermediaries group as secondary stakeholders and (c) the institutions as other stakeholders (Table 3.14).

Other stakeholders mainly include government agencies, officials and functionaries who are accused of being predatory in their own right. There are anomalies in fishing, honey, timber and *golpata (Nypa fruticans)* collection, protecting valuable species and monitoring the activities of pirates (Table 3.15). The traditional resource

Table 3.14 Local stakeholders of the Sundarbans

S. No	Stakeholder name	Description of stakeholders	Type of stake
Primary stakeholders			
01	*Bawalis*	Poor local and indigenous people (male)	Collecting woods
02	*Golpata* collectors	Poor local and indigenous people (male)	Collecting *golpata*
03	*Mouals*	Poor local and indigenous people (male)	Collecting honey/wax
04	Traditional fishers (occasional and subsistence fishers)	Occasional: poor local and indigenous people (mainly male; sometimes female and children) Subsistence: poor local and indigenous people (male and female)	Catching fishes
05	Post Larvae (PL) collectors	Poor local and indigenous people (male, female and children)	Collecting PL of *Golda* and *Bagda* shrimp
Secondary stakeholders			
01	*Mahajans*	Rich local people, influential persons, have monetary power	Lend money, purchase products
02	*Gher* owners and fish culturist	Rich, influential and powerful persons	Purchase shrimp PL from PL collectors, cultivate fish
Other stakeholders (Institutions)			
01	Department of Environment	Government body	In charge of resource management
02	Department of Fisheries	Government body	In charge of fisheries management
03	*Upazila* Administration	Local government	Management of *khas jalmahal* and leasing
04	*Union Parishad*	Local government	Management of *khas jalmahal* (small size) and leasing
05	Forest Department	Government body	Forest management and protection; biodiversity conservation and revenue collection
06	Police Department	Government body	Ensuring livelihood security
07	Bangladesh Navy and Coast Guard	Government body	Securing waterways

Source Prepared based on BFD (2010)

collectors, for example, must obtain access rights (e.g., BLC—Boat Licence Certificate) from FD in order to enter the forest. This process costs money but the collectors most often have to pay extra tolls in the form of bribes. Sometimes, they also have to pay money claimed by the dacoits. To cover such extra costs, they are compelled to collect resources more than they are permitted to, exceeding the sustainable limit, which has a negative impact on the forest's reproduction capacity.

Overall, the politically powerful ones have been able to encroach into the forest illegally with the direct collaboration of forest officials through bribery and other illegal means such as embezzlement and power abuse. Local people portray the

Table 3.15 Irregularities in resource collection and management

Dimensions of irregularities	Description
Irregularities in fishing	• Collection of extra toll along with regular fees of BLC • In exchange of bribe, allowing to catch fish within the prohibited areas such as wildlife sanctuaries, permitting to collect shrimp fries and promoting poison fishing
Irregularities in honey collection	• The FD is supposed to collect certain fees for certain amount of honey from the honey collectors; but in reality, collect a fixed amount as government fees without measuring the quantity • Vicious nexus of middle men, money lenders, corrupt retailers and big city-based honey exporters enjoy a major share of the big profit and IPLCs are being deprived
Irregularities in timber and *golapata* collection	• Log businessmen almost regularly traffic in valuable trees, including the *sundari,* with the help of the FD officials, local politicians and officials of local administration • *Golpata* collectors are bound to collect extra leaves for paying fees and bribe money
Irregularities regarding auction	• Different forest offices arrange need-based auctions of confiscated logs every month. Corrupt FD officials often hide the original quality and quantity of logs and thus help buyers to get logs illegally at a cheaper rate
Illegal tiger and deer poaching	• Both the tiger and the spotted deer are poached in the Sundarbans by professional as well as opportunistic poachers having coalition at the institutional level
Activities of the pirates	• It is nearly impossible to collect forest produce without paying tolls to these pirate groups • Pirates sought to maintain good relations with the local state machinery

Source Prepared by the authors based on Khoda (2008), Basu and Cetzal-Ix (2018)

relationship among the (illegal) business community, politicians and government agencies as a triangular-shaped power-sharing arrangement (Titumir et al., 2019).

Exploitation in terms of access to the credit market exists between the primary and secondary stakeholders. The intermediaries exploit the poor resource collectors by either lending money or providing working opportunities. The poor resource collectors need to depend on this informal credit source for borrowing money even at a higher interest rate as they have limited access to the formal ones. These moneylenders are also the ones who provide the primary stakeholders with the access to information, credit and the market. Thus, the intermediaries, both in money and organisational terms, use their power to make maximum profit from the collection of forest resources by exploiting the traditional forest community. This also has an indirect effect on the sustainability of the biodiversity resources as such exploitations pursue the poor resource collectors to extract large amounts of resources, although they do not want to, for compensating their monetary loss.

Overall, the discussion has revealed that under an unstable property rights structure in the transitional economy, the constellation of individuals and interest groups belonging to relatively influential classes with political and commercial involvement at local, national and international levels use alternative mechanisms in distinct ways to gain access to the resource land and acquire profits. The Forest Department is the most powerful authority whereas the local forest-dependent people have marginal or no power. Moreover, there are some middlemen who also use their monetary and political power to gain benefits by exploiting the resource-dependent communities and by extracting the forest resources. These powerful groups direct the pricing strategy of the resources and regulate the market structure (Titumir et al., 2019) which ultimately alienates the local people from having access to the biodiverse region and causes the degradation of biodiversity resources.

3.9 Pricing, Rent and Extraction of Forest Resources

The goods of the Sundarbans under the monetary framework are broadly categorised as (a) direct use goods (timber, non-timber, aquatic, etc.) and (b) indirect use goods (support for biodiversity, protection from cyclones, maintenance of environmental quality, etc.). There is no comprehensive study that accounts for all types of goods to determine the actual monetary value of this ecosystem. However, some studies measured the value of particular types of goods or services in the forest. According to one study, the economic value of the Sundarbans ecosystem in terms of provisioning, cultural and regulatory functions is approximately \$43 million per year (Uddin, 2011). Another conservative estimation suggests that all the ecosystem services and goods of this forest are worth between \$500 and \$1200 per hectare per year (Haque & Aich, 2014).

In particular, from 2001 to 2010, the amount of timber production was approximately 32,100 m^3 at an average rate of approximately 3567 m^3/year, which included seized timbers or damaged parts of trees resulting from natural catastrophes (Uddin,

2011). According to a previous study, the annual value of wood products removed from the Sundarbans Reserve Forest was $100 million and the value of standing timber was $2.09 billion (Islam & Gnauck, 2009). Timber collection, however, has been banned since 1991 following a moratorium issued by the Forest Department (FD) in 1989 in order to augment the density of trees in the Sundarbans (IUCN Bangladesh, 2012) and later due to some severe cyclones (*Sidr* and *Aila*) the commercial logging ban was in place up to 2015 (BFD, 2010).

Non-Timber Forest Products (NTFPs) including golpata (*N. fruticans*), malia grass (*Phragmites karka*), sun grass, goran, fuel wood, honey and wax are also allowed for collection under permission. The value of these products is measured in different studies (Table 3.16). About 16,000 maunds (1 maund = 37.32 kg) of honey and honeycombs (BFD, 2018a) and 50 tonnes of beeswax are collected from the Sundarbans annually under the management of FD (Gani, 2001) although the collection of honey has fallen over the last few years (Fig. 3.16). In the fiscal year 2019–20, however, the production of honey increased significantly because of low human movement during the COVID-19 pandemic. On a total basis, 1,220.50 quintals of honey and 366.15 quintals of wax were extracted (BFD, 2020).

The major aquatic resources of the forest are fish, larvae, crabs, etc. The annual value of fish caught is $304 million, which is three times larger than the annual value of other forest products (Islam & Gnauck, 2009). Moreover, on an average, $66,000 and $0.177 million annual revenue are collected from dry fish and crab, respectively (Kabir & Baten, n.d; Uddin, 2011).

Table 3.16 Major NTFPs of the Sundarbans, their use and monetary value

NTFPs	Use	Annual revenue and economic value
Golpata	Thatching material for making roof	Revenue in 2003–04 to 2007–08: $83,975 (Kabir & Baten, n.d)
Sun grass	Thatching material for temporary housing	Economic value (annual): $11,995,104 (Kabir & Baten, n.d)
Goran	Raw material for char-coal industry, tannin making and boat building	Economic value (annual): $0.18 million (Kabir & Baten, n.d.)
Fuel wood (e. g., *Hental, Kakra*)	Fuelling	Annual average revenue: $0.06 million (Uddin, 2011)
Honey	Food and medicine	Annual revenue in 2019–20: BDT915,375 (BFD, 2020); $53 per ha/household (Rahman et al., 2018)
Beeswax	Manufacturing, fragrance	Annual revenue in 2019–20: BDT366,150 (BFD, 2020)

Source Authors' compilation from different sources

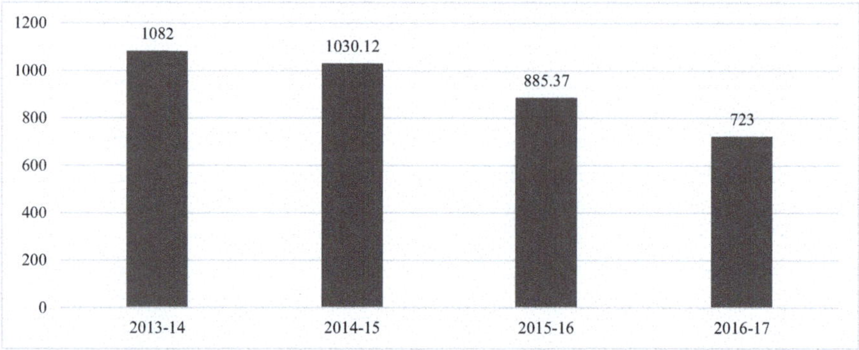

Fig. 3.16 Declining trend in honey collection from the Sundarbans (in quintal). *Source* BFD (2018a)

The commercialisation of the forest area has also been prompted through the nature recreation process by the tourism facility under the advent of globalisation. The Sundarbans, being a beautiful site for nature recreation, attracts tourists—both national and international. The most commonly visited sites in the Sundarbans are *katka*, *hiron point*, *dublar char* and *tiger point* (*kachi khali*). The number of tourists has increased considerably, from about 50,000 in 2001–2002 to about 180,000 in 2018–2019 (Fig. 3.17).

Tourism is also an important source of revenue for the Forest Division of the Sundarbans and the amount of revenue has been increasing gradually (CEGIS, 2016). Nevertheless, the unplanned and unrestricted influx of tourists is putting additional stress on the sensitive ecosystem (Basu & Cetzal-Ix, 2018).

The mangrove ecosystem of the Sundarbans is much more sensitive to the introduction of alien floral species for commercial profits, which threatens the existence of the native species. The government, as per the suggestion of donor agencies, has introduced some exotic species in the adjacent areas of the Sundarbans under

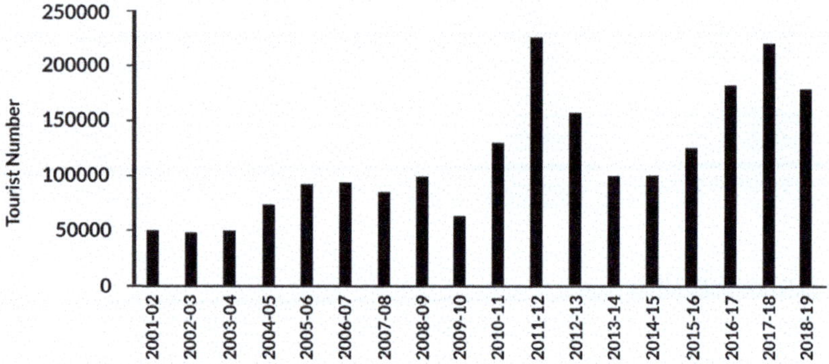

Fig. 3.17 Annual tourists visiting the Sundarbans. *Source* Mahmood et al. (2021)

the social forestry programme (Baten & Kumar, 2010). Around 23 invasive species were recorded in the Sundarbans that include 1 fern, 3 grasses, 4 shrubs, 2 epiphytes, 6 climbers, 2 herbs and 5 trees (Mahmood et al., 2021). The exotic species do not conform to the natural mangrove ecosystem and subsequently, alter the ecological process of the forest (Baten & Kumar, 2010).

3.9.1 Commercialisation and Unequal Rent Distribution

The discussion examines the process through which the commercialised forest products are being channelled into the market using a value chain analysis. Data for value chain analysis has been collected from a government report, a USAID report and a study conducted under *Unnayan Onneshan* (BFD, 2010; Islam, 2010a, Kabir & Baten, n.d). Firstly, the major actors in the resource collection process from the Sundarbans Reserve Forest (SRF) area under two different categories—in general and in terms of specific products—are summarised (Fig. 3.18). In addition, the classification of the commercialised forest products is presented thereafter (Fig. 3.19).

Availability of all the yields derived from the mangrove forest varies between seasons. The most common products of the Sundarbans are presented in a tabular form along with their harvesting period (Table 3.17). It is obvious that the pick extraction season of most of the mangrove products lies between May and December.

In a given year, the total number of collectors is estimated to be approximately 10.8 lakh. The resource collectors, however, mainly (almost 95%) work for other actors for wages. The majority of the collectors work for *boro mahajans* (43.4%), followed by *choto mahajans* (38.3%), *aratdars* (11.6%) and *farias/beparis* (4.7%) (BFD, 2010; Islam, 2010a). The marketing and distribution system of SRF products is actually a complex one but a basic structure can be presented in a simplified manner, as is usually the case with different products (Fig. 3.20).

Value addition statistics shows that the collectors add the highest percentage of value to different forest products but are always deprived in terms of net returns (Table 3.18). The exploiting actors, however, vary from product to product and are the lowest value-adding actors.

First, *golpata* collectors provide the highest value addition (49.7%) of the total price followed by retailers and the lowest value is added by *boro mahajans* (1.5%). The drastic opposite scenario can be depicted in terms of net returns. The *boro mahajans* have the highest share of net returns (36.8%) per month, followed by *aratdars* (33.5%) and *choto mahajans* (9%). The collectors, on the contrary, have net returns of only around 3% which is the lowest. However, in terms of net return as a percentage of the Working Circle (WC) is the highest for the *majhis* in this case.

Second, in the case of honey also, collectors add the highest value, accounting for roughly three-fifths (60%) of the total price. Other high value-adding actors are retailers, *majhis/beparis, boro mahajans*, wholesalers and *choto mahajans*. There appear to be no *aratdars* in the honey value chain, but wholesalers often act as

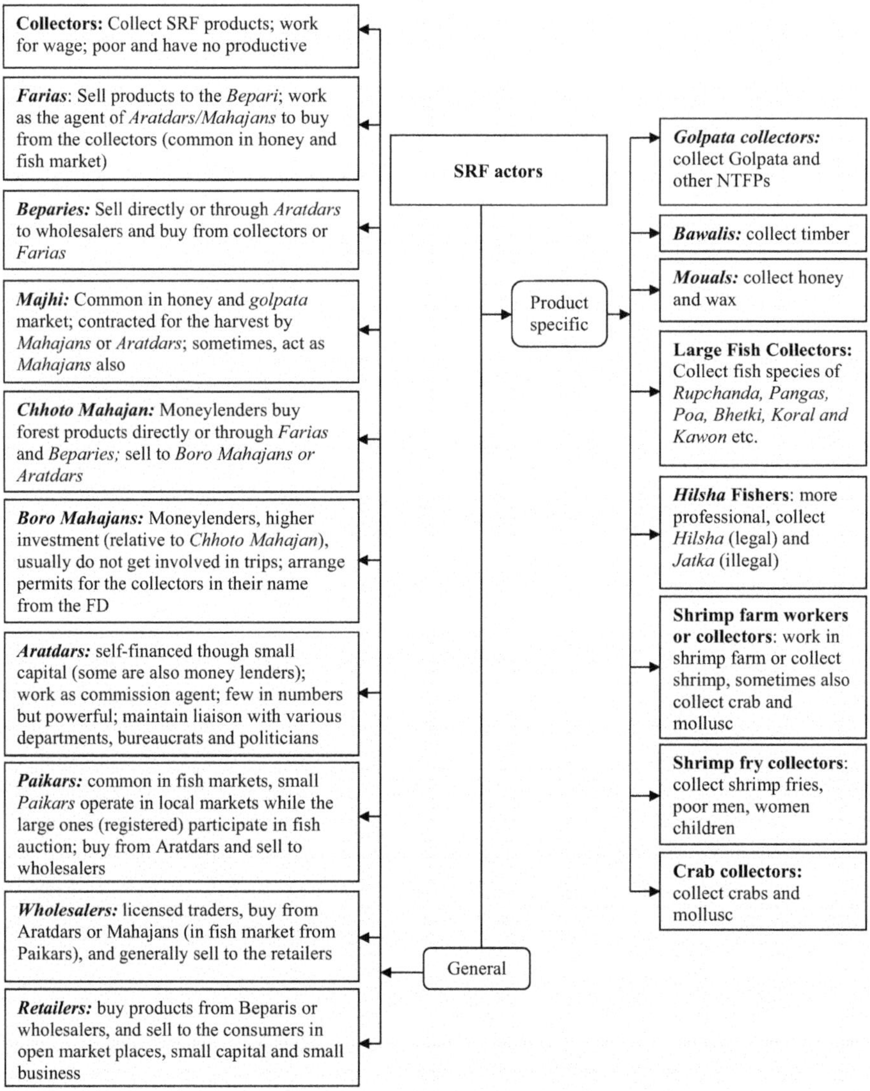

Fig. 3.18 SRF actors under value chain analysis. *Source* Prepared by the authors based on of BFD (2010)

aratdars. The wholesalers have the highest proportion of gross or net returns (28.4%) on a comparative basis, followed by *boro mahajans* (26.3%). The collectors have the lowest proportions also in this case in terms of net return per month (6.7%). Importantly, in terms of return as a percentage of WC, the proportion, however, is the highest for the collectors.

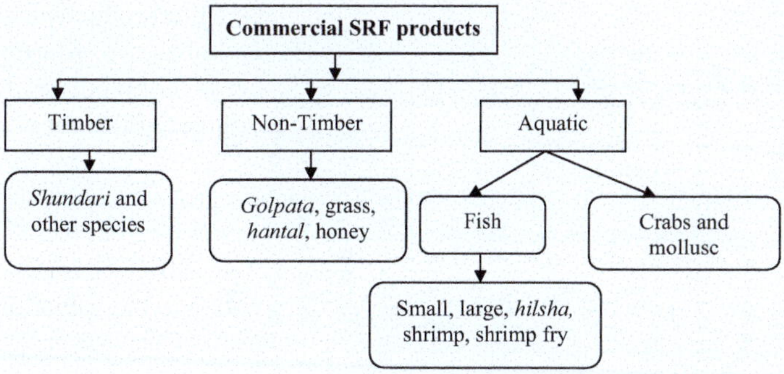

Fig. 3.19 SRF products under value chain analysis. *Source* Prepared by the authors based on BFD (2010)

Table 3.17 Sundarbans' non-timber and aquatic resources extraction calendar

Resources of Sundarbans	Jan	Feb	Mar	Apr	May	Jun	Jul	Aug	Sep	Oct	Nov	Dec
Golpata	⇑	⇑⇑	⇑⇑	⇓							⇓	⇑
Fish	⇑	⇓	⇓	⇓	⇓	⇑	⇓	⇓	⇑	⇑	⇑	⇑
Hilsha						⇓	⇓	⇑	⇑	⇑	⇑	⇓
Shrimp	⇓	⇓	⇓	⇓	⇑	⇑	⇓	⇓	⇑	⇑	⇑	⇑
Shrimp fry	⇓	⇑	⇑	⇑	⇑	⇑	⇓	⇑	⇑	⇑	⇓	⇓
Crab	⇑	⇑	⇓	⇓	⇓	⇓	⇓	⇓	⇓	⇑	⇑	⇑
Honey and wax			⇑	⇑⇑	⇓							

Note Double arrows indicate the dramatic increase or decrease of extraction;
Major extraction ⇑
Reduced extraction ⇓
Source Kabir and Baten (n.d)

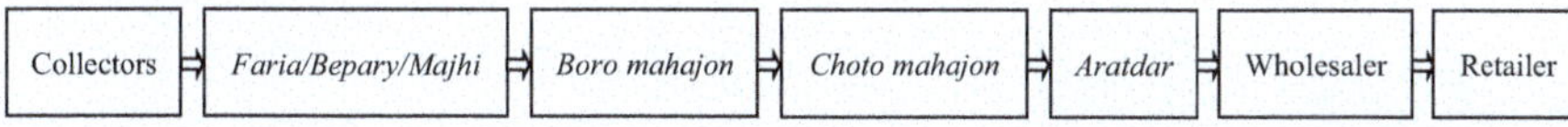

Fig. 3.20 Simplified and typical SRF marketing system and value chain of the actors. *Source* Prepared by the authors

Third, crab collectors, again, provide the highest value addition (50% of the total price). *Majhi/farias* produce the next highest value addition, followed by *choto mahajans, aratdars* and others. Accordingly, *aratdar* ranks third in terms of adding value (8.3%), but has the highest monthly return (29%) and lowest return again is for the collectors. Fourth, in the small fish market, comparatively the *aratdars* have the maximum net returns (59.4%), followed by wholesalers, retailers and *choto mahajans.* In absolute terms, collectors or *beparis* have gross or net returns of only about 5–6%. Fifth, in the large fish market, collectors, obviously, provide the highest value addition (63% of the total price). Leaving collectors aside, like in the case of *gura*

Table 3.18 Value additions and returns for SRF products (in %)

Actors	*Golpata*			Honey			Crab			Small fish		
	Price value addition	Net return (in month)	Net return as % of WC	Price value addition	Net return (in month)	Net return as %of WC	Price value addition	Net return (in month)	Net return as % WC	Price value addition	Net return (in month)	Net return as % WC
Collector	49.7	2.7	–	60.0	6.7	119.4	50.0	4.1	158.2	64.6	3.8	72.4
Majhi/Bepari/Faria	11.2	4.2	122.0	12.0	12.9	64.8	17.6	6.3	27.0	9.2	6.6	12.9
Choto Mahajan	12.7	9.0	22.7	1.3	17.8	29.3	13.8	12.1	17.6	1.5	8.8	10.9
Boro Mahajan	1.5	36.8	23.3	6.7	26.3	12.4	6.9	21.5	4.6	–	–	–
Aratdar	6.1	33.5	25.1	–	–	–	8.3	29.0	24.6	4.6	59.4	11.1
Wholesaler	5.1	8.2	7.5	3.3	28.4	8.9	3.4	26.9	5.3	7.7	12.2	9.1
Retailer	13.7	5.5	12.7	16.7	7.8	18.5	–	–	–	12.3	9.2	78.7
Total	100.0	100.0	–	100.0	100.0	–	100.0	100.0	–	100.0	100.0	–

Large fish			*Hilsha*			Shrimp (*Bagda*)		Shrimp fry (*Bagda*)	
Price value addition	Net return (in month)	Net return as % WC	Price value addition	Net return (in month)	Net return as % WC	Price value addition	Net return (in month)	Price value addition	Net return (in month)
62.5	4.6	239.4	63.3	4.2	–	66.7	5.5	64.1	16.8
11.5	5.8	56.0	10.0	7.3	91.2	6.7	8.2	19.2	22.1
1.0	7.2	66.6	8.3	21.0	59.8	4.4	8.6	–	–
1.0	39.8	45.4	1.0	31.3	21.3	4.4	10.9	–	–
4.0	21.4	6.4	2.7	23.1	12.3	3.6	37.8	16.7	61.1
4.5	11.7	12.0	2.3	8.7	–	3.1	19.3	–	–
15.5	9.5	103.4	12.3	4.5	–	11.1	9.7	–	–

(continued)

Table 3.18 (continued)

Large fish			*Hilsha*			Shrimp (*Bagda*)		Shrimp fry (*Bagda*)	
Price value addition	Net return (in month)	Net return as % WC	Price value addition	Net return (in month)	Net return as % WC	Price value addition	Net return (in month)	Price value addition	Net return (in month)
100.0	100.0	–	100.0	100.0	–	100.0	100.0	100.0	100.0

Note Please see BFD (2010) for details on primary data collection methodology
Source Prepared by the authors based on BFD (2010)

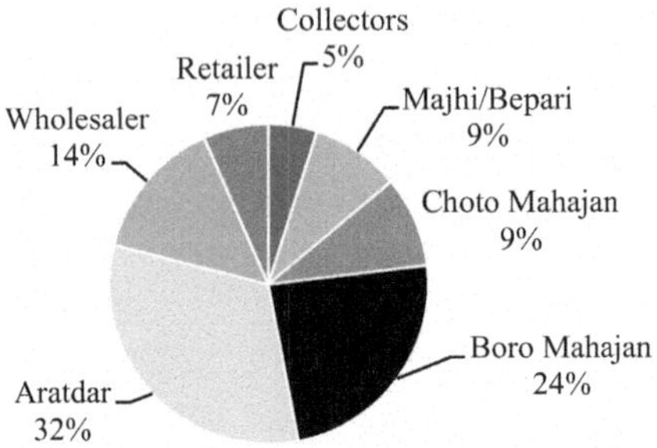

Fig. 3.21 Annual income level of SRF actors (%). *Source* Prepared by the authors based on BFD (2010)

fish, retailers receive the highest value addition, followed by *farias* (who are also frequently involved in collection), *aratdars*, wholesalers and *choto mahajans*. The *boro mahajans*, in this case, have the highest gross or net returns (39.8%).

In the same way, *hilsha* fish, shrimp and shrimp fry collectors have added the highest value (63.3%, 66.7%, and 64.1%, respectively). On the other hand, the highest net return earning actors are *boro mahajan* for hilsha fish (31.3%), *aratdar* for shrimp (37.8%), and shrimp fry (61.1%). Overall, the collectors are the ones who produce price value addition by as high as 50–75% and earn a net return in the range of only 3–7%.

The existing unequal benefit sharing in terms of net return through the extraction of forest products also deepens the degree of the poverty level. It signifies the widening gap between the actors in terms of income level. When all SRF products are considered together, the average income earned by an *aratdar* or a *mahajan* is found to be almost five to seven times of the collector's income. In terms of the total income share, the income of collectors constitutes only 5%, whereas the *aratdars* are the highest income earning actors (32%) followed by *boro mahajans*, wholesalers, *choto mahajans* and *majhi/bepari* (Fig. 3.21).

The existing inequality can also be demonstrated in terms of the Gini coefficient—a measure of inequality (value 0 signifies perfect equality and 1 perfect inequality). The estimated value of 0.52 in terms of all products signifies quite high inequality (Fig. 3.22).

The value in the case of specific products shows, additionally, that the highest inequality exists in the small fish market (0.53), followed by crab (0.52), and *golpata* (0.51) markets. Correlatively, Gini coefficients for large fish and shrimp (0.44) and for *hilsha* (0.48) are a bit lower compared to other SRF products. The lowest degree of inequality can be observed in the case of honey (0.39). These variations clearly signify that in each type of market structure for different forest products of the Sundarbans, there is the existence of inequality among different agents, especially between the powerful groups and resource-dependent communities.

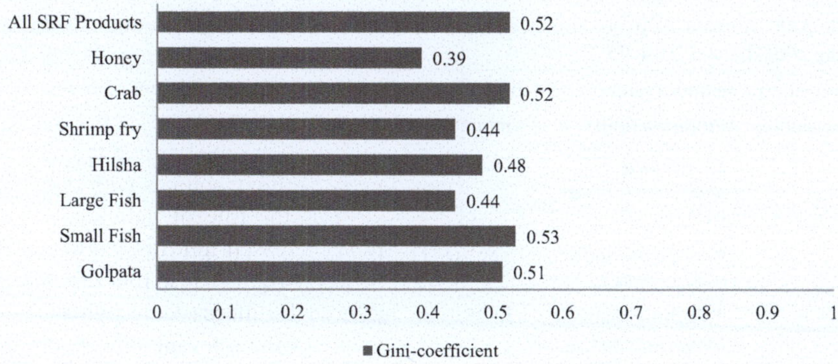

Fig. 3.22 Inequality based on occupational pattern in terms of products. *Source* Prepared by the authors based on BFD (2010)

3.9.2 Loss of Forest Revenue: Evidence of Rent Dissipation

The majority of resource collections from the Sundarbans have decreased over the years, resulting in a loss of revenue for the government as well (Table 3.19). It becomes visible thus that the extraction of the most of the forest resources has increased. As a result, the producer surplus of the forest has reduced and consumers are getting fewer resources day by day.

The overall growth of the real forest revenue is also found to be waning over the years. It has often been claimed that the growth of income from revenue collection by the forestry sector has increased over the years, which is actually factual in nominal terms (Fig. 3.23a). The trend in real value signifies the decreasing trend (Fig. 3.23b). The Sundarbans, as the single most important forest area, contributed about 41% of the total forest revenue (BFD, 2010). Hence, the declining trend of forest revenue indicates that the revenue earned from this forest has also reduced.

Then, year-to-year revenue growth in real terms exhibits a fluctuating trend from 1996–97 to 2015–16 (Fig. 3.24). In recent times, however, the revenue has declined sharply as the data indicates.

The trend in revenue thus reveals that the natural resource rent is being dissipated and absorbed into the non-market transactions arena because of unequal power-sharing arrangements and the existing nature of political settlement in the transitional economy of Bangladesh. The ultimate result is that neither can the resources be conserved at a sustainable limit nor can the economy benefit from revenue earning through the sustainable utilisation of the resources.

Overall, the discussion demonstrates that turning the 'forest resources' into 'commodities' through monetary valuation induces the powerful agents to extract the resources. Accordingly, the intermediaries in the market system intend to exploit the poor local communities to fulfil their goal of earning more profit. Such valuation through market penetration does not count the social benefits and promotes inequity and anti-competitive behaviour. The persistent and existing nature of inequality in

Table 3.19 Amount of major resources and respective revenue earnings from the Sundarbans during 2001–02 and 2014–15

Types of ecosystem services		2001–02		2014–15	
Provisioning	Produces (unit)	Amount	Revenue ($)	Amount	Revenue ($)
	Excoecaria agallocha (*Gewa*) (ft^3)	84,630	33,187	6026	3894
	Ceriops decandra (*Goran*) (no.)	15,865 (MT)	47,742	118,451 (no.)	7520
	Thatching material *Nypa fruticans* (*Golpata*) (MT)	17,525	33,123	16,868	57,338
	Thatching material grass (MT)	3621	790	668	225
	Phoenix paludosa (Hental)	543 (MT)	348	19,761 (no.)	1044
	Fuel wood (ft^3)	69,370	47,523	14,455	10,190
	Honey (MT)	84	7970	67	24,048
	Wax (MT)	23	1665	63	8108
	Fish (MT)	2061	58,374	3432	158,368
	Crab (MT)	123	2148	1123	52,026
	Dry fish (MT)	1095	18,998	2773	179,761
Cultural	Tourist (No.)	59,169	14,588	100,817	144,832

Source BFD (2015)

case of income and return distribution, therefore, compels both the local poor and the powerful ones to extract the resources as much as possible. The poor ones are forced to extract more resources to compensate for their loss in the case of having returns from their effort. The powerful, on the other hand, knowing that the cost of production for them is too low is always inclined to capture the resources without any concern for conservation. The national economy is ultimately losing its revenue over the years as the resources are degrading. Naturally, the overall commercialisation of resources results in the destruction of biodiversity resources and the dissipation of rent.

3.10 Traditional Knowledge and Cooperation for Sustainable Management of Biodiversity Resources

The relationship of the forest-dependent people with forests is intimate, reciprocal and spiritual. This is the key to sustainable conservation and utilisation of biodiversity resources of the Sundarbans. This mutuality between humans and nature is ensured in

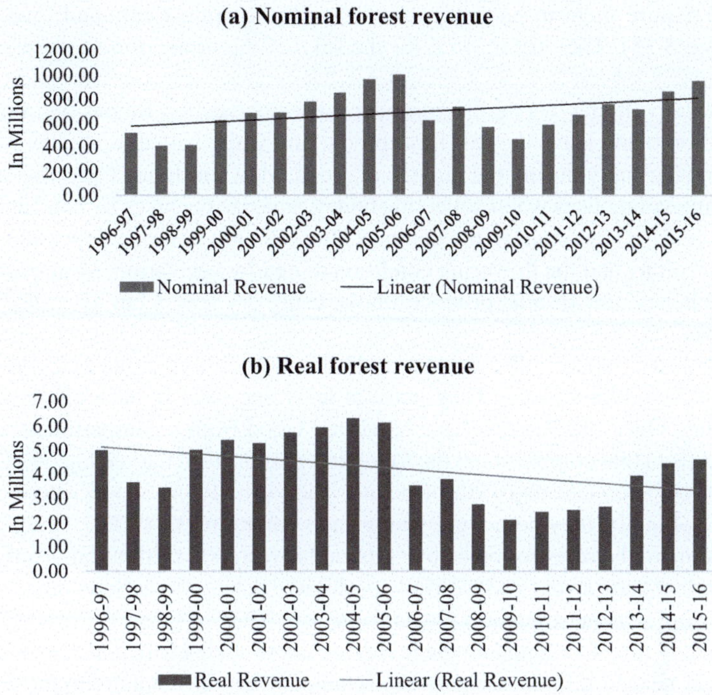

Fig. 3.23 Nominal and real forest revenue earned by the FD. *Source* Author's calculation based on data obtained from BFD (2018b)

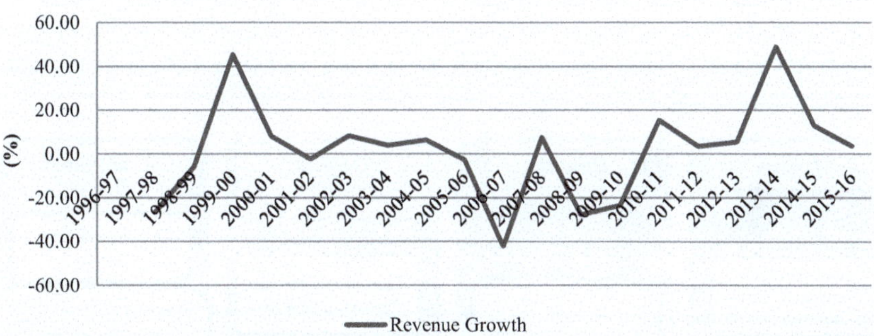

Fig. 3.24 Revenue growth in real term (year-to-year basis). *Source* Author's calculation based on data obtained from BFD (2018b)

one way in which the local and indigenous people apply their traditional knowledge to sustainably utilise the resources.

The IPLCs of the Sundarbans sensibly believe that as the forest provides their livelihood, it must be safeguarded from all forms of overuse and abuse for both current and future generations (Titumir et al., 2020). Therefore, they follow some rules that require them to harvest resources with the utmost care and love for the nature (Fig. 3.25). They try to do it for the sake of the protection of themselves in the future.

The *Mouals* (honey/wax collectors) cut a specific section (about two thirds) of the honeycomb and leave the rest for reproduction while collecting honey from the honeycombs, usually during the months of April, May and June. They also strive to avoid killing young bees when collecting honey, squeeze beehives by hand and never use metal tools. They return to the colonies after a month or more, depending on the size of the colony and the flowering condition of nearby vegetation. Moreover, while collecting honey, the *Mouals* produce smoke using dry leaves but never use fire on the beehive.

The *Bawalis* (wood collectors) in a similar way follow some rules to ensure sustainable wood harvesting. They leave at least one stem in each clump of trees after cutting. Once the *Bawalis* have harvested wood from a compartment, they do not collect from that compartment the following year, instead harvesting on a cyclical basis to allow adequate regrowth of plants. They usually cut wood where there is plenty of it. Finally, they do not cut young and straight trees.

According to the rules followed by *Golpata* harvesters, exploitation in any area is only permitted once a year and is strictly prohibited during June–September, when it is the growing season of *golpata (Nypa fruticans).* They only cut the leaves that are approximately nine feet long and the leaves are cut in such a way that the central leaf and the leaf next to it in each clump are retained. They also adhere to the rule that the flowers and fruits must not be disturbed in any way when cutting leaves. They also maintain that young plants with only one utilisable leaf should not be pruned.

The *Jele* (traditional fishermen) follows several rules while catching fish. Because the fishermen are aware that catching fish fry will eventually reduce the number of fish in the water bodies, they avoid doing so. They do not use '*jal*' net (very small-meshed net) usually. They use nets such as *behundi jaal* (bag net) or *char-paataa*

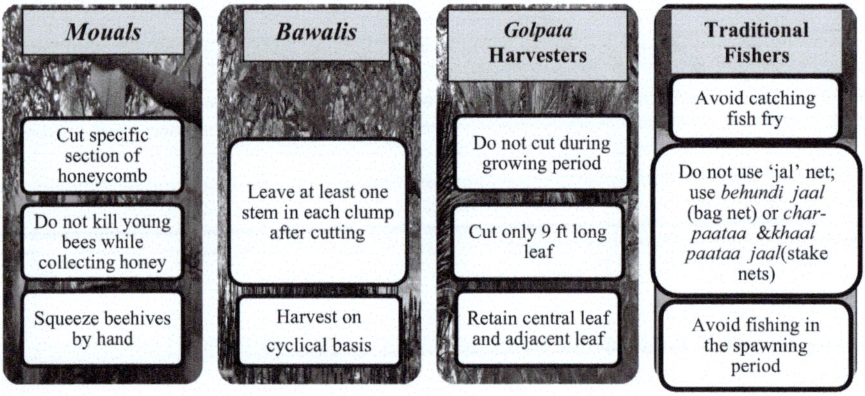

Fig. 3.25 Traditional rules and practices followed by IPLCs occupational groups at a glance. *Source* Prepared by the authors

and *khaal-paataa jaal* (stake nets), which have been innovated and customised to benefit the distinct waterscape of the Sundarbans. They use big-meshed net for rivers and small-meshed net for closed water bodies. They do not catch all fish species and also avoid fishing during the spawning season. They make their boats with *goran* and *sundari* mostly (see Titumir et al., 2019 for details on differences between traditional and non-traditional rules followed by fishers).

The IPLCs believe that the Sundarbans as "a tidal forest is a sacred place and the Creator washes the forest twice a day and maintains its sanctity and therefore tries to maintain sustainable use of the forest" (Titumir et al., 2020, p. 81). Irrespective of religion, they believe in the existence of *Bonbibi* (the main goddess of the Sundarbans) and other gods and goddesses. The Hindus, among them, however, also believe that there is another special god for honey whose name is *Dakshina Roy*. Thus, they consider the forest region to be a holy place because of the existence of gods and goddesses and therefore, try to keep the forest safe and pure. *Bawalis,* in particular, feel that their minds become cool when they stay in the forest. On the other hand, to the *Mouals*, honey is sacred food and therefore, they remain careful not to pollute while collecting it. Above all, they value the Sundarbans as their 'mind' denoting that it is equal to their life (Titumir, 2022).

In addition to the above-mentioned traditional rules, practices and cultural beliefs that have been passed down through generations, the IPLCs have recently diversified their livelihood options by innovating diverse production methods and techniques. They are doing so in response to the ongoing deterioration of their livelihood opportunities due to anthropogenic pressures (e.g., degradation of forest resources, loss of agricultural land) and natural drivers (e.g., climate change). These techniques are novel because the IPLCS developed them to improve their adaptation capacity to the changing circumstances.

Small farmers in the area have developed some innovative agricultural techniques that are adaptable to local biophysical conditions while also maintaining environmental sustainability. They are now cultivating rice seedlings on raised land to limit the risk of saline water contamination and ensure maximum survival in the face of climate change. These seedlings are then transplanted into the main agricultural land. For example, to respond to higher salinity in soil and water, they harvest rice plants at 8–12 inch high from the ground. Practically, this saline contaminated rice straw is decomposed within a very short time if these are used as roofing materials. They, therefore, allow those to be decomposed in the field, which in turn adds organic matter, mainly nitrogen, to the soil and also reduces the saline intensity, both of which are conducive to the growth of their next crop. Furthermore, those who do not have access to land, grow vegetables on sheds or roofs as well as in the front or back yards of their homes (Baten & Kumar, 2010; Titumir et al., 2020).

Then, they have innovated Community-based Mangrove Agro Aqua Silvi (CMAAS) culture as an alternative to Commercial Shrimp (CS) culture (Titumir et al., 2020). The CMAAS culture refers to the practice of integrated cultivation of some mangrove faunal species—crabs, oysters or fish (e.g., shrimps, bhetki [*Latescalcarifer*], etc.) and floral species—golpata (*Nypa fruticans)*, keora (*Sonerati aapetala*), goran (*Ceriops decandra*) etc. at the same time on any swampy land of brackish water.

In addition, integrated cultivation of some mangrove floral species like *golpata* and a few faunal species like tengra (*Mystus tengara*), baila (*Awaous guamensis*), tilapia (*Tilapia nilotica*), etc., are practised in a freshwater swampy land. The CMAAS culture has been found to be profitable (Table 3.20).

The CMAAS culture has little or no negative impact on the Sundarbans ecosystem. On the contrary, the commercial shrimp cultivation has had a significant adverse ramification on the Sundarbans, as has already been stated in previous discussions. Here, a comparative analysis of these two types of culture is provided in summary (Table 3.21). In this case, the Cost–Benefit Analysis (CBA) method was used to compare the economic returns. In terms of Net Present Value (NPV) and Net Benefit (NB), CMAAS culture looks more profitable than Commercial Shrimp (CS) culture. However, the scenario is quite different when considering Benefit–Cost Ratio (BCR). The BCR scenario implies that the cost effectiveness of CS culture is comparatively higher. Shrimp cultivation, therefore, is unquestionably profitable. However, the beneficiaries are the powerful group of people, and regrettably, it has had a negative impact on the livelihoods of the landless and marginal farmers.

Moreover, the ecological comparison (Table 3.22) validates that the CS culture is extremely harmful to the environment, whereas the CMAAS culture has a negligible or no adverse impact on the environment. The ecological benefits of the CMAAS culture clearly show that it protects land and soil from erosion, ensures better utilisation of fallow land, protects the environment from pollution, helps conserve biodiversity resources of the Sundarbans and, most importantly, provides IPLCs with alternative and sustainable livelihood opportunities.

The CMAAS culture is thus a unique method aimed at adapting to climate change in the coastal region. By inventing this method, the local communities have displayed a strong sense of ownership and a scope for scalability.

Thus, the IPLCs, living adjacent to the forest area, have been practising several unique production methods based upon their traditional knowledge and experience that can significantly contribute to the sustainable management and conservation

Table 3.20 Economic return of CMAAS culture

CMAAS culture		
Economic return (benefits > cost)	Mangrove cultivation (flora)	Mangrove aqua farming (fauna)
	Total income (per *Bigha*/per year): BDT 56,250	Total income (per *Bigha*/per year): BDT 183,000
	Total cost (per *Bigha*/per year): BDT 1800	Total Cost (per *Bigha*/per year): BDT 14,750
	Net benefit: BDT 54,450	Net benefit: BDT 173,250
	Cost–benefit ratio: 1:32	Cost–benefit ratio: 1:12

Note A *bigha*, a unit of land measurement, is 1600 yd^2 (0.1338 ha or 0.3306 acre) and often interpreted as being 1/3 acre (it is precisely 40/121 acre). In metric units, a bigha is hence 1333 m^2
Source Prepared based on findings of the research by *Unnayan Onneshan* (2010b)

Table 3.21 Value of CBA measures of CMAAS and CS culture

Measures of CBA	CMAAS culture (BDT/*bigha*/yr)	CS culture (BDT/*bigha*/yr)
Present Value of Costs (PVC)	16,550.00	8860.00
Present Value of Benefits (PVB)	217,500.00	177,272.72
Net Present Value (NPV)	202,454.54	169,218.18
Net Benefit (NB)	200,950.00	168,412.72
Benefit–Cost Ratio (BCR)	13.00	20.00

Source Prepared based on findings of the research by *Unnayan Onneshan* (2010b)

Table 3.22 Ecological comparison between CMAAS and CS cultures

Criteria	CMAAS culture	CS culture
Salinity	No use of saline water; no salinity intrusion	Increases salinity in soil (in farmland and adjacent land)
Use of land	Homestead adjacent fallow lands are used and no conversion of forest land into cultivation land	Used ponds exhaust usefulness within three to six years of construction. So, destruction of mangroves occurs to make room for more ponds
Use of chemical fertiliser, pesticides, insecticides	No usage of chemical fertiliser or insecticides, natural feeding, and therefore, no pollution	Chemical fertiliser, insecticides, etc. are used, causing pollution
Impact on agricultural productivity	Does not affect agricultural productivity	Restricts crop production in agricultural land (by increasing salinity of land) and conversion of agricultural land to shrimp farming ponds reduces land availability
Impact on the Sundarbans (in particular)	Eases and reduces the increasing anthropogenic pressures, making an alternative source of livelihood for the local people who are dependent on the Sundarbans	Eradication of natural mangrove vegetation, and pollution of aquatic resources (negative)
Adaptation to climate change	An innovative adaptation method to climate change for the vulnerable	Increases the vulnerability to climate change

Source prepared based on findings of the research by *Unnayan Onneshan* (2010b)

of biodiversity resources through symbiotic human–nature relationships (Titumir et al., 2020). Accordingly, these unique informal institutions match the theory of sustainability.

Overall, there is a coexistence of customary and official systems in the management process of the Sundarbans, though the customary rights are not effective fully. The problem actually lies in the lack of appropriate negotiation between customary and legal management systems. If the customary and legal rights are juxtaposed, it can be found that they are sometimes complementary and mostly contradictory (Table 3.23). The Sundarbans is a state-owned forest but by tradition, it belongs to the local communities.

The exclusionist approach by the Forest Department has wielded negative effects on the resource in the long run. The solutions used over the past several years are based on modern science and technology while ignoring the traditional knowledge of the IPLCs. The ultimate result is the wastage of resources and increased vulnerability of the locals. Therefore, integrating the local people's traditional knowledge into the formal governance framework may encourage the sustainable recovery of the degraded ecosystem of the Sundarbans.

Human beings are social beings as the central thesis of this book implies and therefore, act in a reciprocal manner. Whenever their rights over resources and protection strategies are recognised fully, they would behave in a positive way to protect the resources. In the same way, it can be expected that the IPLCs would also play a reciprocal role by utilising their traditional knowledge fully, which can be the best possible strategy to tackle the ongoing problem of resource degradation in the biodiverse-rich region of the Sundarbans. Overall, recognition of the traditional rights of forest resource users, indigenous knowledge, traditional cultural practices, beliefs, histories, philosophies and customary use of forest resources ensure the sustainable use of mangrove biodiversity resources. Their knowledge base, therefore, is to be comprehended, adopted and applied.

3.11 Concluding Remarks

This chapter scrutinises the way in which the biodiversity resources of the Sundarbans are being managed, utilised or degraded in the particular context of a transitional economy in Bangladesh on the basis of both theoretical discussion and empirical evidence and offers a new understanding that can ensure the sustainability of those resources.

There are several anthropogenic pressures that cause the degradation of biodiversity resources in Bangladesh in general and of the Sundarbans in particular. They also hamper the balanced relationship between the *biotic* and the *abiotic* components. The anthropogenic pressures have primarily exaggerated since the advent of neo-liberalism as the exclusive strategy of accumulation of wealth, with profits being prioritised over the intrinsic ecological value of biological resources through the commercialisation of forest products (Titumir et al., 2020).

Table 3.23 Resource rights over the Sundarbans

Customary rights	Legal rights	Status	Entity
Bawalis cut wood where there is abundance, they do not cut young and straight trees, ultimately they manage the forest sustainably	After the moratorium of 1988 no trees would be cut in the SRF	Contradictory/de facto	User right
Exploitation is not allowed in any area more than once a year and it is not allowed during the months of June–September	The *Golpata* collectors get permission in mid-November to mid-March. Sometimes *Bawalis* cut *golpata* by getting permission	Complementary but somehow contradictory due to activities as paid labour/de facto	User right
While collecting honey, *mouals* produce only smoke rather fire. During the collection, they always ensure that no young bees are killed *Mouals* do not enter the forest on Fridays because they believe that the goddesses are busy with their prayers that day	According to Forest Act 1927 (Section 32h) fire protection is ensured in protected forest When the *mouals* enter the forest from mid-March to mid-June, they have to work the whole week following the decision of the boat owner (*Mahajan*)	Complementary with the Forest Act, 1927 but contradictory as they have to work the whole week for collecting honey/de facto	Access right
The traditional fishermen go to the forest to fish every day from midnight to 4 am and return between 10 am and 12 noon	Forest Act 1927 empowers the FD to manage the inshore and offshore fisheries of the Sundarbans *Khal (Bengali name of canal)* closure regulation, 1989 closed 18 *khals* for fishing for safe fish breeding	Contradictory/de facto	Access right
Traditional forest users always see the forest as their way of life. Most of them have no boat of their own, so they have to work under boat owner	According to the Forest Department criteria, traditional forest users are those who have a boat	Contradictory/de facto	Boat ownership

Source Adapted from data reservoir of *Unnayan Onneshan*

The incessant encroachment into the forest region, the conversion of mangrove forestland into commercial shrimp culture farms and the marginalisation of IPLCs all point to institutional fragility in the management of the Sundarbans. The fragility is obvious through several legal and quasi-legal intrusions by a number of powerful entities. Both the formal and informal powerful agents are found to be highly organised in the case of extracting the biodiversity resources and most often, enjoy political patronage as well as administrative support. The powerful class is in fact inclined to capture the resource rent as the biodiversity resources are turned into valuable goods due to the commercialisation of resources. The ultimate outcome, however, is the dissipation of rent, persistent poverty of the local communities and the continuous degradation of biodiversity resources.

By employing such state-of-the-art analyses, the chapter, then, illustrates the livelihood strategies followed by the resource-dependent communities of the Sundarbans. The discussion shows in this regard that those strategies (both traditional practices and innovative tools) are essentially effective and conducive to the protection, conservation and management of the Sundarbans mangrove ecosystem. Such unique production methods, as practised by the local communities living near the Sundarbans, can make a substantial contribution to the revitalisation and sustainable management of forest resources. Moreover, they value the resources as blessings going beyond the monetary framework of valuation. Overall, they seek to achieve a 'holistic well-being', which means the well-being of both nature and people together through the balancing of nature, reciprocity and solidarity (Swiderska, 2021). Nonetheless, their knowledge has frequently been overlooked by the formal institutional management structure.

As a result, the future challenge will be to comprehend the changing nature of formal and informal institutions to design a comprehensive strategy for biological diversity, coalescing conservation, sustainable use and benefit sharing as its fundamental pillars while also incorporating these livelihood strategies and innovations, emanating from traditional knowledge and practice, into such a strategy (Titumir & Afrin, 2017). Alternatively, it is required mainstreaming biodiversity and nature's contributions to people in policy framework in a stronger and systematic manner, which is also recognised in the CBD's post-2020 agenda (Sandström, 2023). In fact, the transition has become critical for both the survival of the IPLCs and the health of people and the planet (Forest People's Programme et al., 2020). Overall, a collaborative management approach must be promoted through the development of novel institutional arrangements and livelihood options based on both formal and informal institutions.

References

Adhikari, B., & Lovett, J. C. (2005). Transaction costs and community-based natural resource management in Nepal. *Journal of Environmental Management, 78*(1), 5–15. https://doi.org/10.1016/j.jenvman.2005.04.005

Adhikari, K. P., & Goldey, P. (2010). Social capital and its "downside": The impact on sustainability of induced community-based organizations in Nepal. *World Development, 38*(2), 184–194. https://doi.org/10.1016/j.worlddev.2009.10.012

Adnan, S. (2013). Land grabs and primitive accumulation in deltaic Bangladesh: Interactions between neoliberal globalization, state interventions, power relations and peasant resistance. *The Journal of Peasant Studies, 40*(1), 87–128. https://doi.org/10.1080/03066150.2012.753058

Adnan, S. (2016). Alienation in neoliberal India and Bangladesh: Diversity of mechanisms and theoretical implications. *South Asia Multidisciplinary Academic Journal, 13*, 1–20. https://doi.org/10.4000/samaj.4130

Alam, M. (2009). Evolution of forest policies in Bangladesh: A critical analysis. *International Journal of Social Forestry (IJSF), 2*(2), 149–166.

Alam, M., Furukawa, Y., Sarker, S. K., & Ahmed, R. (2008). Sustainability of *sal* (*Shorea robusta*) forest in Bangladesh: Past, present and future actions. *The International Forestry Review, 10*(1), 29–37.

Altrell, D., Saket, M., Lyckeback, L., Piazza, M., Ahmad, I. U., Banik, H., Hossain, M. A. A., & Chowdhury, R. M. (2007). *National forest and tree resources assessment Bangladesh 2005–2007*. Bangladesh Forest Department, Ministry of Environment and Forest, Bangladesh Space Research and Remote Sensing Organization, Ministry of Defence & Food and Agriculture Organization of the United Nations.

Ancrenaz, M., Dabek, L., & O'Neil, S. (2007). The costs of exclusion: Recognizing a role for local communities in biodiversity conservation. *PLOS Biology, 5*(11), e289. https://doi.org/10.1371/journal.pbio.0050289

Awty-Carroll, K., Bunting, P., Hardy, A., & Bell, G. (2019). Using continuous change detection and classification of Landsat data to investigate long-term mangrove dynamics in the Sundarbans region. *Remote Sensing, 11*(23), 2833. https://doi.org/10.3390/rs11232833

Aziz, A., & Paul, A. R. (2015). Bangladesh Sundarbans: Present status of the environment and biota. *Diversity, 7*(3), 242–269. https://doi.org/10.3390/d7030242

Bang, A., & Khadakkar, S. (2020). Biodiversity conservation during a global crisis: Consequences and the way forward. *PNAS, 117*(48), 29995–29999. https://doi.org/10.1073/pnas.2021460117

Banglapedia. (2021). *Bay of Bengal*. National Encyclopedia of Bangladesh. Retrieved June 10, 2021 https://en.banglapedia.org/index.php/Bay_of_Bengal

Bank Track. (2015). Equator principles analysis of the Rampal coal-fired power plant project, Bangladesh. Available at https://www.banktrack.org/download/rampal_equator_principles_full_analysis_pdf/rampal_equator_principles_full_analysis.pdf

Barbier, E. B. (2007). Valuing ecosystem services as productive inputs. *Economic Policy, 22*, 177–229.

Barbier, E. B. (2009). Ecosystems as natural assets. *Foundations and Trends Microeconomics, 4*(8), 611–681. https://doi.org/10.1561/0700000031

BARC. (2016). *State of biodiversity for food and agriculture in Bangladesh*. Bangladesh Agricultural Research Council, Ministry of Agriculture, Government of the People's Republic of Bangladesh & Food and Agriculture Organization of the United Nations.

Basu, R. (2016). A power plant like Rampal would never be allowed in India. *The Dhaka Tribune*. http://www.dhakatribune.com/bangladesh/2016/08/23/power-plant-like-rampal-never-allowed-india/

Basu, S. K., & Cetzal-lx, W. (2018). Traditional honey collectors in the Sunderbans region and their impact on the local mangrove ecosystem and biodiversity: A case study with particular reference to human–animal conflict. *Biodiversity, 19*(3–4), 248–254. https://doi.org/10.1080/14888386.2018.1508365

Baten, M. A., & Kumar, U. (2010). *Responses to the changes in the Sundarbans*. Paper presented at International Conference on Biological and Cultural Diversity, UNCBD, 3–5 May, Montreal, Canada.

Battersby, S. (2017). Can humankind escape the tragedy of the commons? *PNAS, 114*(1), 7–10. https://doi.org/10.1073/pnas.1619877114

BBS. (2001). *Population and housing census 2001*. Bangladesh Bureau of Statistics, Government of the People's Republic of Bangladesh.

BBS. (2011). *Population and housing census 2011*. Bangladesh Bureau of Statistics, Government of the People's Republic of Bangladesh.

Behera, M. D., & Haider, M. S. (2012). Situation analysis on biodiversity conservation. In *Ecosystem for life: A Bangladesh-India initiative*. International Union for Conservation of Nature (IUCN).

Benjaminsen, T. A., & Bryceson, I. (2012). Conservation, green/blue grabbing and accumulation by dispossession in Tanzania. *Journal of Peasant Studies, 39*(2), 335–355. https://doi.org/10.1080/03066150.2012.667405

BFD. (2010). Integrated resources management plans for the Sundarbans (2010–2020). In *Nishorgo network* (Vol. 1). Bangladesh Forest Department, Ministry of Environment and Forest, Government of the People's Republic of Bangladesh.

BFD. (2015). *Forest production and revenue generation from the Sundarbans from 2001–02 to 2014–15 (Unpublished Report)*. Department of Forest, Ministry of Environment and Forests, Government of the People's Republic of Bangladesh.

BFD. (2016a). *Bangladesh forestry master plan 2017–2036*. Bangladesh Forest Department, Ministry of Environment and Forests, Government of the People's Republic of Bangladesh.

BFD. (2016b). *Forest fact sheet: National tree planting campaign and tree fair*. Bangladesh Forest Department, Ministry of Environment and Forests, Government of the People's Republic of Bangladesh.

BFD. (2017). *Forest investment programme*. Bangladesh Forest Department, Ministry of Environment and Forests, Government of the People's Republic of Bangladesh.

BFD. (2018a). Forest produces latest. In *The Dhaka tribune. Honey collection declines in the Sundarbans*. https://www.dhakatribune.com/climate-change/2018a/07/13/honey-collection-declines-in-the-sundarbans

BFD. (2018b). *Forest revenue earned from 1996 to 2016 (Unpublished Report)*. Ministry of Environment and Forests, Government of the People's Republic of Bangladesh.

BFD. (2019). *Status of tiger in Bangladesh Sundarbans 2018: Second phase*. Bangladesh Forest Department, Ministry of Environment and Forests, Government of the People's Republic of Bangladesh.

BFD. (2020). Forest produces latest. In *The Dhaka tribune. Beekeepers in good spirit as Sundarbans sees honey boom*. https://www.dhakatribune.com/bangladesh/nation/2020/09/23/beekeepers-in-good-spirit-as-sundarbans-sees-honey-boom

BFD. (2021). *Forest assessment by Forest Department (Last Updated Data on July, 2021)*. Bangladesh Forest Department, Government of the People's Republic of Bangladesh. Retrieved August 12, 2021 from http://www.bforest.gov.bd/

Bhattacharya, P., Pradhan, L., & Yadav, G. (2010). Joint forest management in India: Experiences of two decades. *Resources, Conservation and Recycling, 54*(8), 469–480. https://doi.org/10.1016/j.resconrec.2009.10.003

Bieg, C., McCann, K. S., & Fryxell, J. M. (2017). The dynamical implications of human behaviour on a social-ecological harvesting model. *Theoretical Ecology, 10*(3), 341–354. https://doi.org/10.1007/s12080-017-0334-3

Bowles, S. (2008). Policies designed for self-interested citizens may undermine "the moral sentiments": Evidence from economic experiments. *Science, 320*(5883), 1605–1609. https://doi.org/10.1126/science.1152110

Bräuer, I. (2003). Money as an indicator: To make use of economic evaluation for biodiversity conservation. *Agriculture, Ecosystems and Environment, 98*, 483–491. https://doi.org/10.1016/S0167-8809(03)00107-5

Byomkesh, T., Nakagoshi, N., & Shahedur, R. M. (2009). State and management of wetlands in Bangladesh. *Landscape and Ecological Engineering, 5*(1), 81–90. https://doi.org/10.1007/s11355-008-0052-5

CEGIS. (2007). Effect of cyclone *Sidr* on the Sundarbans: A preliminary assessment. Center for Environmental and Geographic Information Services.

CEGIS. (2016). *Joint landscape narrative by India and Bangladesh*. Unpublished Report. Center for Environmental and Geographic Information Services.

Chacraverti, S. (2014). *The Sundarbans fishers: Coping in an overly stressed mangrove estuary. Samudra Monograph*. International Collective in Support of Fish workers.

Chee, Y. E. (2004). An ecological perspective on the valuation of ecosystem services. *Biological Conservation, 120*(4), 549–565. https://doi.org/10.1016/j.biocon.2004.03.028

Chen, H. L., Lewison, R. L., An, L., Tsai, Y. H., Stow, D., Shi, L., & Yang, S. (2020). Assessing the effects of payments for ecosystem services programs on forest structure and species biodiversity. *Biodiversity and Conservation, 29*, 2123–2140. https://doi.org/10.1007/s10531-020-01953-3

Choudhury, M. S. H., Koike, M., & Muhammed, N. (2009). Embracing collaborative protected area management for conservation: An analysis of the development of the forest policy of Bangladesh. *The International Forestry Review, 11*(3), 359–374.

Corbera, E., Kosoy, N., & Martínez-Tuna, M. (2007a). The equity implications of marketing ecosystem services in protected areas and rural communities: Case studies from Meso-America. *Global Environmental Change, 17*, 365–380. https://doi.org/10.1016/j.gloenvcha.2006.12.005

Corbera, E., Brown, K., & Adger, W. N. (2007b). The equity and legitimacy of markets for ecosystem services. *Development and Change, 38*(4), 587–613. https://doi.org/10.1111/j.1467-7660.2007.00425.x

Costanza, R., d'Arge, R., de Groot, R., Farberk, S., Grasso, M., Hannon, B., Limburg, K., Naeem, S., O'Neill, R. V., Paruelo, J., Raskin, R. G., Sutton, P., & van den Belt, M. (1997). The value of the world's ecosystem services and natural capital. *Nature, 387*, 253–260. https://doi.org/10.1038/387253a0

Costanza, R., de Groot, R., Braat, L., Kubiszewski, I., Fioramonti, L., Sutton, P., Farber, S., & Grasso, M. (2017). Twenty years of ecosystem services: How far have we come and how far do we still need to go? *Ecosystem Services, 28*, 1–16. https://doi.org/10.1016/j.ecoser.2017.09.008

Dasgupta, P. (2021). *The economics of biodiversity: The Dasgupta review*. HM Treasury.

Deacon, R. T., & Mueller, B. (2003). *Political economy and natural resource use*. Departmental Working Paper. Department of Economics, UC Santa Barbara.

de Freitas, C. T., Lopes, P. F. M., Campos-Silva, J. V., Noble, M. M., Dyball, R., & Peres, C. A. (2020). Co-management of culturally important species: A tool to promote biodiversity conservation and human well-being. *People and Nature, 2*(1), 61–81. https://doi.org/10.1002/pan3.10064

de Groot, R., Brander, L., van der Ploeg, S., Costanza, R., Bernard, F., Braat, L., Christie, M., Crossman, N., Ghermandi, A., Hein, L., Hussain, S., Kumar, P., McVittie, A., Portela, R., Rodriguez, L. C., ten Brink, P., & van Beukering, P. (2012). Global estimates of the value of ecosystems and their services in monetary units. *Ecosystem Services, 1*(1), 50–61. https://doi.org/10.1016/j.ecoser.2012.07.005

Díaz, S., Demissew, S., Carabias, J., Joly, C., Lonsdale, M. … Zlatanova, D. (2015). The IPBES conceptual framework-connecting nature and people. *Current Opinion in Environmental Sustainability, 14*, 1–16.https://doi.org/10.1016/j.cosust.2014.11.002

DoE. (2010). *Fourth national report to the convention on biological diversity: Biodiversity national assessment and programme of action*. Department of Environment, Ministry of Environment and Forest, Government of the People's Republic of Bangladesh.

DoE. (2015). *Fifth national report to the convention on biological diversity: Biodiversity national assessment and programme of action 2020*. Department of Environment, Ministry of Environment and Forest, Government of the People's Republic of Bangladesh.

FAO. (2011). *Bangladesh forestry outlook study. Asia-Pacific forestry sector outlook II*. Working paper series, Paper no. APFSOS II/WP/2011/33, pp. 101. Food and Agriculture Organization of the United Nations.

Forest Peoples Programme, International Indigenous Forum on Biodiversity, Indigenous Women's Biodiversity Network, Centres of Distinction on Indigenous and Local Knowledge and Secretariat of the Convention on Biological Diversity. (2020). *Local biodiversity outlooks 2: The contributions of indigenous peoples and local communities to the implementation of the strategic plan for biodiversity 2011–2020 and to renewing nature and cultures*. A complement to the fifth edition of Global Biodiversity Outlook. Forest Peoples Programme.

Gagné, S. A., Bryan-Scaggs, K., Boyer, R. H. W., & Xiang, W. (2019). Conserving biodiversity takes a plan: How planners implement ecological information for biodiversity conservation. *Ambio, 49*, 1490–1505.https://doi.org/10.1007/s13280-019-01281-z

Gani, M. O. (2001). The giant honey bee (Apis dorsata) and honey hunting in Sundarbans reserved forests of Bangladesh. In *Proceedings of the 37th International Apicultural Congress, 28 October–1 November, Durban, South Africa.*

Giri, C., Pengra, B., Zhu, Z., Singh, A., & Tieszen, L. L. (2007). Monitoring mangrove forest dynamics of the Sundarbans in Bangladesh and India using multitemporal satellite data from 1973 to 2000. *Estuarine Coastal and Shelf Science, 73*, 91–100. https://doi.org/10.1016/j.ecss.2006.12.019

Giri, C., Long, J., Abbas, S., Murali, R. M., Qamer, F. M., Pengra, B., & Thau, D. (2015). Distribution and dynamics of mangrove forests of South Asia. *Journal of Environmental Management, 148*, 101–111. https://doi.org/10.1016/j.jenvman.2014.01.020

Gómez-Baggethun, E., & Muradian, R. (2015). In markets we trust? Setting the boundaries of market-based instruments in ecosystem services governance. *Ecological Economics, 117*, 217–224. https://doi.org/10.1016/j.ecolecon.2015.03.016

Gómez-Baggethun, E., & Ruiz-Pérez, M. (2011). Economic valuation and the commodification of ecosystem services. *Progress in Physical Geography: Earth and Environment, 35*(5), 613–628. https://doi.org/10.1177/0309133311421708

Gopal, B., & Chauhan, M. (2006). Biodiversity and its conservation in the Sundarbans mangrove ecosystem. *Aquatic Sciences, 68*, 338–354. https://doi.org/10.1007/s00027-006-0868-8

Gowdy, J. M. (1997). The value of biodiversity: Markets, society and ecosystems. *Land Economics, 73*(1), 25–41.

Grandcolas, P., & Justine, J. (2020). Covid-19 or the pandemic of mistreated biodiversity. *The Conversation.* https://theconversation.com/covid-19-or-the-pandemic-of-mistreated-biodiversity-136447

Greenberg, J. B., & Park, T. K. (1994). Political ecology. *Journal of Political Ecology, 1*, 1–12. https://doi.org/10.2458/v1i1.21154

Hahn, T., McDermott, C., Ituarte-Lima, C., Schultz, M., Green, T., & Tuvendal, M. (2015). Purposes and degrees of commodification: Economic instruments for biodiversity and ecosystem services need not rely on markets or monetary valuation. *Ecosystem Services, 16*, 74–82. https://doi.org/10.1016/j.ecoser.2015.10.012

Hakim, S. (2007). *Ecotourism and the Sundarbans: Alternative policies for development.* University Grants Commission of Bangladesh.

Haque, A. K. E., & Aich, D. (2014). Economic valuation of ecosystem services. In M. Z. Hussain (Ed.), *Bangladesh Sundarban delta vision 2050: A first step in its formulation—Document 2: A compilation of background information* (pp. 55–63). International Union for Conservation of Nature, Bangladesh Country office.

Haroon, A. K. Y., & Kibria, G. (2017). Wetlands, biodiversity and livelihood values and significance with special context to Bangladesh. In B. A. K. Prusty, R. Chandra & P. A. Azeez (Eds.), *Wetland science* (pp. 317–346). Springer. https://doi.org/10.1007/978-81-322-3715-0_17

Harvey, D. (2003). *The new imperialism.* Oxford University Press.

Harvey, D. (2005). *A brief history of neoliberalism.* Oxford University Press.

Helm, D., & Hepburn, C. (2012). The economic analysis of biodiversity: An assessment. *Oxford Review of Economic Policy, 28*(1), 1–21. https://doi.org/10.1093/oxrep/grs014

Higgins, P., Short, D., & South, N. (2013). Protecting the planet: A proposal for a law of ecocide. *Crime, Law and Social Change, 59*(3), 251–266. https://doi.org/10.1007/s10611-013-9413-6

Hill, R., Adem, C., Alangui, W. V., Molner, Z ... Xue, D. (2020). Working with indigenous, local and scientific knowledge in assessments of nature and nature's linkages with people. *Current Opinion in Environmental Sustainability, 43*, 8–20.https://doi.org/10.1016/j.cosust.2019.12.006

Hossain, M. K. (2009). Remnant *sal* (*Shorearobusta*) forest of norther Bangladesh. In IUCN, UNEP & UNU (Eds.), *Biodiversity conservation and response to climate vulnerability to community level*. International Union for Conservation of Nature, United Nations Environmental Programmes & United Nations University.

Hossain, M. S. (2001). Biological aspects of the coastal and marine environment of Bangladesh. *Ocean & Coastal Management, 44*(3–4), 261–282. https://doi.org/10.1016/S0964-5691(01)00049-7

Hossain, M. S., Khan, Y. S. A., Chowdhury, S. R., Kashem, M. B., & Jabbar, S. M. A. (2004). Environmental and socio-economic aspects: A community based approach from Chittagong coast, Bangladesh. *Jahangirnagar University Journal of Science, 27*, 155–176.

Hussain, M. Z. (2014). *Bangladesh Sundarban delta vision 2050: A first step in its formulation—Document 2: A compilation of background information*. International Union for Conservation of Nature (IUCN), Bangladesh Country Office, viii+192 pp.

Iftekhar, M. S., & Islam, M. R. (2004). Degeneration of Bangladesh's Sundarbans mangroves: A management issue. *The International Forestry Review, 6*(2), 123–135. https://doi.org/10.1505/ifor.6.2.123.38390

IPBES. (2019). Summary for policymakers of the global assessment report on biodiversity and ecosystem services. In S. Díaz, J. Settele, E. Brondízio, H. T. Ngo, M. Guèze, J. Agard, A. Arneth, P. Balvanera, K. Brauman, S. Butchart, K. Chan, L. A. Garibaldi, K. Ichii, J. Liu, S. M. Subramanian, G. F. Midgley, P. Miloslavich, Z. Molnár, D. Obura, A. Pfaff, S. Polasky, A. Purvis, J. Razzaque, B. Reyers, R. Roy Chowdhury, Y. J. Shin, I. J. Visseren-Hamakers, K. J. Willis, C. N. Zayas (Eds.), *Intergovernmental science-policy platform on biodiversity and ecosystem services*. IPBES Secretariat.

Irfanullah, H. M. (2020). Will nature conservation remain a priority in post-corona Bangladesh? *The Daily Star*. https://www.thedailystar.net/opinion/news/will-nature-conservation-remain-priority-post-corona-bangladesh-1904146

IRP. (2019). Global resources outlook 2019: Natural resources for the future we want. In B. Oberle et al. (Eds.), *A report of the international resource panel*. United Nations Environment Programme.

Islam, K. M. N. (2010a). *A study of the principal marketed value chains derived from the Sundarbans Reserved Forest*. International Resources Group (IRG), USAID/Bangladesh. https://www.preventionweb.net/files/18616_18616nabiulislamsrfreportsummaryfin.pdf

Islam, M. N., & Al-Amin, M. (2019). The Rampal power plant, ecological disasters and environmental resistance in Bangladesh. *International Journal of Environmental Studies, 76*(6), 922–939. https://doi.org/10.1080/00207233.2019.1662183

Islam, M. S. (2003). Perspectives of the coastal and marine fisheries of the Bay of Bengal, Bangladesh. *Ocean & Coastal Management, 46*(8), 763–796. https://doi.org/10.1016/S0964-5691(03)00064-4

Islam, M. S. N., & Gnauck, A. (2009). Threats to the Sundarbans mangrove wetland ecosystems from transboundary water allocation in the Ganges basin: A preliminary problem analysis. *International Journal of Ecological Economics & Statistics, 13*(9), 64–78.

Islam, M. T. (2014). Vegetation changes of Shundarbans based on Landsat imagery analysis between 1975 and 2006. *Landscape & Environment, 8*(1), 1–9.

Islam, S. N. (2010b). Threatened wetlands and ecologically sensitive ecosystems management in Bangladesh. *Frontiers of Earth Science in China, 4*(4), 438–448. https://doi.org/10.1007/s11707-010-0127-0

Ituarte-Lima, C., Schultz, M., Hahn, M. C., & Cornell, S. (2014). Biodiversity financing and safeguards: Lessons learned and proposed guidelines. In *Information Document UNEP/CBD/COP/12/INF/27 for the 12th Conference of the Parties of the Convention on Biological Diversity in Pyeongchang Korea*. SwedBio/Stockholm Resilience Centre at Stockholm University.

IUCN Bangladesh. (2012). *Environmental management and biodiversity conservation plan for Bangladesh Sundarbans' biodiversity*. Draft Final Report, Vol. 1. International Union for Conservation of Nature, Bangladesh Country Office.

IUCN Bangladesh. (2014). *Bangladesh Sundarban delta vision 2050: A first step in its formulation—Document 1: The Vision*. International Union for Conservation of Nature, Bangladesh Country Office, vi + 23 pp.

Jaureguiberry, P., Titeux, N., Wiemers, M., Bowler, D. E., Coscieme, L., Golden, A. S., Guerra, C. A., Jacob, U., Takahashi, Y., Settele, J., Díaz, S., Molnár, Z., & Purvis, A. (2022). The direct drivers of recent global anthropogenic biodiversity loss. *Science Advances, 8*(45). https://doi.org/10.1126/sciadv.abm9982

Kabir, D. M. H., & Hossain, J. (2008). *Resuscitating the Sundarbans: Customary use of biodiversity and traditional cultural practices in Bangladesh*. Unnyan Onneshan, BELA, Forest peoples Programme and Nijera Kori.

Kabir, M. H., & Baten, M. A. (n.d). *The economics of the biodiversity and ecosystem services of the Sundarbans*. Unpublished Research Report. Unnayan Onneshan.

Kelly, A. (2011). Conservation practice as primitive accumulation. *Journal of Peasant Studies, 38*(4), 683–701. https://doi.org/10.1080/03066150.2011.607695

Khan, M. H. (2010). *Political settlements and the governance of growth-enhancing institutions*. School of Oriental and African Studies, University of London.

Khan, N. A. (2001). *Regional study on forest policy and institutional reform: Final report of the Bangladesh case study*. Asian Development Bank (ADB).

Khoda, M. E. (2008). *Transparency and accountability in forest conservation and management: Problems and way out*. Transparency International Bangladesh.

Khor, M. (2011). *Risks and uses of the green economy concept in the context of sustainable development, poverty and equity*. Research Paper No. 40. The South Centre.

Kibria, G., & Haroon, A. K. Y. (2017). Climate change impacts on wetlands of Bangladesh, its biodiversity and ecology, and actions and programs to reduce risks. In B. A. K. Prusty, R. Chandra & P. A. Azeez (Eds.), *Wetland science* (pp. 189–204). Springer. https://doi.org/10.1007/978-81-322-3715-0_10

Kohlar, F., & Brodizio, E. S. (2019). Considering the needs of indigenous and local populations in conservation programs. *Conservation Biology, 31*(2), 245–251. https://doi.org/10.1111/cobi.12843

Kosoy, N., & Corbera, E. (2010). Payments for ecosystem services as commodity fetishism. *Ecological Economics, 69*, 1228–1236. https://doi.org/10.1016/j.ecolecon.2009.11.002

Kumar, P. (2005). *Market for ecosystem services*. International Institute for Sustainable Development (IISD).

Kumar, P. (Ed.). (2010). *The economics of ecosystems and biodiversity: Ecological and economic foundations*. United Nations Environment Programme. Earthscan.

Lovera, S. (2020). Introduction: On forests, finance and perverse partnerships. In O. Munnion & S. Lovera (Eds.), *Circular economy or vicious cycle?: How corporate capture of policy-making and perverse incentives are driving deforestation*. JG Utrecht & Asunción: Global Forest Coalition.

Lovera, S., Gupta, J., & van Ros-Tonen, M. (2015). *Forests, finance and fairy tales: The economic inefficiency of REDD+*. Paper presented at the XIV World Forestry Congress, September, Durban.

Lu, D. (2021). Covid-19 hit biodiversity across the globe. Here's how to fix things. *New Scientist*. https://www.newscientist.com/article/mg24933252-700-covid-19-hit-biodiversity-across-the-globe-heres-how-to-fix-things/

Luck, G., Chan, K., Eser, U., Gómez-Baggethun, E., Matzdorf, B., Norton, B., & Potschin, M. (2012). Ethical considerations in on-ground applications of the ecosystem services concept. *BioScience, 62*(12), 1020–1029.

Mahmood, H., Ahmed, M., Islam, T., Uddin, M. Z., Ahmed, Z. U., & Saha, C. (2021). Paradigm shift in the management of the Sundarbans mangrove forest of Bangladesh: Issues and challenges. *Trees, Forests and People, 5*(100094). https://doi.org/10.1016/j.tfp.2021.100094

Martínez-Alier, J. (2002). *The environmentalism of the poor: A study of ecological conflicts and valuation*. Edward Elgar.

Marx, K. (1976). *Capital: A critique of political economy* (Vol. 1). Vintage Books.

McAfee, K. (1999). Selling Nature to Save it? Biodiversity and the Rise to Green Developmentalism. *Environment and Planning D: Society and Space, 17*(2), 133–154.

McAfee, K. (2012). The contradictory logic of global ecosystem services markets. *Development and Change, 43*(1), 105–131. https://doi.org/10.1111/j.1467-7660.2011.01745.x

McAfee, K., & Shapiro, E. N. (2010). Payments for ecosystem services in Mexico: Nature, neoliberalism, social movements, and the state. *Annals of Association of American Geographers, 100*(3), 579–599. https://doi.org/10.1080/00045601003794833

McCauley, D. J. (2006). Selling out on nature. *Nature, 443*(7107), 27–28. https://doi.org/10.1038/443027a

MEA (Millennium Ecosystem Assessment). (2005). *Ecosystems and human well-being: Biodiversity synthesis*. World Resources Institute.

Minar, M. H., Hossain, M. B., & Shamsuddin, M. D. (2013). Climate change and coastal zone of Bangladesh: Vulnerability, resilience and adaptability. *Middle-East Journal of Scientific Research, 13*(1), 114–120. https://doi.org/10.5829/idosi.mejsr.2013.13.1.64121

Mishra, M., Acharyya, T., Santos, C. A. G., da Silva, R. M., Kar, D., Kamal, A. H. M., & Raulo, S. (2021). Geo-ecological impact assessment of severe cyclonic storm Amphan on Sundarban mangrove forest using geospatial technology. *Estuarine, Coastal and Shelf Science, 260*(107486). https://doi.org/10.1016/j.ecss.2021.107486

MoEF. (2001). *State of environment Bangladesh*. Ministry of Environment and Forests, Government of the People's Republic of Bangladesh. Retrieved January 10, 2020 from: http://old.moef.gov.bd/html/state_of_env/state_of_env.html

Mukhopadhyay, A., Payo, A., Chanda, A., Ghosh, T., Chowdhury, S. M., & Hazra, S. (2018). Dynamics of the Sundarbans mangroves in Bangladesh under climate change. In R. Nicholls, C. Hutton, W. Adger, S. Hanson, M. Rahman, M. Salehin (Eds.). *Ecosystem services for well-being in deltas*. Palgrave Macmillan. https://doi.org/10.1007/978-3-319-71093-8_26

Muradian, R., & Gómez-Baggethun, E. (2013). The institutional dimension of "market-based instruments" for governing ecosystem services: Introduction to the special issue. *Society and Natural Resources, 26*(10), 1113–1121. https://doi.org/10.1080/08941920.2013.829380

Muradian, R., Arsel, M., Pellegrini, L., Adaman, F., Aguilar, B., Agarwal, B., Corbera, E., Ezzine de Blas, D., Farley, J., Froger, G., Garcia-Frapolli, E., Gómez-Baggethun, E., Gowdy, J., Kosoy, N., Le Coq, J. F., Leroy, P., May, P., Méral, P., Mibielli, P., … Urama, K. (2013). Payments for ecosystem services and the fatal attraction of win-win solutions. *Conservation Letters, 6*(4), 274–279. https://doi.org/10.1111/j.1755-263X.2012.00309.x

Mustafa, M. M. (2002). *A review of forest policy trends in Bangladesh—Bangladesh forest policy trends*. Policy Trend Report 2002: 114–121. Institute for Global Environmental Strategies (IGES).

Neuteleers, S., & Engelen, B. (2015). Talking money: How market based valuation can undermine environmental protection. *Ecological Economics, 117*, 253–260. https://doi.org/10.1016/j.ecolecon.2014.06.022

Newmann, R. P. (2017). Commoditization, primitive accumulation and the spaces of biodiversity conservation. In S. Chari, S. Freidberg, V. Gidwani, J. Ribot & W. Wolford (Eds.), *Other geographies: The influences of Michael Watts*. Wiley.

Nyborg, K., et al. (2016). Social norms as solutions. *Science, 354*(6308), 42–43.https://doi.org/10.1126/science.aaf8317

Ostrom, E. (2000). *Private and common property rights*. Center for the Study of Institutions, Population and Environmental Change, Indiana University.

Otero, I., Farrell, K. N., Puyeo, S., et al. (2020). Biodiversity policy beyond economic growth. *Conservation Letters, 13*(4), e12713. https://doi.org/10.1111/conl.12713

Pearce, D. (2001). Valuing biological diversity: Issues and overview. In OECD (Ed.), *Valuation of biodiversity benefits: Selected studies*. Organisation for Economic Co-operation and Development. https://doi.org/10.1787/9789264195844-en

Perrings, C., Folke, C., & Mäler, K. (1992). The ecology and economics of biodiversity loss: The research agenda. *Ambio, 21*(3), 201–211.

Purushothaman, S., Thomas, B. K., Abraham, R., & Dhar, U. (2013). Beyond money metrics: Alternative approaches to conceptualising and assessing ecosystem services. *Conservation and Society, 11*(4), 321–325.

Quader, O. (2010). Coastal and marine biodiversity of Bangladesh (Bay of Bengal). In: *Proceedings of International Conference on Environmental Aspects of Bangladesh (ICEAB10)*, September 2010, Japan.

Rabbani, G., Rahman, S. H., & Faulkner, L. (2013). Impacts of climatic hazards on the small wetland ecosystems (ponds): Evidence from some selected areas of coastal Bangladesh. *Sustainability, 5*(4), 1510–1521. https://doi.org/10.3390/su5041510

Rahman, M. M. (1995). *Wetlands and biodiversity: A case study of common property resources in Bangladesh.* Center for Natural Resource Studies.

Rahman, M. M., Jiang, Y., & Irvine, K. (2018). Assessing wetland services for improved development decision-making: A case study of mangroves in coastal Bangladesh. *Wetlands Ecology and Management, 26*, 563–580. https://doi.org/10.1007/s11273-018-9592-0

Rahman, M. M., Rahman, M. M., & Islam, K. S. (2010a). The causes of deterioration of Sundarbans mangrove forest ecosystem of Bangladesh: Conservation and sustainable management issues. *Aquaculture, Aquarium, Conservation & Legislation International Journal of the Bioflux Society (AACL Bioflux), 3*(2), 77–90.

Rahman, M. M., Rahman, M. M., Guogang, Z., & Islam, K. S. (2010b). A review of the present threats to tropical moist deciduous *sal* (*Shorearobusta*) forest ecosystem of central Bangladesh. *Tropical Conservation Science, 3*(1), 90–102.

Rahman, M. R., & Asaduzzaman, M. (2010). Ecology of Sundarbans, Bangladesh. *Journal of Science Foundation, 8*(1&2), 35–47.

Rahman, M. R., & Hossain, M. B. (2015). Changes in land use pattern at Chakaria Sundarbans mangrove forest in Bangladesh. *Bangladesh Research Publications Journal, 11*(1), 13–20.

Rashid, S. M. A. (2019). *Coastal biodiversity—A review.* Report prepared for Long Term Monitoring Research and Analysis of Bangladesh Coastal Zone.

Rasul, G. (2005). *State policies and land use in the Chittagong Hill Tracts of Bangladesh.* International Institute for Environment and Development (IIED).

Rasul, G., Chettri, N., & Sharma, E. (2011). *Framework for valuing ecosystem services in the Himalayas.* ICIMOD Technical Report. The International Centre for Integrated Mountain Development.

Reddy, C. S., Pasha, S. V., Jha, C. S., Diwakar, P. G., & Dadhwal, V. K. (2016). Development of national database on long-term deforestation (1930–2014) in Bangladesh. *Global and Planetary Changes, 139*, 173–182. https://doi.org/10.1016/j.gloplacha.2016.02.003

Robbins, J. (2012). The ecology of disease. *The New York Times.* https://www.nytimes.com/2012/07/15/sunday-review/the-ecology-of-disease.html

Rode, J., Gómez-Baggethun, E., & Krause, T. (2014). Motivation crowding by economic incentives in conservation policy: A review of the empirical evidence. *Ecological Economics, 117*, 270–282. https://doi.org/10.1016/j.ecolecon.2014.11.019

Rodríguez-Labajos, B., & Martínez-Alier, J. (2013). The economics of ecosystems and biodiversity: Recent instances for debate. *Conservation and Society, 11*(4), 326–342.

Roy, A. K. D., & Alam, K. (2012). Participatory forest management for the sustainable management of the Sundarbans mangrove forest. *American Journal of Environmental Science, 8*(5), 549–555. https://doi.org/10.3844/ajessp.2012.549.555

Roy, A. K. D., Alam, K., & Gow, J. (2011). A review of the role of property rights and forest policies in the management of the Sundarbans mangrove forest in Bangladesh. *Forest Policy and Economics, 15*, 46–53. https://doi.org/10.1016/j.forpol.2011.08.009

Rubino, M. C. (2000). Biodiversity finance. *International Affairs, 76*(2), 223–240. https://doi.org/10.1111/1468-2346.00131

Sadmo, A. (2015). The early history of environmental economics. *Review of Environmental Economics and Policy, 9*(1), 43–63. https://doi.org/10.1093/reep/reu018

Salam, M. A., Noguchi, T., & Koike, M. (1999). The causes of forest cover loss in the hill forests in Bangladesh. *GeoJournal, 47*(4), 539–549. https://doi.org/10.1023/A:1006947203052

Sandström, C., Ring, I., Olschewski, R., Simoncini, R., Albert, C., Acar, S., Adeishvili, M., Allard, C., Anker, Y., Arlettaz, R., Bela, G., Coscieme, L., Fischer, A., Fürst, C., Galil, B., Hynes, S., Kasymov, U., Marta-Pedroso, C., Mendes, A., . . . Pergl, J. (2023). Mainstreaming biodiversity and nature's contributions to people in Europe and Central Asia: insights from IPBES to inform the CBD post-2020 agenda. *Ecosystems and People, 19*(1). https://doi.org/10.1080/26395916.2022.2138553

Sarker, S. K., Reeve, R., Thompson, J., Paul, N. K., & Matthiopoulos, J. (2016). Are we failing to protect threatened mangroves in the Sundarbans world heritage ecosystem? *Scientific Reports, 6*(21234). https://doi.org/10.1038/srep21234

SCBD. (2022). Decision adopted by the Conference of the Parties to the Convention on Biological Diversity: Kunming-Montreal Global Biodiversity Framework. Secretariat of the Convention on Biological Diversity. https://www.cbd.int/doc/decisions/cop-15/cop-15-dec-04-en.pdf

Schill, C. (2017). *Human behaviour in social-ecological systems: Insights from economic experiments and agent based modelling*. Doctoral Thesis in Sustainability Science. Stockholm Resilience Centre, Stockholm University.

Schlager, E., & Ostrom, E. (1992). Property rights regimes and natural resources: A conceptual analysis. *Land Economics, 68*(3), 249–262. https://doi.org/10.2307/3146375

Sen, S. G. (2010). *Conservation of the Sundarbans in Bangladesh through sustainable shrimp aquaculture*. Harvard Kennedy School, Harvard University.

Sharda, J., & Buckley, T. (2016). *Risky and over-subsidized: A financial analysis of the Rampal power plant*. Institute of Energy Economics and Financial Analysis.

Singh, R., & Singh, G. S. (2017). Ecosystem services: A bridging concept of ecology and economics. *Ecological Questions, 25*, 95–101. https://doi.org/10.12775/EQ.2017.008.s

Swiderska, K. (2021). *Here's why indigenous economics is the key to saving nature*. International Institute for Environment and Development. Retrieved August 20, 2021, from https://www.iied.org/heres-why-indigenous-economics-key-saving-nature

Terborgh, J., & Peres, C. A. (2017). Do community-managed forests work? A biodiversity perspective. *Land, 6*(2), 22. https://doi.org/10.3390/land6020022

Titumir, R. A. M. (2021). COVID-19 and climate change: Double jeopardy for traditional resource users in the Sundarbans. In *CEESP News*. International Union for Conservation of Nature (IUCN). Retrieved January 13, 2021 from https://www.iucn.org/news/commission-environmental-economic-and-social-policy/202101/covid-19-and-climate-change-double-jeopardy-traditional-resource-users-sundarbans

Titumir, R. A. M., & Afrin, T. (2017). Complementarities of human-nature well-beings: A case illustrated through traditional forest resource users of Sundarbans in Bangladesh. In *Satoyama initiative thematic review (SITR)* (Vol. 3). United Nations University Institute for the Advanced Study of Sustainability (UNU-IAS).

Titumir, R. A. M., Paran, M. S., & Pash, M. W. (2019). "The Sundarbans is our mind": An exploration into multiple values of nature in conversation with traditional resource users. In *Satoyama initiative thematic review (SITR)* (Vol. 5). United Nations University Institute for the Advanced Study of Sustainability (UNU-IAS).

Titumir, R. A. M., Afrin, T., & Islam, M. S. (2020). Traditional knowledge, institutions, and human sociality in sustainable use and conservation of biodiversity of the Sundarbans of Bangladesh. In O. Saito, S. M. Subramanian, S. Hashimoto & K. Takeuchi (Eds.), *Mapping socio-ecological production landscapes, stakeholders, policy and action*. Science for Sustainable Societies, Integrated Research System for Sustainability Science (IR3S), University of Tokyo.

Titumir, R. A. M. (2022). Introduction: The Sundarbans as our mind. In R. A. M. Titumir (Ed.), *Sundarbans and its Ecosystem Services: Traditional Knowledge, Customary Sustainable Use and Community Based Innovation* (pp. 1–20). Palgrave Macmillan. https://doi.org/10.1007/978-981-19-3000-3_1

The Daily Star. (2018). Licence to harm Sundarbans. https://www.thedailystar.net/frontpage/licence-harm-sundarbans-1558918

The Daily Star. (2019). Unesco labels Sundarbans 'World Heritage in Danger'. https://www.thedailystar.net/country/news/unesco-labels-sundarbans-world-heritage-danger-1756996

The Daily Star. (2021). 2.57 lakh acres of forest occupied: Forest department tells JS body that 1.6 lakh people, entities grabbed the land. https://www.thedailystar.net/frontpage/news/257-lakh-acres-forest-occupied-2040877

TIB. (2015). *Rampal and Matarbari power projects: Governance challenges in environmental impact assessment and land acquisition*. Transparency International Bangladesh.

Tilman, A. R., Watson, J. R., & Levin, S. (2017). Maintaining cooperation in social-ecological systems: Effective bottom-up management often requires sub-optimal resource use. *Theoretical Ecology, 10*(2), 155–165.

Turnhout, E., Waterton, C., Neves, K., & Buizer, M. (2013). Rethinking biodiversity: From goods and services to "living with." *Conservation Letters, 6*(3), 154–161. https://doi.org/10.1111/j.1755-263X.2012.00307.x

Uddin, M. S. (2011). *Economic valuation of the Sundarbans mangrove ecosystem services—A case study in Bangladesh*. Master's Thesis. UNESCO-IHE Institute for Water Education.

Uddin, M. S., Shah, M. A. R., Khanom, S., & Nesha, M. K. (2013). Climate change impacts on the Sundarbans mangrove ecosystem services and dependent livelihoods in Bangladesh. *Asian Journal of Conservation Biology, 2*(2), 152–156.

United Nations. (1992). *Convention on biological diversity*. United Nations.

Unnayan Onneshan, U. (2010a). Agriculture biodiversity and food security: two sides of a coin. In: M. A. Baten & R. A. M. Titumir (Eds.), Unnayan Onneshan.

Unnayan Onneshan, U. (2010b). *Community based mangrove agro aqua silvi (CMAAS) culture: Promoting as a community adaptation tool and an alternative to commercial shrimp culture*. Unpublished Research report. Unnayan Onneshan.

Unnayan Onneshan, U. (2020). *Biodiversity, climate change and traditional resource users in the Sundarbans: An exploration through public participation geographic information system (PPGIS)*. Unnayan Onneshan.

Vahabi, M. (2008). *Protection costs, transaction costs, and economic theory*. MPRA Paper No. 17648. University Library of Munich.

Vatn, A. (2000). The environment as a commodity. *Environmental Values, 9*(4), 493–509. https://doi.org/10.3197/096327100129342173

Vatn, A. (2010). An institutional analysis of payments for environmental services. *Ecological Economics, 69*(6), 1245–1252. https://doi.org/10.1016/j.ecolecon.2009.11.018

Vidal, J. (2020). 'Tip of the iceberg': Is our destruction of nature responsible for Covid-19? *The Gurdian*. https://www.theguardian.com/environment/2020/mar/18/tip-of-the-iceberg-is-our-destruction-of-nature-responsible-for-covid-19-aoe

Walters, B. B. (2004). Local management of mangrove forests: Effective conservation or efficient resource exploitation? *Human Ecology, 32*, 177–195. https://doi.org/10.1023/B:HUEC.0000019762.36361.48

Williams, P. A., Sikutshwa, L., & Shackleton, S. (2020). Acknowledging indigenous and local knowledge to facilitate collaboration in landscape approaches—Lessons from a systematic review. *Land, 9*(9), 331. https://doi.org/10.3390/land9090331

World Bank. (2015). World development report 2015: Mind, society, and behavior. The World Bank.

World Bank. (2018). Review of key environmental issues in Bangladesh. The World Bank.

WWF. (2020). Living planet report 2020: Bending the curve of biodiversity loss. In R. E. A. Almond, M. Grooten, T. Petersen (Eds.), World Wide Fund for Nature.

Chapter 4
Water Resources: Provision, Distribution and Sustainable Production

4.1 Introduction

Water is a natural compound that is crucial for all forms of life on earth to exist. It covers the majority of the earth's surface, exists in all three states in nature—liquid, solid and gas—and maintains an ecological balance by following a unique hydrological cycle. Despite the fact that the earth has water in abundance, a small proportion of it is directly usable. Water covers approximately 71% of the earth's surface and has a total volume of approximately 1.4 billion cubic kilometres, but about 96.5% of the total amount of water is salt water, whereas 3.5% is freshwater. Of this proportion of freshwater, 3.2% is unavailable and only 0.3% is relatively easily accessible (Mucke, 2019). The current estimate shows that some 1.1 billion people lack access to water and another 2.7 billion would experience water shortage at least for one month in a year (WWF, 2021a). The available amount of usable water, however, is enough for sustaining human lives, but the disparity in terms of access and usage, the human control over the natural flow of the resource and the unavailability of technology are posing a key challenge to ensure water security for all. Nevertheless, the demand for water is expected to increase to the extent that over the next 30 years, it is unlikely to be met by natural supplies in some regions of the world (Piesse, 2020). Besides, water resources, such as fish, wetland species and other biological resources residing in water are under the greatest threat due to the increasing anthropogenic pressures such as pollution, waste disposal and overexploitation.

Given water's outstanding significance, the likelihood of 'water wars' has even been discussed in international politics from time to time (Leithead, 2018; Rahaman, 2012; Wolf, 1999), which, however, has been claimed by others as misleading (Thielbörger, 2019). Overall, water governance becomes a highly contested issue both in policy and research domains. This chapter, therefore, aims to examine the loopholes in the existing water governance regimes that result in the unsustainable use and inequitable distribution of such resources. The chapter, in this regard, develops

R. A. M. Titumir et al., *Natural Resource Degradation and Human-Nature Wellbeing*,
https://doi.org/10.1007/978-981-19-8661-1_4

an alternative framework based on the human–nature nexus to comprehend the sustainable governance of water and water resources.

Water resources specifically comprise two aquatic ecosystems. They are (a) freshwater ecosystem and (b) marine ecosystem. The freshwater ecosystem represents the entire drainage basins where water flows from land and groundwater runoff to streams and river channels as well as to recipient lakes or reservoirs (Wetzel, 2001). Thus, it includes Transboundary Rivers and their distributaries, wetlands and groundwater, among others. Conversely, the marine ecosystem is separable from the freshwater ecosystem because it is distinctively rich with *abiotic* and *biotic* factors, contains salt water and plays a crucial role in dissolving carbon dioxide (CO_2). Both ecosystems provide human beings with valuable service, contributing to food supply, economic and socio-cultural well-being.

Regarding the Transboundary River system, interstate and regional conflicts are emerging from the contestation over the control of resources in the system. Simultaneously, natural rivers are decreasing and getting blocked for multi-purpose uses that are disrupting both the flow of water and the ecosystems at the same time. On the contrary, the excessive withdrawal of groundwater, which accounts for approximately 99% of all liquid freshwater on earth (UNESCO, 2022), in the face of a spiralling demand due to a growing population is causing disputes in ensuring equitable distribution across the globe and within a particular state. On a total basis, about 20% of the world's groundwater has already been overexploited and the majority of the world's aquifers would be destroyed by 2050 as the global water demand is expected to rise by more than 1% a year in the upcoming days (Piesse, 2020; UNESCO/UN-Water, 2019).

Water, however, is not only a substance for consumption but it constitutes a unique ecosystem that houses a particular stock of resources. Wetlands are those ecosystems the value of which, however, has been degrading gradually, primarily with the advent of excessive extraction. The world has already lost 85% of its natural wetlands (Carlson, 2020) and the rate of declination appears to be alarming in recent years, as during 1970–2008, about 25–30% of the wetlands had disappeared worldwide (IPBES, 2015). In such a scenario, sustainable production of wetland resources has also become a growing concern in the realm of water governance. Simultaneously, 66% of the marine ecosystem has been experiencing increasing cumulative impact because of multiple human actions (IPBES, 2019).

Furthermore, the suboptimal production of water resources (e.g., fish) from a variety of sources such as rivers, wetlands and oceans has become a common phenomenon. The estimated global fish production has reached to approximately 179 million tonnes with an increasing trend indicating excessive pressure on fishery resources (FAO, 2020). Of all known fish species, 51% are found in freshwater (WWF, 2021b), and since the 1970s, migratory species have declined by an estimated 76% on average (Deinet et al., 2020) while the percentage (94%) for mega species is more catastrophic (He et al., 2019). Marine fish stocks have also continued to decline, with the proportion of fish stocks within a biologically sustainable level falling from 90 to 65.8% between 1990 and 2017 (FAO, 2020). Overall, 33% of the total fish stocks are estimated to be overexploited (IPBES, 2019). Besides overexploitation, there is a growing concern about inequality in the production and utilisation of marine fisheries.

The effects of climate change have aggravated the situation further by challenging the sustainability of water resources (UNESCO/UN Water, 2020). Recently, the stress on water has multiplied with the advent of the COVID-19 pandemic. There is no accurate conclusion regarding the net outcome (positive or negative) of the impact of the pandemic on water resources (Balamurugan et al., 2021), but it is evident that there are some obvious impacts, both direct and indirect. Increasing demand for Water, Sanitation and Hygiene (WASH) facilities and the disposal of medical waste (e.g., masks, personal protective equipment) were increasing the pressure on the water reservoirs (Shivakumar, 2020). Overall, the intense pressure on available water and its resources is directly affecting agricultural production, ecosystem services, biodiversity and the livelihoods of millions.

The current governance mechanism of water resources can be explained with the help of market-centric approaches such as neo-classical economics and new institutional economics. The neo-classical economics has introduced the idea of market for water by putting a price tag on it (e.g., Bakker, 2004; Dinar & Subramanian, 1997; Dinar et al., 2015; Fridman, 2015; Grafton et al., 2014; Griffin, 2006; Olmstead, 2010; Sjödin et al., 2016; Zetland, 2021). It considers the problems related to water as a scarcity issue by arguing that water, which is no longer abundant, necessitates being tackled by creating a market for aquatic resources in order to increase efficiency. The school of New Institutional Economics (NIE) has extended the market-centric argument by stating that the market fails largely due to the absence of stable institutions (North, 1991; Ostrom, 1990; Ostrom et al., 1994; Williamson, 2002). The implication of NIE in the case of water is that it reveals the importance of reforming water institutions (IBRD/World Bank, 2004; Menard & Saleth, 2012; Saleth, 2018) and defines the nature of ownerships of some types of water resources. In particular, the Integrated Water Resource Management (IWRM) system can be identified as a specific type of institutional arrangement, which suggests that water can be managed at the river basin level by considering diverse sectors related to it. However, some argue that market-centric propositions lead to negative ramifications, including unequal provisioning, intensifying rent-seeking behaviour in the complex market of water and the heavy transaction costs arising from privatisation (Bayliss, 2016; Molle et al., 2008). In particular, complexities have multiplied due to the involved political settlement in the transitional economies (Allouche, 2016; Benson et al., 2015; Clement et al., 2017). The core problem of market-centrism is that it has transformed the understanding of water governance from 'right-based access' to 'commodity-based treatment' (Haren, 2007).

This chapter challenges the neo-classical tradition by arguing that water-related problems are not emerging from scarcity but from inequitable distribution and unsustainable production. However, it considers the significance of stable water institutions but contends that the factors that cause fragility in institutional arrangements need to be identified right at the onset. Therefore, political-economic factors are crucial here, and this is where the political economy approach comes up with thought-provoking explanations. However, the political economy approach alone could not adequately capture the significance of human nature nexus in the realm of water governance. Overall, the existing varieties of theoretical positions are not sufficient in explaining the complexities of water resource governance, particularly in the context of a transitional economy. The chapter accordingly develops an alternative understanding of the water governance regime that may help in devising appropriate policy prescriptions to ensure the sufficient provision, equitable distribution and sustainable production of water and aquatic resources.

It requires articulating here that the chapter examines the problems of water governance by considering the context of Bangladesh, as the country is not an exception against the backdrop of the global crisis of water resource governance, and in some cases, the problems are more acute because of the country's particularities as a transitional economy. Given the complexity of water-related problems, however, the chapter has used some case studies to present various levels of water governance failure. It has probed the provisioning problem to describe the inequality in access to groundwater resources facilitated by the commodification process. The exploration of the depletion of wetland resources has demonstrated the failure of water governance under the conditions of institutional fragility and power politics. Furthermore, the case of governance disputes over transboundary water has highlighted the political contestation over water rights. Finally, the challenges of governing marine (e.g., fisheries) resources sustainably have been examined, taking the factors of institutional fragility and technological incapacity into account. Overall, the chapter, through theoretical and empirical analyses, validates that the central causes of unequal distribution and overexploitation of water and aquatic resources are widespread politicisation, application of neo-liberal policies and the segregation of human and nature (water) mutuality.

4.2 State of Water Resources in Bangladesh

Being a riverine country, Bangladesh has an abundance of water resources. Given the heterogeneity in the features of water, the chapter interprets the country's water resources by the 'stock and flow analysis' (Fig. 4.1). The first level of analysis is concerned with the state of the groundwater. It elucidates that rising demand is putting an additional strain on the stock of groundwater. At the second level, the focus is given to the river basin to illustrate the current state of transboundary water. The exploration of wetlands at the third level displays the degradation of resources provided by them. The state of marine resources has been examined at the fourth level. Lastly, the fisheries resource is the overlapping one, and the discussion focuses on it while examining the state of the wetlands and marine ecosystems.

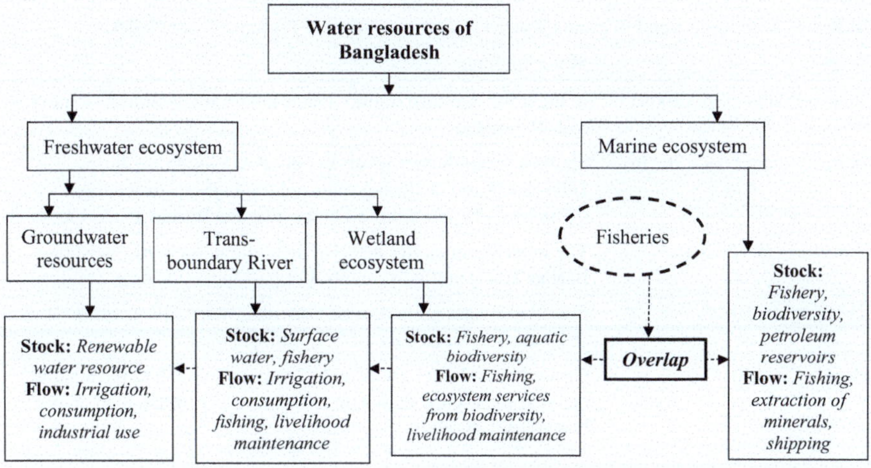

Fig. 4.1 Water resources in Bangladesh under stock and flow analysis. *Source* Prepared by the authors

4.2.1 Groundwater Resources

Groundwater is the most important source of water in Bangladesh for domestic consumption, irrigation and industrial use. The country is considered rich in groundwater resources, but there is no recent data on total groundwater storage. National Water Plan Phase-II estimated the average groundwater to be 21 km^3 in 1991 (FAO, 2014a). However, the little amount of accessible surface water available for consumption has been mounting pressure on groundwater levels over the years.

The total amount of groundwater withdrawn is estimated to be 90% for irrigation and 10% for domestic and industrial purposes combined, which is equivalent to approximately 4% of global groundwater withdrawal (Shamsudduha et al., 2019). About 73.09% of the total cultivated area is irrigated by groundwater, while the remaining (26.91%) is irrigated by surface water (BADC, 2020). The area irrigated by surface water decreased from 76% in 1981 to 23% in 2012, whereas groundwater-irrigated area soared to 80% from 16% during the same period (Qureshi et al., 2014). Water usage for industries is also rapidly increasing and causing significant overuse from groundwater sources in the Dhaka, Mymensingh and Chittagong areas. In particular, approximately 0.2 Bm^3, 0.04 Bm^3, 5.4 Bm^3 and 0.5 Bm^3 of groundwater are used normally for paper mills, textile mills, leather factories and others, respectively (WARPO, 2017).

Apart from irrigation in rural areas and industrialisation in urban areas, a significant amount of groundwater is withdrawn for domestic consumption in both areas. About 90% of rural people rely on groundwater for drinking while in Dhaka city, 95% of the total water supply comes from groundwater alone and the rest is provided by treated surface water (Banglapedia, 2021a). On an overall basis, groundwater provides 98% of the drinking water supplied in Bangladesh (Shamsudduha et al.,

Table 4.1 Declining trend in groundwater level across different regions of Bangladesh

Study	Major findings
Shamsudduha et al. (2009)	In the north-western region, water tables are declining steadily (0.1–0.5 m/year)
Shamsudduha et al. (2011)	Shallow groundwater levels declined at a high rate during 1985–2005 and declining rates are the highest (exceeding 0.5 m/year) in and around Dhaka city and Barind Tract region
Qureshi et al. (2014)	The most significantly affected areas lie in the north-west (e.g., Barind Tract) and north-central (e.g., Madhupur Tract) regions
Khaki et al. (2018)	Country's water table dropped drastically by 32% from 2003 to 2013 at an average pace of 8.73 mm per year
Mojid et al. (2019)	Groundwater tables are found to remain below 6 m throughout the year in Bogura, Rajshahi, Naogaon, Joypurhat and Chapai Nawabganj districts

Source Authors' compilation from different sources

2019). Consequently, the groundwater level of the country has been dwindling over the years as is widely documented (Table 4.1).

The future scenario can be depicted through the estimation of the sector development plan (2011–2025), which shows that the domestic water requirement would be 2208 mm^3, the ecological requirement around 2999 mm^3 and the agriculture would require about 19,993 mm^3 by 2025 (Table 4.2). This growing demand represents that 50.8% of the rechargeable groundwater would be depleted. The estimate, regarding regional differences, shows that the north-east and north-west regions will be facing water deficit. The north-east, especially, will become water-zero with a demand of 3498 mm^3 by 2025 (Table 4.2). Overall, heavy withdrawals of water than the recharge rate will lower the water table of each region in the years to come.

Moreover, arsenic contamination of shallow aquifers affects approximately 26% of all tube wells in Bangladesh (Shamsudduha et al., 2019). Water reservoirs in south-western coastal regions are also vulnerable to increasing salinity intrusion (Ayers et al., 2016; Worland et al., 2015). Such contamination is likely to amplify due to climate change in the upcoming years. Overall, a substantial proportion of Bangladesh's land area (5–24% under 'extremely high' to 'high risks') is currently under the combined risk of arsenic contamination, salinity intrusion and groundwater storage depletion (Shamsuddoha et al., 2019). This can affect as few as 6.5 million to 24.4 million people altogether.

In particular, the demand and supply condition of urban areas is at stake. The capital city Dhaka, following an unsystematic urban sprawl, is facing the greatest acuteness of the problem because of the unavailability of alternative sources of water for domestic consumption. The rate of population growth in metropolitan Dhaka is much higher than in the other cities of the country. The total demand in Dhaka city is 2.45 billion litres per day of which 78% is supplied through groundwater abstraction under the administration of Dhaka Water Supply and Sewerage Authority (DWASA, 2020). This dependency augments a very high depletion rate of the groundwater

Table 4.2 Regional estimates of usable recharge and groundwater demand

Region	Gross area (million ha)	Usable recharge (million m^3) (NWP-II*)	Groundwater demands, (million m^3) (NWMP**)				Balance UR-GD*** (million m^3)
			Water supply	Environment	Agriculture	Total	
North-west	3.016	12,100	539	1290	9548	11,377	723
North-east	–	–	222	170	1357	1749	– 3498
North-central	3.569	23,100	566	637	3082	4285	18,815
South-east	3.007	9800	232	149	1158	1584	8216
South-central	1.426	3500	179	88	652	919	2581
South-west	2.562	5600	289	620	4196	5105	495
Eastern Hills	–	–	181	–	–	181	181
Total (mm^3)	13.580	54,100	2208	2999	19,993	25,200	27,513
%			8.80	11.90	79.30	100	50.8

Source MoLGRDC (2011)
*Usable recharge data assessed under National Water Plan Phase II (MPO, 1991)
**Groundwater demand assessed under National Water Management Plan (WARPO, 2001)
***UR-GD: Usable recharge minus groundwater demand

table (Akther et al., 2009; Islam & Islam, 2017; IWM, 2008; *Unnayan Onneshan*, 2011). During the period 1970–2014, there was an increasing trend in groundwater withdrawal in the city, which reached from 0.18 Mm^3/d to 2.5 Mm^3/d while at the same time, the groundwater table lowered significantly (Fig. 4.2).

This is inevitable as the installation of tube wells has increased sharply over the years reaching 896 wells in 2020 (Fig. 4.3) although the DWASA envisaged that dependence would be more on purifying surface water under the Water Supply Master Plan for Dhaka City-2015.

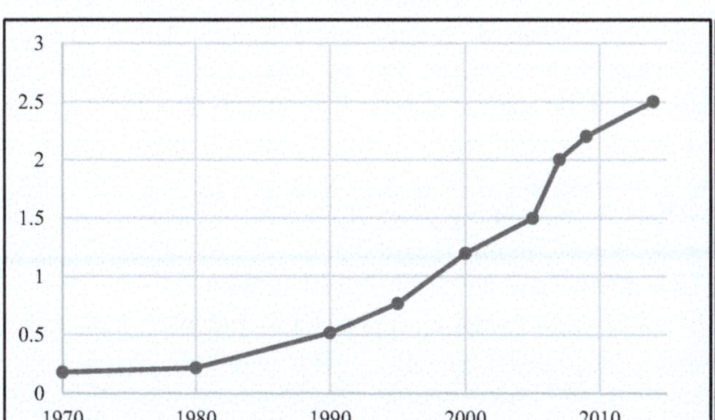

Year	Depth to water table (m)
1970	0.5-10.5
1980	0.5-9
1990	6-22.5
1995	12.5-32
2000	19-41.5
2005	19-54
2007	19-67
2009	21-67
2014	24-71

Fig. 4.2 Increase in groundwater withdrawal (Mm^3/d) and lowering of groundwater table (m) in Dhaka city from 1970 to 2014. *Source* Hassan and Zahid (2017), Zahid (2015)

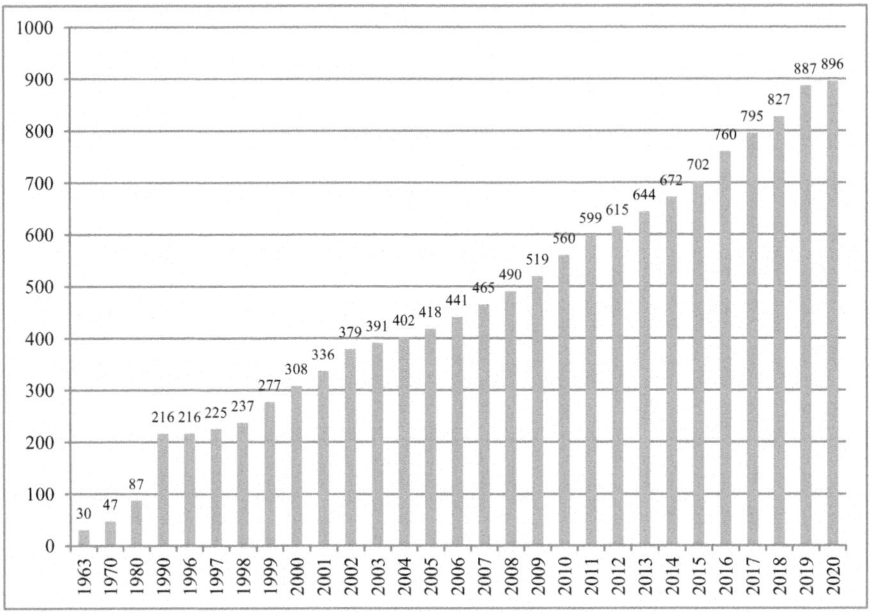

Fig. 4.3 Trend of installation of deep tube wells and bore wells in Dhaka city. *Source* DWASA (2020)

An earlier study by considering the depletion rate of 2.81 m per year projected that the groundwater level would go down to 120 m by 2050 (Fig. 4.4) (*Unnayan Onneshan*, 2011). The study also showed that the potential groundwater recharge of Dhaka city was only about 1.33 m/y, implying that the city had been experiencing an annual groundwater recharge deficit of about 1.48 m/y. The projected depletion trend is also congruent with a recent study, which predicts that the table will go down to about 110–115 m by 2050 (Islam & Islam, 2017). According to the '2030 Water Resources Group', however, Dhaka's groundwater will fall to 100–150 m by 2050 (Islam, 2021). An estimation of annual decline by another study reveals that it would be 5.1 m per year by 2030 (approximately 70% higher than the existing rate) (BWP, 2019). Although various studies projected different scenarios, there is consensus that the Dhaka city is going to face a severe challenge in provisioning adequate water to city dwellers in the coming years if preventive measures are not taken. Iftekhar and Islam (2022) noted in this regard that the city has all elements to be water sensitive, but it lacks investment and institutional structure.

Poor people, mostly living in slum areas, are deprived of potable water under authorised sources. Therefore, they depend typically on the unauthorised sources of water created by affluent residents and water vendors. Provisioning through informal sources would be more inevitable in future as demand would exceed the production capacity of DWASA then (Fig. 4.5). The projection indicates that there will always

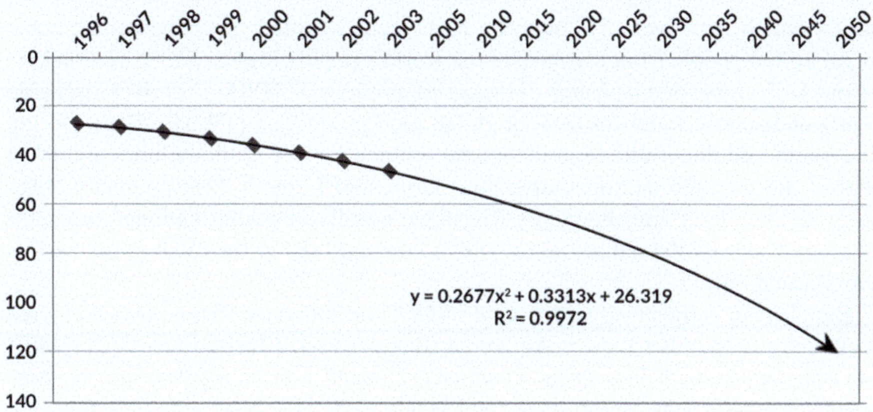

Fig. 4.4 Projection of groundwater depletion in Dhaka city. *Source Unnayan Onneshan* (2011)

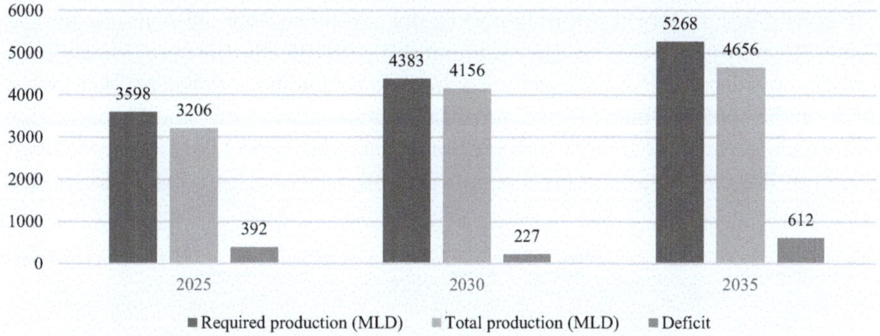

Fig. 4.5 Projected gap in required production and total production capacity of DWASA (in million litres per day-MLD). *Source* DWASA and IWM (2014)

be a deficit in supplying required water to the city across different years and the highest gap is observable in 2035.

In such a scenario, the state of groundwater in city areas indicates that on the one hand, the amount of water has been decreasing and the inequality in terms of water provisioning widening on the other.

4.2.2 Transboundary Rivers

The river basin, among the other options, is one of the vital sources of freshwater. It denotes a watershed area, which encompasses the entire land surface crisscrossed by many streams and creeks, emanating from headwater that usually originates in the upland and descends to the plains (Finger et al., 2006). Approximately, 286 river basins, which span over 151 countries, have been identified. These basins are home to 2.8 billion people (roughly 38% of the world's population) cover 62 million km^2 (about 42% of the total land area of the earth), produce 22,000 km^3 of river discharge yearly and supply water for 60% of global food production (TWAP, 2016). During its courses and the coverage in its natural settings, a river basin includes and often crosses the national administrative boundaries and forms a Transboundary River basin regime. Such Transboundary Rivers are usually treated as national instead of common water by the nation states, which fabricates the governance of the basin a complicated one.

Bangladesh is traversed with a total of 58 Transboundary Rivers, which entered the country from either India or Myanmar (Banglapedia, 2021b). In particular, India and Bangladesh share 54 rivers all of which originated in India and flow into the three major rivers—Ganges, Brahmaputra and Meghna (GBM)—that ultimately flow into the Bay of Bengal. The combined outflow of these rivers—widely known as the GBM river basin—is indeed one of the most intriguing, vigorous and magnificent fluvial common river systems in the world (Map 4.1).

It constitutes the world's third largest freshwater outlet after the Amazon and the Congo River system to the oceans (Kummu et al., 2016). The basin region covers a total area of just over 1.7 million km^2 comprising a number of regions, distributed among India (64%), China (18%), Nepal (8%), Bangladesh (7%) and Bhutan (3%), which is altogether 83% of the South and Southeast Asia (Table 4.3). The entire fluvial system of the GBM basin plays a significant role in modulating the hydroclimate

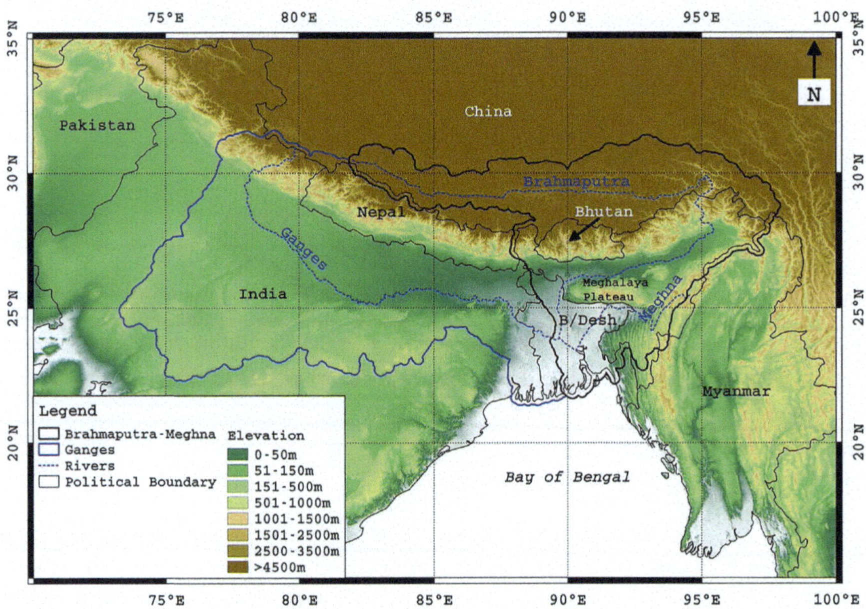

Map 4.1 Ganges–Brahmaputra–Meghna (GBM) basin. *Source* Khandu et al. (2016)

conditions, biological processes and agro-economic activities in the central South Asian region in particular (IUCN, 2014a). The basin also provides a vast reserve of surface water available for both domestic and industrial uses and has substantial hydropower potential. Altogether, it supports more than 9% of the world population (Author's calculation from World Bank Data Portal). The river system thus becomes a part of the lives and livelihoods and the culture of the millions of people in this region (Akter, 2016).

Water resources, however, in the shared region are unevenly distributed. In the upstream, it drains an area of 304.4 thousand km^2 of China, 147.5 thousand km^2 of Nepal, 38.4 thousand km^2 of Bhutan and a large part of India of about 1102 thousand km^2. Bangladesh, being a downstream country, receives the upstream drainage and the basin covers an area of 46.3 thousand km^2 by the Ganges, 39.1 thousand km^2 by the Brahmaputra and an area of 35 thousand km^2 by the Meghna River systems (Table 4.3). In such a scenario, during March–April, the Brahmaputra and the Ganges jointly account for 80% of the total flow measured within the country, while the Meghna contributes 2% (World Bank, 2000). The GBM basin is important for Bangladesh because it is the prime source of freshwater for this country to support its agriculture and consumption purposes as well as to protect the environment from the threat of salinity intrusion. Overall, the country is largely dependent on the transboundary water for its sustenance.

In particular, the total length of *the Ganges* in the Indian part is 2525 km^2 and extends 290 km^2 in the territory of Bangladesh (Table 4.4) before it confluences with the Meghna River. The Ganges basin has a vast reserve of surface water. A summary scenario of available water resources in the Ganges basin, as well as projected average annual per capita water availability in 2025 and 2050, depicts some noteworthy issues (Table 4.5). India's total renewable water resources (TRWR) available are approximately 525 BCM per year, of which 353 BCM are internal renewable surface water resources (IRSWR) and 172 BCM are rechargeable groundwater resources (RGR). On the contrary, Bangladesh has TRWR of 27 BCM, of which 22 BCM are IRSWR and 5 BCM RGR.

The projected available water per person shows that Bangladesh has only 0.41 and 0.37 m^3 of water against 885.27 and 752.48 m^3 available for India in 2025 and 2050, respectively (Table 4.5). Moreover, the Indian part has developed a live storage capacity of 42,060.2 million cubic metres (MCM) and some capacity of 18,600.18 MCM are under construction (Table 4.4).

Then, the Ganges basin is also characterised by its strong seasonal variations governed by the south-west monsoon. The average annual rainfall is 1059.74 mm (mm) (Table 4.4), which increases the volume of flow of water during the wet season (June–October). This increased flow often turns into terrible flooding, particularly in the downstream. Another hydrological feature of the basin is that it transports huge sediment with its current. The estimated annual sedimentation load of the basin is about 1.84 billion tonnes where in Bangladesh, about 316–520 million tonnes have been transported with the discharge of water (Table 4.4). Such huge sedimentation fills up the riverbed and reduces the depth of rivers in the downstream.

Table 4.3 GBM basin area, population and water distribution

Basin	Drainage area (Total)		Countries included	Area of country in basin ('000 km^2)	Total area of the basin (%)	Total area of the country (%)	Arable area ('000 km^2)	Mean annual precipitation (mm)	Surface water availability (km^3)
	'000 km^2	% of South and Southeast Asia							
Ganges	1087.3	52	Bangladesh	46.3	4	32	30	1568	525
			Bhutan	–	–	–	–	–	–
			China	33.5	3	0.3	Negligible	N/A	12.0
			India	860	79	26	602	450–2000	525
			Nepal	147.5	14	100	26	1860	210.2
Brahmaputra	543.4	27	Bangladesh	39.1	7	27	36	2400	537.2
			Bhutan	38.4	7	100	2	500–5000	78.0
			China	270.9	50	3	Negligible	400–500	165.4
			India	195	36	6	55	2500	537.2
			Nepal	–	–	–	–	–	–
Meghna	82	4	Bangladesh	35	43	24	225	3574	48.4
			Bhutan	–	–	–	–	–	–
			China	–	–	–	–	–	–
			India	47	57	1	15	2640	48.4
			Nepal	–	–	–	–	–	–
Total	1712.7	83	Bangladesh	120.4	7	83	291	1568–3574	1110.6
			Bhutan	38.4	3	100	2	500–5000	78.0
			China	304.4	18	3	Negligible	400–500	177.4
			India	1102	64	33	672	450–2640	1110.6
			Nepal	147.5	8	100	26	1860	210.2

Source Prepared by the authors based on Brichieri-Colombi and Brandnock (2003), FAO (2011a), Rahman et al. (2019), Rasul (2015)

Table 4.4 Salient features of the Ganges

Subject	Country	Values
Length of Ganges (km^2)	India	2525
	Bangladesh	290
Average annual rainfall (mm)		1059.74
Average max. temperature (°C)		32.05
Average min. temperature (°C)		18.44
Average annual water flow (BCM)	Nepal to India	210.2
	India to Bangladesh	525.02
Live storage capacity developed by India (MCM)		42,060.2
Live storage under construction by India (MCM)		18,600.18
Est. annual sediment load (billion tonnes)		1.84
Est. annual sediment load in Bangladesh (million tonnes)		316–520
Number of water resources structures	Dam	784
	Barrage	66
	Weir	92
	Anicut	1
	Lift	54

Source Prepared by the authors based on FAO (2011a), India-WRIS (2012), Islam et al. (1999), Lodrick and Ahmad (2018)

Table 4.5 Water availability accounts at Ganges basin

River basin		Bm^3 $year^{-1}$	Estimated population (million)			Estimated per capita average annual water availability (m^3)		
			2010	2025	2050	2010	2025	2050
Ganges (Indian portion)	TRWR [a]	525	494.47	593.04	697.69	1061.74	885.27	752.48
	IRSWR [b]	353				713.90	595.24	505.96
	RGR [c]	172				347.85	290.03	246.53
Ganges (namely Padma, Bangladesh portion)	TRWR [a]	27	55.95*	65.50*	72.54*	0.48	0.41	0.37
	IRSWR [b]	22				0.39	0.34	0.30
	RGR [c]	5				0.09	0.08	0.07

Source Author's calculation from BBS (2015a) and Amarasinghe et al. (2016)
Note [a]TRWR = Total Renewable Water Resources; [b]IRSWR = Internal Renewable Surface Water Resources; [c]RGR = Rechargeable Groundwater Resources; BCM/Bm^3 $year^{-1}$ = Billion Cubic Metre per Year; m^3 = Cubic Metre; *Population estimation years for Bangladesh are 2011, 2026 and 2051, respectively

In the context of Bangladesh, the Ganges or the Padma plays a vital role in the socio-economic structure of the regions of Rajshahi, Pabna, Kushtia, Jessore and Khulna (32% of the total land). Additionally, it works as a shield against salinity intrusion in the coastal region of Bangladesh and as a lifeline for the Sundarbans ecosystem (FAO, 2011a). However, man-made obstacles such as barrages and dams built over the tributaries and distributaries of the river system by the Indian counterpart cause serious disruption in its natural flow. Being a low-riparian country, Bangladesh faces water scarcity, particularly in the dry season (November–May) and the situation has been deteriorating over the years. Specifically, before 1975, the Ganges' water flow during the dry season was more than enough to meet Bangladesh's water requirements. The flow decreased considerably when India commissioned the Farakka barrage in 1975 on the Ganges, 18 km from the western border of Bangladesh. Bangladesh has been experiencing economic, social and environmental challenges because of the reduction of water flow following the construction of the barrage.

The *Brahmaputra*, another mighty river of the GBM basin, crosses three countries—China, India and Bhutan—upstream to reach Bangladesh. The Brahmaputra basin covers a catchment area of about 552,000 km^2, which is almost half the size of the Ganges, but it produces double the flow than that of the latter (IUCN, 2014b). In total, it sustains approximately 80 million people by providing them with water, fish and livelihood options (Pradhan et al., 2021). Of the total population, 41% reside in north-western and north-central Bangladesh (Fig. 4.6). The second highest proportions of the population reside in Assam and West Bengal in India. The remaining 9% of the basin population reside in Tibet, Bhutan and other north-eastern states of India.

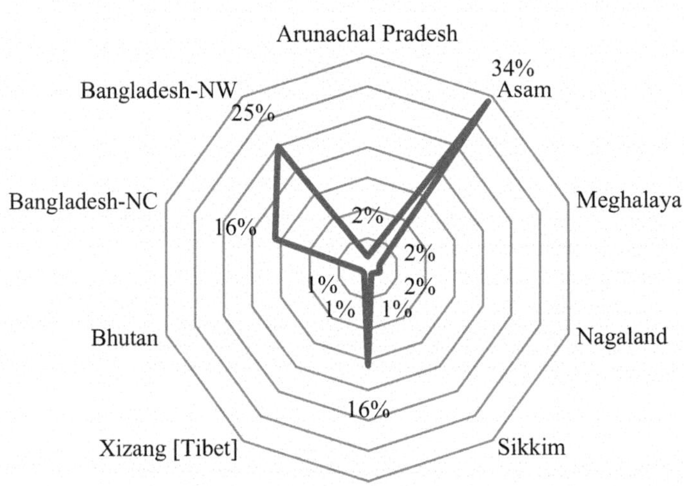

Fig. 4.6 Distribution of population in the Brahmaputra basin. *Source* Prepared by the authors based on IUCN (2014b)

While the Brahmaputra basin has been facing relatively fewer anthropogenic interventions compared to the Ganges, the efforts from upstream to initiate multi-purpose projects on this stream are buzzing. Moreover, there has been very little progress at the regional level to manage this basin. There is no international treaty exists involving all the riparian countries barring a few bilateral agreements or Memoranda of Understanding (MoUs) (Barua, 2018; Liu, 2015).

A river named Teesta, a tributary of the Brahmaputra stream (Jamuna River in Bangladesh), runs through the five northern districts named Gaibandha, Kurigram, Lalmonirhat, Nilphamari and Rangpur, comprising 9667 km^2 and supporting irrigation of 14% of the total cropped area (Asia Foundation, 2013). The areas covered by the Teesta through its course are highly reliant on the water from the river. The water crisis in this river has been an acute bone of contention between India and Bangladesh in recent times. Moreover, both India and China have more plans to build dams and barrages in the Brahmaputra basin, which can further aggravate the water crisis in northern Bangladesh in future.

Finally, the *Meghna* basin remains less discussed when it comes to the discourse on transboundary water governance (Sinha et al., 2018). The basin covers an area of 82,000 km^2, with 47,000 km^2 (57% of the total area) in India and 35,000 km^2 (43% of the total area) in Bangladesh (JRC Bangladesh, 2011). Meghna River was formed in Bangladesh near the Kishoreganj district by the convergence of two rivers named Surma and Kushiyara originating in India as the Barak River. The Meghna flows into the Bay of Bengal after joining the Padma (the combined flow of the Ganges and the Brahmaputra) near Chandpur district in Bangladesh (Baten & Titumir, 2016). On a total basis, more than 50 million people of Bangladesh and India, including several indigenous communities, depend on the ecosystem services provided by the basin (IUCN, 2021). Wetlands, in particular, on which a large number of people rely for their livelihoods, have developed because of the hydrological and topographical characteristics of the Meghna basin. The basin is currently under pressure from climate change, which has direct implications for Bangladesh's food security as the region is considered as the 'rice bowl' of the country (Sinha et al., 2018). Simultaneously, the human-induced pressures are also on the rise. In particular, the proposal by India to build the Tipaimukh dam has sparked controversy recently because it would severely disrupt the natural flow of the Meghna.

Overall, Transboundary Rivers provide Bangladesh with immense benefits such as replenishment of the rivers and water bodies, reduction in salinity intrusion, recharging of groundwater, renewal of lands in floodplains with fresh sediment, fisheries resources and many more. Considering the key types of water resources, the contribution of Transboundary Rivers is the highest bringing about 1010 billion cubic metre (BCM) of water to Bangladesh (Fig. 4.7). These rivers are the lifelines for Bangladesh, but the country is incapable to manage water resources of these rivers on its own, as 92% of the GBM basin is located outside of the country (Faisal, 2002; Wirsing et al., 2013). The country relies upon the co-operative intent of the upstream countries, particularly of India. The non-cooperative action by the Indian counterpart to hold maximum water, however, poses a severe threat to the water security in Bangladesh (mainly in the dry season), as well as a disruption in the ecological balance and system of life in the GBM regions.

The long-term implications of water shortage are more severe for Bangladesh as it leads to significant disruptions in agriculture, fisheries and forestry, which ultimately results in the economic vulnerability of the people (Fig. 4.8).

The scenario is becoming further complicated in the face of climate change, with increased glacier melting in the short term and drought in the long term, which could pose a further challenge to the ecological state of the region (Baten & Titumir, 2016;

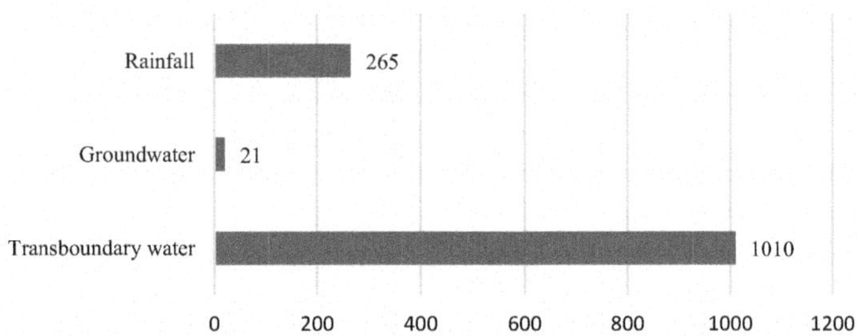

Fig. 4.7 Contribution of Transboundary Rivers in supplying water to Bangladesh (in BCM). *Source* Prepared by the authors based on Ahmed (2013)

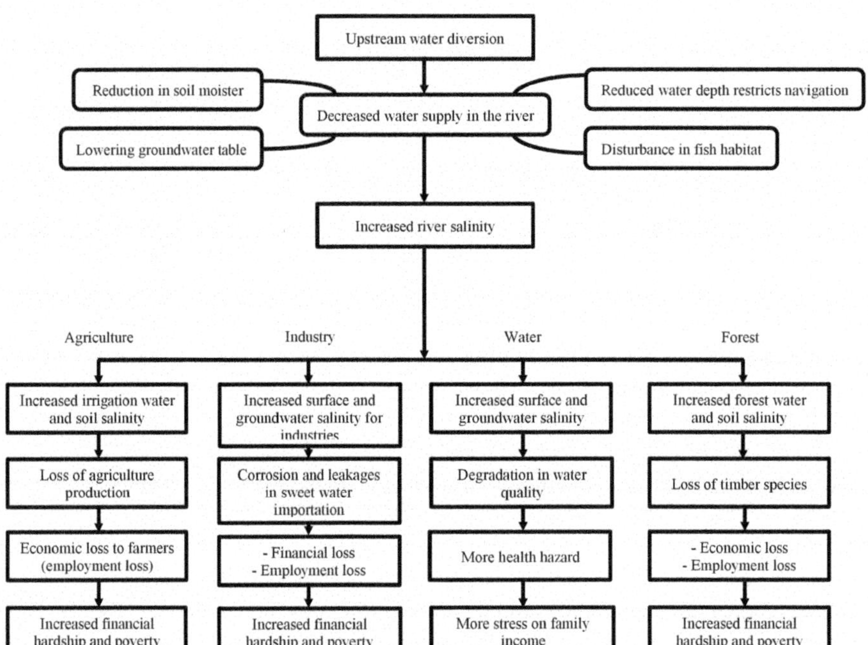

Fig. 4.8 Impacts of water diversion upstream in Bangladesh. *Source* Baten and Titumir (2016)

Immerzeel et al., 2010; Rahman et al., 2019). However, the available amount of water in the basin could meet the demand of all the co-riparian countries. The shortage of water is essentially an outcome of a lack of cooperation among the countries.

4.2.3 Wetland Resources

Wetlands are low-lying land extensions saturated with water either for the entire year or on a seasonal basis. According to the Ramsar Convention, wetlands are characterised as "areas of marsh, fen, peatland or water, whether natural or artificial, permanent or temporary, with water that is static or flowing, fresh, brackish or salt, including areas of marine water the depth of which at low tide does not exceed six meters" (Article 1.1) (Ramsar Convention Secretariat, 2016). The wetland ecosystem has very high ecological, economic, cultural, scientific and recreational values as it provides a number of services vital for human well-being (Keddy et al., 2009; Kumar et al., 2017; Maltby & Acreman, 2011; MEA, 2005; Ramsar Convention Secretariat, 2016). However, this valuable ecosystem continues to be lost and degraded across the world due to the dual effects of anthropogenic pressures and natural factors (Davidson & Finlayson, 2018; Xu et al., 2019). Subsequently, the essential benefits that the wetlands provide to human beings continue to be seriously eroded.

Bangladesh is one of the low-lying river basin countries in South Asia, with the majority of its land formed by river alluvium from the Ganges and Brahmaputra and their tributaries (DoE, 2015) (as discussed in the previous section too), which during its formation produced some pockets known as wetlands that hold water either temporarily or permanently. According to the Ramsar Convention, wetlands comprise two-thirds of the country's total land (Byomkesh et al., 2009). The wetlands of Bangladesh are mainly distributed into (a) the *Haor* basin of the north-eastern region and (b) the lowlands, locally known as *Beel*, of the GBM flood basin (Byomkesh et al., 2009; Debnath, 2016; Huq et al., 2020). Additionally, there is another type of wetland called *Baor* (the ox-bow lakes). It is, therefore, difficult to concentrate on each type to examine the problems of governance affecting the wetland ecosystem. In this backdrop, this chapter focuses on the *haor*-based wetland ecosystem in particular for empirical analysis in a later section because it is the most critically sensitive wetland area in Bangladesh.

The *haors* are swampy lowlands that turn into a vast span of water bodies recharged by surging surface rainfall water during the monsoon. The *haor* areas are located mostly in the north-eastern region in Bangladesh (Map 4.2) covering an area of 1.99 million ha and accommodating around 19.37 million people (CEGIS, 2012).

About 373 *haors* are located in the districts of Sunamganj, Habiganj, Netrakona, Kishoreganj, Sylhet, Moulvibazaar and Brahmanbaria (Table 4.6), covering an area of roughly 858,000 ha (approximately 43% of the total area of the region). Some of the prominent *haors* are Hakaluki haor, Tanguar haor, Hail haor, Matian haor, Pasuar Beel haor, Dekar haor, Baro haor, Gurmar haor, Sonamorol haor, Baram haor, Kalni haor, Kawadighi haor and Pagner haor among which Tanguar Haor and Hakaluki

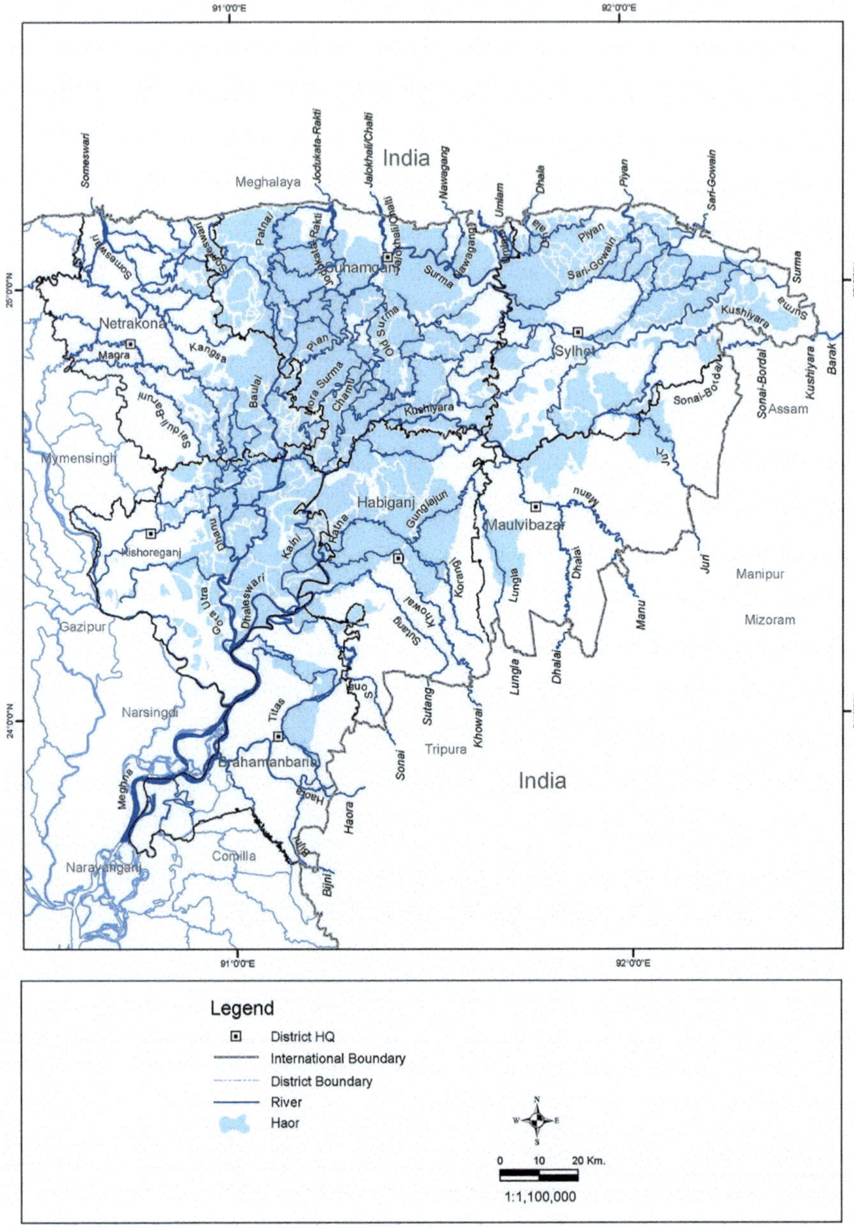

Map 4.2 Location of *haor* region in Bangladesh. *Source* CEGIS (2012)

Haor have been designated as Ramsar Site of ecological importance as well as an Ecologically Critical Area (ECA) (MoWR, 2016).

The wetlands of Bangladesh have immense importance in terms of supporting human lives and that of diverse flora and fauna. Broadly, the values of wetlands are categorised into three types: economic, environmental and social or cultural (see

Table 4.6 *Haor* areas of Bangladesh

District	Upazila	Total area (Hectare)	Haor area (Hectare)	No. of Haors
Sunamganj	Bishwambarpur, Chhatak, Dakshin Sunamganj, Derai, Dharampasha, Dowarabazar, Sulla, Tahirpur, Jagannathpur, Jamalganj, Sunamganj Sadar	367,000	268,531	95
Sylhet	Balaganj, Beani Bazaar, Bishwanath, Zakiganj, Companiganj, Dakshin Surma, Fenchuganj, Golapganj, Gowainghat, Jaintiapur, Kanaighat, Sylhet Sadar	349,000	189,909	105
Habiganj	Ajmiriganj, Bahubal, Baniachong, Chunarughat, Habiganj Sadar, Lakhai, Madhabpur, Nabiganj	263,700	109,514	14
Maulvibazaar	Barlekha, Juri, Kamalganj, Kulaura, Rajnagar, Maulvibazaar Sadar, Sreemangal	279,900	47,602	3
Netrakona	Atpara, Barhatta, Durgapur, Khaliajuri, Madan, Kalmakanda, Kendua, Mohanganj, Purbadhala, Netrakona Sadar	274,400	79,345	52

(continued)

Table 4.6 (continued)

District	Upazila	Total area (Hectare)	Haor area (Hectare)	No. of Haors
Kishoreganj	Austagram, Bajitpur, Bhairab, Hossainpur, Itna, Karimganj, Katiadi, Kishoreganj Sadar, Nikli, Kuliar Char, Mithamain, Pakundia, Tarail	273,100	133,943	97
Brahmanbaria	Akhaura, Banchharampur, Ashuganj, Kasba, Sarail, Brahmanbaria Sadar, Nabinagar, Nasirnagar	192,700	29,616	7
Total		1,999,800	858,460	373

Source CEGIS (2012), MoWR (2016)

Byomkesh et al., 2009; Islam, 2010 for detail classification of values). The *haor* regions in particular are one of the most important economic hubs of Bangladesh. Agriculture and fisheries are the main sources of resource. Millions of people, particularly the marginalised and the poor, depend on the wetland ecosystem for their livelihoods, food and nutrition, fuel and transportation.

In terms of environmental value, the ecosystem supports diverse species of biodiversity resources of which some are threatened as well (Table 4.7). Additionally, it reduces the impact from storm damage and flooding, stabilises climate by storing carbon and helps in recharging groundwater. Overall, the wetlands play a significant role in maintaining the ecological harmony of the country. Finally, the region has social and cultural values as it offers recreation and tourism opportunities.

Table 4.7 State of biodiversity resources of *haors*

Category of species	Number of species	Number of threatened species
Amphibians	9	1
Birds	257 (plus thousands of migratory birds)	18
Fishes	289	–
Mammals	29	13
Reptiles	40	24
Plants	300	–

Source Authors' compilation from Birdlife International (2004), Byomkesh et al. (2009), CEGIS (2012)

The wetlands of Bangladesh, however, have been declining and their resources degrading gradually as is evident from various studies (Byomkesh et al., 2009; Haq, 2013; Haque & Basak, 2017; Hossain et al., 2009; Islam, 2010; Shopan et al., 2013). Several anthropogenic pressures have been identified as the causes of degradation of the wetland ecosystem such as encroachment into wetlands for housing, industry and agricultural purposes; overexploitation of fisheries resources; over-cutting of wetlands' vegetation; shrimp cultivation; hunting of water birds; crude oil discharge; household waste disposal; industrial (e.g., fertiliser, cement) waste disposal; use of chemical fertilisers in agriculture; construction of flood embankments and rural roads; gas and oil exploration, among others. Overall, the degradation of the ecosystem is resulting from the impact of the burgeoning human population and unsustainable usage of their resources.

Specifically, the *haor* area is a vital source of the inland (capture) fisheries sector of the country, and the *haor*-dependent communities primarily rely on this sector for managing their livelihoods. The production of fish from this source has increased from 97,119 MT in FY2017-18 to 108,880 MT in FY2018-19 with a growth rate of 12.11% (DoF, 2019). However, the noticeable fact is that the contribution of the inland (capture) fisheries sector is significantly low compared to that of the inland (culture) fisheries (Table 4.8). The productivity rate of capture fisheries was 318 kg/ha, whereas the number is 3028 kg/ha for culture fisheries. The productivity rate of each of the capture sources, including *haor*, is also lower than that of the culture sources. The success of the fisheries sector, therefore, depends largely on the culture fisheries rather than the capture one.

In the past, however, the key source of fish production in Bangladesh was the inland open water capture fisheries. Inland culture fisheries surpassed the former since the year of 2000 (Alam, 2011). It indicates that the natural inland fish stocks have waned significantly, and the livelihoods of poor fishers have been affected seriously. This scenario is also relevant to the wetlands or *haors*, which are the essential parts of the inland open water system.

The vulnerable state of the wetland ecosystem in Bangladesh thus can be depicted from two perspectives (Fig. 4.9). Apart from the degradation of resources, the condition of livelihood of the wetland-dependent communities is also worsening. Although the country has made significant strides in development over the past few decades, the socio-economic conditions of the *haor* regions have remained stagnant throughout the decades (Alom, 2016; Nahar et al., 2017; Parvin & Akteruzzaman, 2012). On an average, 29.56% of people in the *haor* areas live below the Lower Poverty Line (LPL) (CEGIS, 2012). They have always been deprived of sufficient access to the necessities of life and the condition has become critical over the years due to the degradation of available resources. Therefore, addressing both these challenges is central to the wetlands governance framework.

Overall, the wetland ecosystem of the *haor* regions in Bangladesh is under a severe threat of degradation, which has implications in the context of the economy—both local and national. The identified causes of the degradation of resources in previous studies, however, were external factors only. Critical scrutiny is required to understand the vulnerability by identifying the inherent factors.

Table 4.8 Sectorwise inland fish production in Bangladesh, 2018–19

Inland fisheries	Water area (Ha)	Production (MT)	Productivity (Kg/Ha)
Inland open water (capture)			
1. River and estuary	853,863	325,478	381
2. Sundarbans	177,700	18,282	103
3. Beel	114,161	99,890	875
4. Kaptai Lake	68,800	10,578	154
5. Floodplain (a) *Subsistence fisheries* (b) *Fry released programme* (c) *Haor**	**2,675,758** *2,317,175* *107,346* *251,237*	**781,481** *623,607* *48,994* *108,880*	**292** *269* *456* *433*
Capture Total	**3,890,282**	**1,235,709**	**318**
Inland closed water (culture)			
1. Pond	397,775	1,974,632	4964
2. Seasonal culture water body	144,217	217,340	1507
3. Baor	5671	10,343	1824
4. Shrimp/Prawn farm	258,553	258,039	998
5. Pen culture	6330	12,361	1953
6. Crab production	9377	12,084	1289
7. Cage culture	1.76 lakh cum	3802	22 kg/cum
Culture Total	**821,923**	**2,488,601**	**3028**

Note **Bold** numbers denotes the total area for "floodplain", "capture", and "culture" and *italics* are the sub-categories of floodplain fisheries. Haor* under floodplain is given an asterisk to signify that the discussion mainly focuses on this one category in detail.
Source DoF (2019)

4.2.4 Marine Resources

With 710 kms of coastline, Bangladesh is located at the tip of the Bay of Bengal (BoB). At the end of the ultimate settlement of the maritime border dispute with Myanmar and India in 2012 and 2014, respectively, Bangladesh received entitlement to 118,813 km^2 in the BoB, consisting of its territorial sea and the Exclusive Economic Zone (EEZ) (MoFA, 2014). Bangladesh accordingly possesses 121,110 km^2 of total marine water (Map 4.3) with an increase of maritime boundary by 15–20%, a size equal to more than 80% of the country (Patil et al., 2018). Since then, marine resources

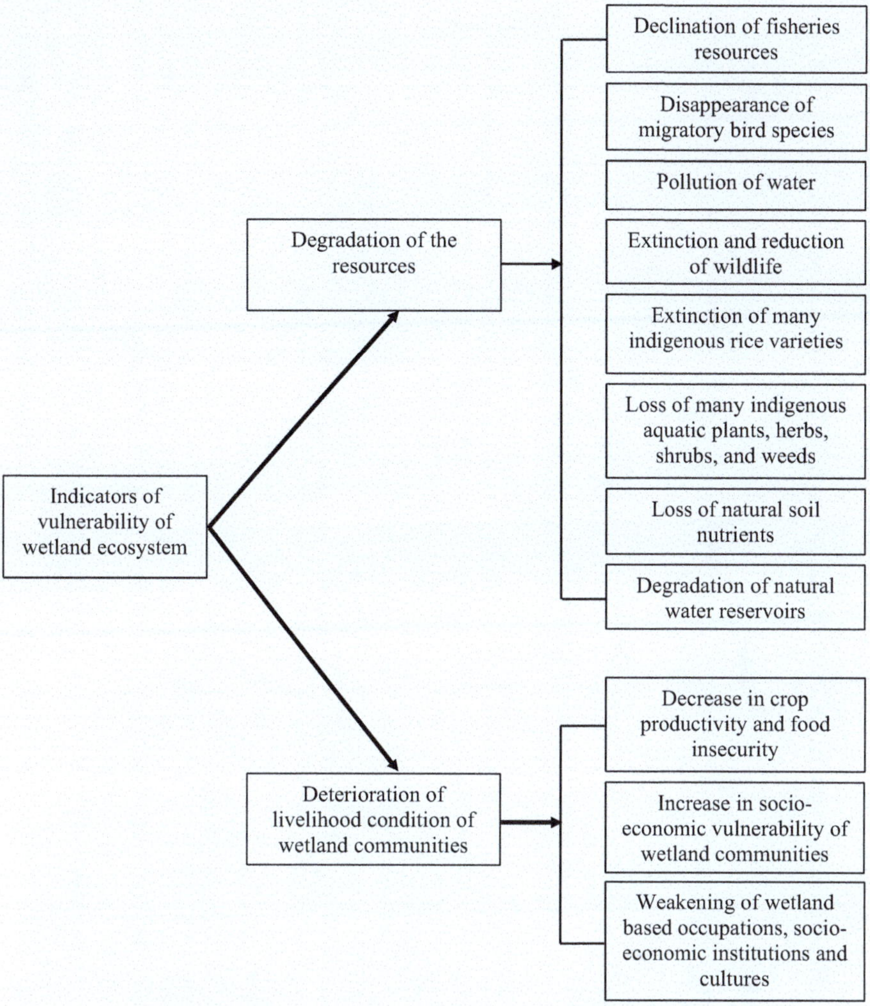

Fig. 4.9 Vulnerability of the wetland (haor) ecosystem. *Source* Prepared by the authors

as another form of vital aquatic resources have gained particular attention, as the marine ecosystem could be the country's major source of natural resource supply in the future.

Both the coastal and marine ecosystems are also a part of the Bay of Bengal Large Marine Ecosystem (BOBLME)—one of the 64 large marine ecosystems in the world (Habib & Islam, 2020; IUCN Bangladesh, 2015). The ecosystems are endowed with a rich reservoir of both living and non-living resources (Shamsuddoha & Islam, 2017). The coastal region of the country has all along contributed to the national economy, but the marine ecosystem added a new dimension to harness the potentiality of the blue economy. The 'blue economy' is an emerging concept in the realm

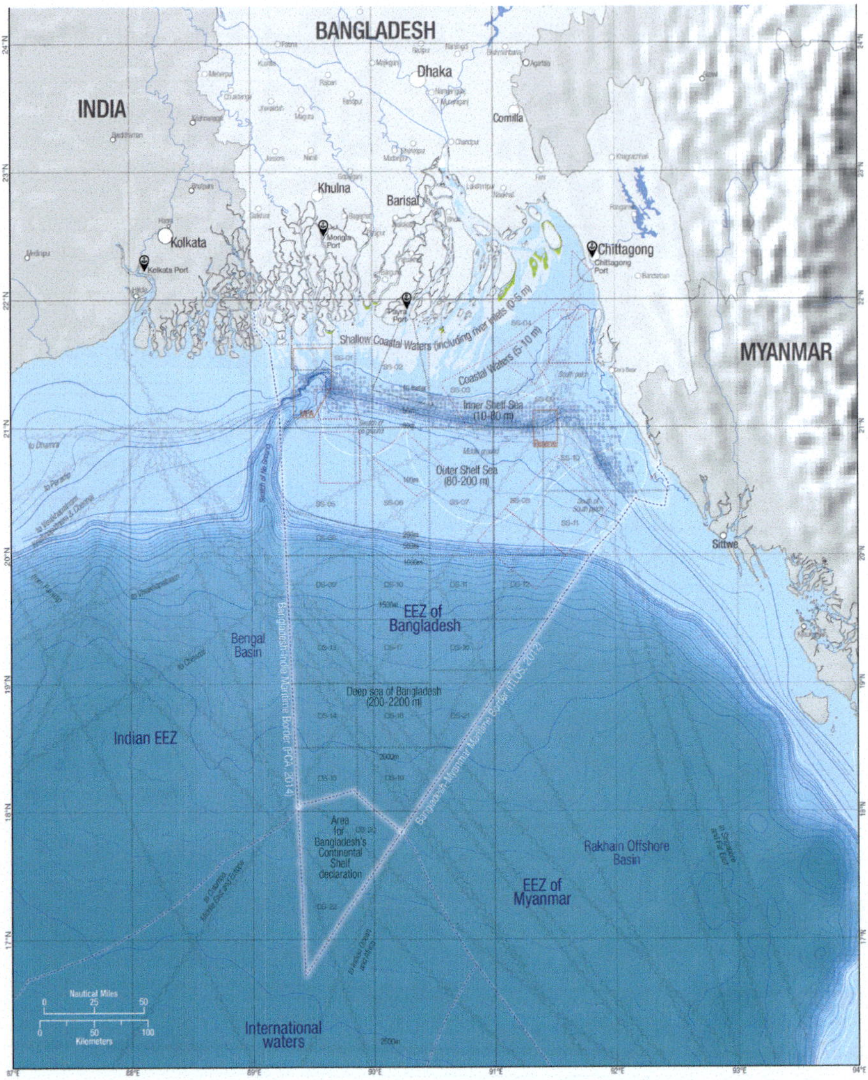

Map 4.3 Bangladesh's maritime entitlements. *Source* Chowdhury (2017)

of economic development that entails using ocean resources sustainably through technological innovation to enhance livelihood opportunities while preserving the ocean ecosystem (Sarker et al., 2018; World Bank, 2017). It covers a wide range of interconnected sectors and economic activities relevant to oceans, seas and coasts (Table 4.9).

Available estimate suggests that the marine resources contributed more than $6 billion annually to the Bangladeshi economy (The Daily Star, 2019), but there is also a claim that there is no accurate estimate of the actual contribution (Alam, 2021).

Table 4.9 Status and future potential of marine-based economic sectors of Bangladesh in brief

Sector		Current status and future potentials
Living resources	Fisheries	(a) An important source of animal protein, income and employment
		(b) Contributes to earn foreign currency
		(c) Marine aquaculture on a commercial basis, marine stock enhancement and sea ranching are yet to be developed
		(d) Technological incapacity to harness the full potentiality
		(e) Most of the commercially important fish stocks are either overexploited or under threat
		(f) Need to develop options to utilise the resources sustainably and efficiently
	Non-conventional fishery items	(a) Different non-conventional fishery items such as seagrass, algae and seaweeds, squid and cuttlefish, jellyfish, oyster, crabs, snails, shellfish, octopuses also hold promise
	Marine biotechnology	(a) Biotechnology has not yet been utilised for any marine living resources
		(b) Scope remains to apply marine biotechnological tools for developing pharmaceutical drugs, chemicals and other products
Non-living resources	Oil, gas, minerals, renewable resources	(a) Despite Bangladesh issuing 26 blocks for oil and gas exploration, no offshore commercial discoveries have been made till date
		(b) 22 sites have been identified for onshore coastal wind power generation
		(c) True potential of its offshore gas prospects is yet to assess
	Salt production	(a) The goal is to meet the domestic demand only; discouragement of import (National Salt Policy 2011)
		(b) Creates over 5 million employment opportunities
		(c) Contributes to the livelihood of coastal communities

(continued)

Table 4.9 (continued)

Sector		Current status and future potentials
Others	Shipping and transport; shipbuilding	(a) Remains of fundamental importance in terms of connecting Bangladesh to the global economy
		(b) Sea born trade and fleet use need to be expanded
		(c) More than 100 shipyards engaged in traditional and domestic shipbuilding while more than 10 engaged in building ships of international quality
		(d) Exportation of ships has started, but further scope remains
	Tourism and recreation	(a) Tourism sector is less developed compared to neighbouring countries
		(b) Sustainable tourism can create new employment opportunities and reduce poverty
	Coastal protection and carbon sequestration	(a) Ecological significance in addressing climate change

Source Prepared by the authors based on Mannan et al. (2020), Patil et al. (2018), Sarker et al. (2018), Sanaullah (2021)

Among the diverse sectors of the marine ecosystem, however, the most commercially viable one is the fishery and it can directly be attributed to the interpretation of water resources in the context of this book. Therefore, a special attention has been given to marine fisheries.

The fisheries sector contributes 3.5% to the national GDP and 25.72% to the total agricultural GDP in Bangladesh and more than 12% of the population is directly or indirectly involved in various fisheries-related activities for maintaining their livelihoods (DoF, 2019). The total production of fish has increased over the years from 28.99 lakh MT in 2009–10 to 43.85 lakh MT in 2018–19 (DoF, 2019). In particular, marine fisheries have been playing an important role in fulfilling the domestic demand for protein and in earning foreign currency through export. The harvest of marine fish was 5.89 MT in 2012–13. It ramped up to 6.60 lakh MT in 2018–19 and was projected to reach 6.78 lakh MT in 2019–20 (MoF, 2020).

Between the 1970s and 1980s, a number of surveys examined the state of marine fisheries resources (Table 4.10), but there is no current or complete information on the fisheries stocks of the marine ecosystem (Shamsuzzaman et al., 2017).

Hilsha (*tenualosa ilisha*)—a Geographical Indication (GI) product—as a single species contributes to around 12% of the country's total fish production and its contribution to the GDP is more than 1% (GED, 2020). Around 50–60% of the global *Hilsa* catch occurs in Bangladeshi coastal and marine waters, followed by 20–25% in Myanmar, 15–20% in India, and the remaining 5–10% in other countries

Table 4.10 Standing stock (in tonnes) of demersal fish, pelagic fish and shrimp of the Bay of Bengal during the1970s and 1980s

Demersal fish	Pelagic fish	Shrimp	Reference
264,000–373,000	–	9000	West (1973)
160,000	90,000–160,000	–	Saetre (1981)
200,000–250,000	160,000–200,000	4000–6000	Penn (1983)

Source FAO (2014b)

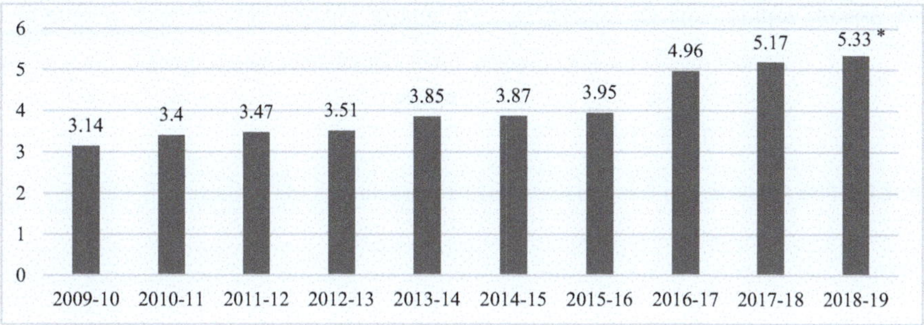

Note: *projected

Fig. 4.10 Production of Hilsha fish from 2009–10 to 2018–19 (DoF, 2019)

(Hussain, 2019). In 2017–18, a total of 5.17 lakh MT *Hilsha* was caught, up from 3.14 lakh MT in 2009–10 (Fig. 4.10).

However, the rate of expansion of production of marine fisheries has been very slow although absolute production is increasing. As a result, marine fisheries have seen a decline in their share of total fish production—shrinking from nearly 18% in 2009–10 (GED, 2020). In addition, the production of marine fisheries has remained stagnant compared to that of inland (capture) and inland (culture) fisheries over the years (Fig. 4.11). In concrete terms, the marine fisheries have the least contribution to the total fish production of the country. In 2018–19 FY, the production of marine fisheries was 6.6 lakh MT, whereas the numbers were 12.36 lakh MT and 24.89 lakh MT for inland (capture) and inland (culture) fisheries, respectively. As a result, the country has failed to utilise the fisheries resources from the BoB at the full scale.

Moreover, Bangladesh earned BDT 4250.31 crore by exporting 73,171.32 metric tonnes of fish and fishery products in FY2018-19. The overall export scenario, however, shows that the amount of total fish and fish products exports have remained relatively stagnant over the years (Fig. 4.12). If utilised appropriately, the recently settled maritime boundary has enormous potential for marine fish, which could contribute to an exponential rise in the country's export revenues from the fisheries sector. Overall, the state of the marine ecosystem shows that there is a great potential to increase revenue using marine resources, especially the fisheries resources, which are currently underutilised and undervalued.

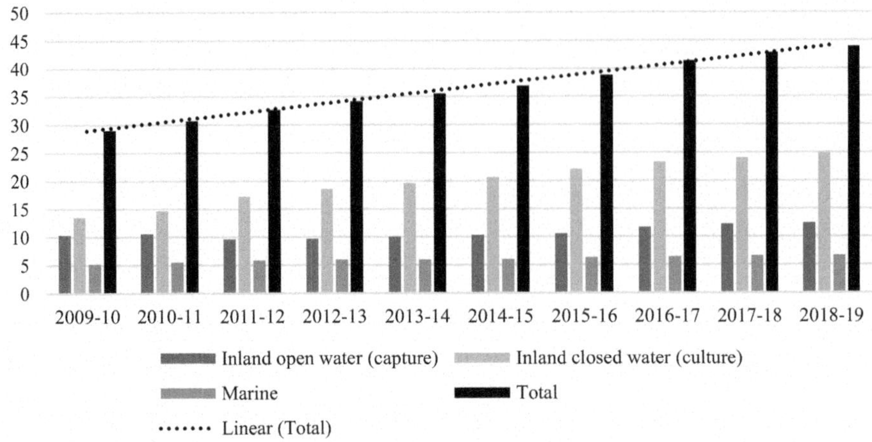

Fig. 4.11 Fish production in Bangladesh in last 10 years (in lakh MT). *Source* DoF (2019)

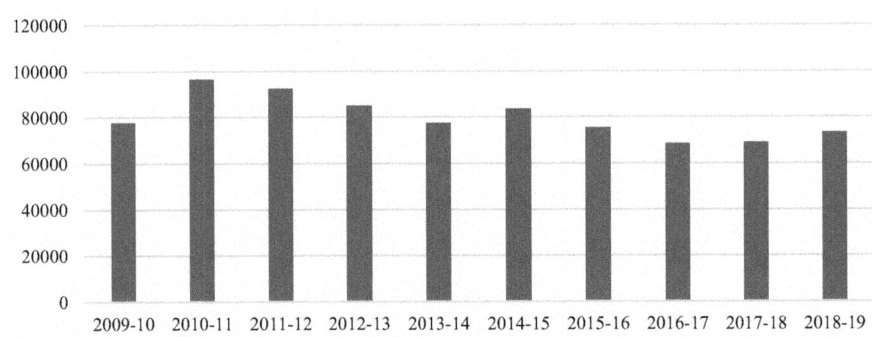

Fig. 4.12 Trend in export of fish and fish products (Amount in MT). *Source* GED (2020)

Finally, the marine fisheries sector faces challenges in terms of several anthropogenic pressures and natural drivers. Ocean pollution, degradation of marine habitats such as coral reefs, mangroves, estuaries, spawning and nursing grounds under coastal development projects seriously affect the productivity of the sector (GED, 2020; Patil et al., 2018). Moreover, ocean warming, sea level rising, ocean acidification, etc. resulting from climate change have been seriously altering the lifecycle of marine species (IPCC, 2019). Being a climate-vulnerable country, the marine ecosystem of Bangladesh is facing similar threats.

4.3 Water Under Market: Scarcity, Pricing and Institutions

The market-centric approach advocates using the market as an allocation mechanism to solve water governance problems, including environmental externalities through pricing and establishing private property rights (Anderson & Leal, 1988; Bakker, 2004, 2005, 2007; Dinar & Subramanian, 1997; Dinar et al., 1997, 2015; Fridman, 2015; Grafton et al., 2014; Griffin, 2001, 2006; Lee, 1999; Olmstead, 2010; Olmstead & Stavins, 2007; Sjödin et al., 2016; Zetland, 2021). The aim is to portray water as an increasingly scarce resource and, accordingly, it suggests that pricing is the best instrument to allocate it efficiently. In view of this, it relies on profit-maximising private entities to distribute water at the socially optimal level. The accountability of private entities to their customers and shareholders is argued to be more direct and effective than that of the political representatives to the citizens (Rogers et al., 2002; Winpenny, 1994). The underlying implication is that the state has failed to allocate water resources among the citizens efficiently because of the politicisation of resources and regulatory capture (Bakker, 2007; Parker & Kirkpatrick, 2005).

The marketisation of water, however, portrays multiple meanings such as introducing market principles to public sector management, the extension of private property rights, the introduction of new technologies, liberalisation, deregulation, reregulation, the creation of water markets, private sector partnerships, decentralisation and many more (Furlong, 2010). Overall, there are conflations as well as overlapping in defining water governance under the neo-classical framework. Specific policy instruments are also difficult to categorise. Some of the instruments are raw water abstraction charges, household water tariffs, irrigation tariffs, tradable water permits, re-selling of piped waters, etc. (Sjödin et al., 2016).

Such commercialisation of water is promoted with the advent of neoliberal policy reforms since the mid-1980s under 'market environmentalism' (Bakker, 2004). In particular, the journey began with the proclamation of water as an economic good in the Dublin Principles—adopted by the United Nations at the 1992 International Conference on Water and Environment (Figueres et al., 2003; Rogers et al., 2002; World Bank, 2004), though there is a long historical evolution (Furlong, 2010). The reform initiatives have had a profound impact on different types of water resources ranging from piped water, raw water, watershed management and transnational water (Furlong, 2010). In the context of this book, the actual implications of pricing water would be explained concerning the distribution of piped or groundwater for everyday consumption in an urban context.

Neo-classical economics' prescriptions, however, rarely deliver the predicted results and fail to offer sustainable solutions to the complex, prolonged and geographically diverse problems of water governance. Instead, the branch offers a very straightforward suggestion considering the water as similar to any other private commodity and thus could not capture the multiple values of water. The critical review on 'market environmentalism' signifies the negative impacts of neo-liberal reform measures in terms of both environmental degradation and distributional implications in various forms of capital accumulation (Heynen et al., 2007; Mansfield, 2004; McCarthy & Prudham, 2004). This has acute implications in the context of water governance too (Allouche & Finger, 2001; Barlow & Clarke, 2003; Hukka & Katko, 2003; McDonald & Ruiters, 2005; Prudham, 2004). The market creates barriers for agents to have access to water; in this process, the poor are often excluded, as their purchasing power may not be adequate to afford the water with a price tag. Thus, the profit motives of private firms lead to higher prices resulting in a loss of consumer surplus and marginalisation of the poor.

The facts that policy objectives are not clearly stated and that policymakers frequently attempt to address multiple purposes with a single water price are additional reasons for the poor performance of water pricing instruments (Sjödin et al., 2016). Then, large-scale investment is often required for implementing the market instruments through improved infrastructure, but empirical evidence shows that water infrastructure suffered from underinvestment even in the industrialised countries (Gandy, 1997). Simultaneously, large water projects such as the construction of dams are questioned because of their social and environmental impacts (Conca, 2006). Some other critics by identifying water as a non-substitutable resource argue for recognising access to water as a 'human right' (Morgan, 2004; Trawick, 2003).

The human rights concept, however, has several drawbacks, but the fundamental weak point is that the human right to water could not exclude the 'commodification of water' (Furlong, 2010). The inherent reason is that the basis of the 'right' doctrine has emerged from individualistic libertarian philosophy (Mutua, 2002), which also promotes the commodification of water. Above all, the success of the market mechanism often depends on the existing nature of a particular social and political setting. Considering this, it can be argued that in a transitional economy, the pricing of water can have serious implications, especially in the capturing of rent by powerful entities, which has not been addressed in the neo-classical framework.

The New Institutional Economics (NIE), which argues for a stable institutional arrangement for governing resources sustainably (North, 1991; Ostrom, 1990), is relevant here as it reveals the importance of reforming water institutions (IBRD/Menard & Saleth, 2012; Saleth, 2018; World Bank, 2004). Moreover, water resources are sometimes managed as a common pool resource under robust community-controlled mechanisms (Berkes, 1989; Mehta, 2003). The community ownership or collective action is said to be successful in managing the commons resources efficiently by incorporating obligation-based property rights distribution (Ostrom, 1990, 2005; Schlager & Ostrom, 1992). The argument is that it is not possible to manage water wisely without community involvement as water is tightly bound to communities of a particular local context (Shiva, 2002). However, the inequitable power structure within and outside the community can distort this mechanism (McCarthy, 2005; Mehta, 2001). The heterogeneity in the group behaviour increases the transaction cost that breaks the cooperative action. Alternatively, cooperative water management may be difficult to implement in practice due to stakeholders' strategic behaviour and the high transaction costs of organising collective action (Kahil et al., 2016). Specifically, the common water resources cannot accrue efficiency in transitional economies because of heterogeneous groups and the persistence of weak enforcement of institutional arrangements under the existing nature of the political settlement. Furthermore, the powerful groups often violate the existing nature of property rights under the same condition. The NIE branch could not put their argument to describe such factors behind the institutional fragility.

In particular, understandings of the NIE School are also relevant to marine resource governance as the debate regarding the distribution of property rights over marine resources among different states continues. Individual states are declaring the exclusive economic zones for governing the marine resources but there is a chronic failure, which in large part is attributable to distributive bargaining problems (Alcock, 2002). The developing countries are primarily deprived of utilising the resources efficiently because of a lack of power in negotiating with the powerful countries as well as of technological backwardness. These factors are beyond the institutional economics framework.

Finally, the Integrated Water Resources Management (IWRM) system can be recognised as an institutional arrangement that reflects upon the need to strike a balance between economic efficiency, social equity and environmental sustainability by considering water as an economic good (Lenton, 2011). Water management at the river basin scale, public participation in decision-making and setting pre-determined goals or outcomes are the policy suggestions under IWRM for addressing water crises (Young et al., 1994). It is, however, raising the question of how participatory a planning process can be if the goals are pre-determined by IWRM constructs (Lautze et al., 2011). Moreover, IWRM only considers the physical volumes of water rather than the various values derived from their uses (Roy et al., 2011). Consequently, water governance to meet human demands has come at the expense of freshwater biodiversity and the integrity of wetland ecosystems (Gilman et al., 2004). Therefore, the sustainability arguments remain contentious when it theoretically considers water as a commodity (Anokye & Gupta, 2012). Finally, it methodologically allows economic and political interests, more than IWRM principles, to work more in decision-making (Godinez-Madrigal et al., 2019). In this process, it disregards the traditional management of water resources.

Overall, the market-centric approach and the concomitant reform initiatives have several implications for water resources although it is difficult to pinpoint the impacts of a particular policy given the fact that water resources can be of diverse types. However, in the context of this chapter, there are some major implications of this approach. First, the chapter challenges the neo-classical proposition of water pricing in monetary terms by arguing that valuation through pricing could not capture the inherent value of water. From a political ecological standpoint, the existing crises and conflicts in the water sector can be attributed to the fact that the water governance framework values water as a commodity while national and international policies defined it as a right (Johnston, 2003).

The chapter, however, does not place emphasis on the 'human right' perspective of water either. Rather, it argues that the valuation of water needs to be reframed under the human sociality framework. There is a historical nexus between humans and water, which needs to be revitalised, and once that is done, then human beings themselves could ensure the equitable distribution and creative conservation of water resources. Second, the chapter emphasises the importance of stable institutional arrangements to govern common water resources such as wetlands, Transboundary Rivers, and marine resources in a sustainable manner. However, the political-economic factors are more significant in this regard.

4.4 Politico Economy of Commodification, Exchange and Accumulation

The consideration of the politico economy branch in understanding the problems associated with water governance has focused primarily on the Marxian doctrine. The branch finds neo-liberalisation or privatisation of water resources to be the newer form of accumulation under capitalism. To some extent, it framed the market environmentalism as the 'green imperialism' in which particular instances of environmental deterioration are mobilised as opportunities for continued profit (O'Connor, 1996; Pratt & Montgomery, 1997).

Furthermore, there are also distributional implications in the various forms of accumulation of capital, which can be related to Harvey's (2003, 2004) concept of 'accumulation by dispossession' (Glassman, 2006). In this process, various forms of property rights are converted into exclusive private property rights, which is a form of expropriation of resources (Swyngedouw, 2005). Overall, commodification is a just strategy for expropriating water resources, which is inevitable under capitalism. Capitalism requires expanded reproduction for accumulation, where a natural resource like water is a potential avenue for reproduction under the neo-liberal policy regime.

The Marxian concept of 'unequal ecological exchange' can also be related to the water governance regime, specifically to the issue of commodification of water and governance of transboundary water. First, markets for water are, in fact, politically produced as foreign governments and international financial institutions have crucial influence over the creation of such markets in developing countries (Haughton, 2002). It signifies that international stakeholders are creating scope to promote capitalism at the expense of the degradation of water resources in developing countries. Secondly, the existing nature of the international political economy causes unequal distribution of transboundary water flow among various nation states where the powerful states are accruing more benefits. Thus, water has become a political good that is connected to contested relationships of power and authority (Bakker, 2012).

Moreover, political economy variables such as power and political settlement are also relevant to the governance of water and in this regard, the propositions of political ecology in particular are relevant. The political ecological school argues that the failures of the market and institution in governing water are stemming from the power-laden socio-ecological relationship in society (Andreucci et al., 2017; Harris et al., 2012; Lebel et al., 2005; Swyngedouw, 2009). The imbalance of power among agents can also distort the governance system and create an unequal exchange of water (Boelens et al., 2016; Mirumachi & Allan, 2007; Tiwary, 2006). Moreover, the political ecology doctrine challenges the conservationist approach of ecological economics, which prescribes enclosure to protect aquatic resources (Kelly, 2011). The argument is that enclosure is the means of primitive accumulation (Harvey, 2004) and it cuts into the rights of people entitled to access the resources.

Overall, the chapter considers the political economy approach in order to understand why the existing forms of property rights over water resources are not enforced appropriately in a given context and the reasons behind the existing inequality in distributing water resources.

4.5 An Alternative Framework

By critically examining the market and non-market comprehensions of water-related problems (Table 4.11), the chapter comes up with its own explanation of water resources degradation and inaccessibility to water resources. The discussion states that the neoliberal tenets cannot explain the problem of water resources and the market solution they offer is not efficient. The policies are in favour of water pricing, deregulation of water resource regimes and integrated water resource management that has worsened the crisis other than reduced it. Alternatively, the politico economy tenets need revision, which only argue that the power makes the institutions fragile but could not offer any solution.

Therefore, the new framework, on the one hand, considers the political economy approach to examine the failure of market-centric measures and on the other, tries to develop an alternative understanding for creative conservation of water and sustainable solution to the complex problems associated with water governance based on human sociality.

4.5.1 Proposition 1: Provisioning and Access

Water, if provisioned through the market, there is a possibility of excludability in the forms of access leading to unequal distribution, distortionary price and price discrimination due to imperfect market with entry barriers and high transaction costs.

This proposition implies that the commodification of water transforms the use value of water into the exchange value. The exchange increases the price and reduces the consumer surplus. Because of price and the reduction of consumer surplus, access to water largely correlates with the socio-economic hierarchy of households. The formation of class and exclusion by the socio-economic hierarchy make two forms of water markets that are formal and informal. As the lower socio-economic class does not get access to the formal channel, they largely depend on informal sources of water. The informal channel dictates by the powerful agents of the society and

Table 4.11 Underlying gaps in approaches regarding water

Branches name	Arguments for water	Underlying gaps
Neo-classical economics	(a) Creation of water market is necessary to allocate it to its most valuable uses as water is scarce in nature (b) Privatisation to create proper incentives for management and cost control (c) Full cost pricing to improve allocation and reduce wastage of water, as well as improved infrastructure and service delivery (d) Reduction of subsidies in the water sector for sending proper price signals for improved resource allocation (e) Less government intervention in managing water resources	(a) Water as an essential component of life on earth cannot be left to the vagaries of the market (b) Enclosure of common water bodies results in marginalised access where the poor can be the sufferer (c) Detrimental to poor users (d) Ignores the weak governance and the political transaction cost in developing countries (e) Political feasibility of market-based instruments is questionable; sometimes unrealistic, given the investment required (f) Treating water as a marketable commodity is fallacious as it ignores the inherent values of it (g) Water is a natural monopoly for which competition is limited
Institutional economics	(a) Emphasises institutional stability for ensuring sustainable management and equitable distribution of water resources (b) Reformation of existing legislation and regulations (deregulation and reregulation) (c) Policy tools are (re) allocation of property rights (e.g., declaration as protected regions)	(a) Could not incorporate distributional aspects (b) Gap in explaining the political economy factors (power and political settlement for example) behind unstable nature of institutions (c) Human sociality aspect is missing; less emphasises informal institutional set-up
Political economy	(a) Address the equity and distributional aspect (b) Identify the expropriation of water as the means of capital accumulation which leads to degradation of water resources	(a) Could not explicitly internalise the 'human–nature' mutuality into the sustainable utilisation of resources framework (b) Does not provide any measures but a broad understanding of the contributing elements of the degradation of water resources

Source Prepared by the authors

creates opportunity to generate rent. As the demand for water is inelastic, the agents enjoy the monopoly power in determining the price. Thus, by imposing higher prices compared to the formal channel the agents secure a bulk amount of rent.

When the natural resources face market intervention, the use value changes into exchange value (see Chap. 2 for details). Certainly, in this case, the demand pattern for water changes in response to the price. In Panel-A, the demand changes from the perfectly elastic condition to the inelastic condition at D_2 from D_1 (Fig. 4.13). The demand curve in a dynamic situation is inelastic because the users require basic minimum access to water. The event changes the price also from P_1 to P_2. The consequences of this price effect are immense as the consumer surplus reduces from BP_1A to BP_2E; the cost per unit increases from P_1 to P_2, the total rent generates equal to P_2P_1HE and the society losses EHA amount as consumption inefficiency (deadweight loss).

The price affects not only the demand but also the supply of water in a dynamic situation. The supplies of water (formal and informal) then follow the supply law, which makes elastic supply curve S_2 shown in Panel-B. Both the inelastic demand for water and the elastic supply of water create scope for the trader to make economic rent. In Panel-B, the dynamic equilibrium condition will be E^* representing the price at P^* and the quantity demanded at Q^*. The higher price, then, reduces the

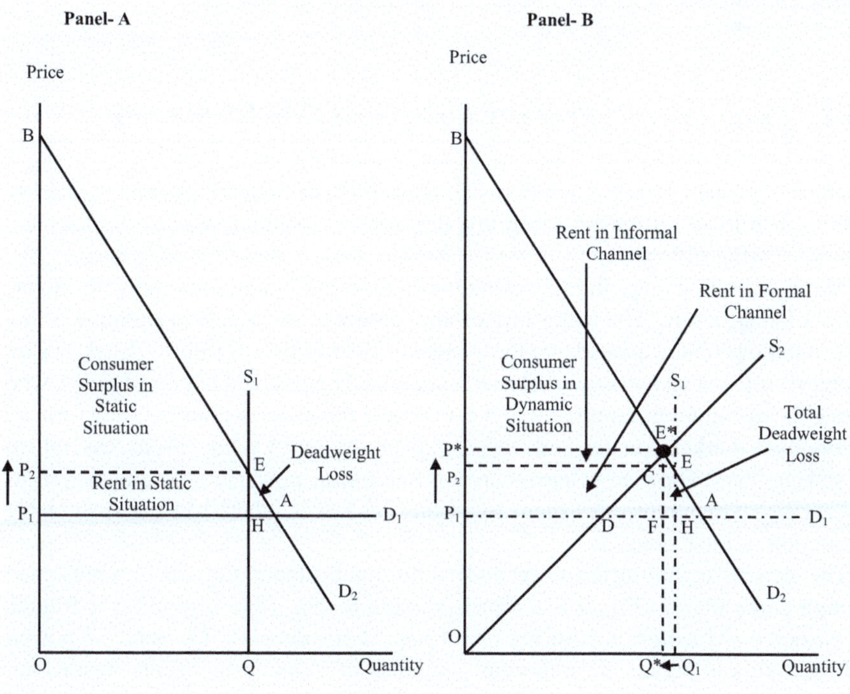

Fig. 4.13 Dynamics of pricing mechanism in access to water. *Source* Prepared by the authors

consumer (the household) surplus from BP_2E to BP^*E^* and increases the producer (the agents) surplus from P_2OC to P^*OE^*. A significant portion that society loses due to increasing deadweight loss arises from production inefficiency which is E^*DF and from consumption inefficiency which is E^*FA. The society thus loses a total amount of E^*DA. Overall, the loss of welfare of society is the outcome of commodification.

The water market comprises both formal and informal channels. The coexistence of both of the channels systematically generates economic rent. Panel-A indicates the P_2P_1HE is the rent for the formal channel where the supply of water is fixed. The supply is fixed because the piped and other improved sources in a formal channel supply water at a constant rate. In the other case (Panel-B), where both the formal and informal channels operate, the supply is no more fixed but price responsive. The traders (powerful agents) of the informal sector monopolise the good (supply of water) and demand a higher price than the formal channel. The graph shows that under a dynamic situation of the market, the total rent of $P^*E^*DP_1$ is shared by both the formal sector represented by P_2CDP_1 and the informal sector represented by $P^*E^*CP_2$. In this way, the poorer income group faces tremendous disparity in the distribution of and access to water largely constrained by the power in the economisation process of water.

The hypothesis is that under the pricing mechanism, the household-level access to water correlates with the socio-economic positions of the households and the informal channel is constrained by power operated in the market.

4.5.2 Proposition 2: Property Rights and Benefits Sharing

Absence of property rights, including claimant rights of stakeholders and regulatory regimes, results in institutional fragility, depletion of available water resources and unequal sharing of benefits.

At the institution level, the arrangements of water governance resemble the imperfect market structure. The water estates are subject to the regulatory regime of the state, but the problem arises from the system of distribution of rights. The rights are given to a form of an association through auction. The political and local elites, who dictate the association, capture the rights to access the water resources. They form an oligopolistic market structure and receive high premiums by selling the access right to the stakeholders. The market barrier creates rent for the power holders and the profit-seeking behaviour motivates them to extract resources other than the sustainable production of the resources.

The scenario regarding the power and institutional vulnerability can be represented through game theory (Fig. 4.14). Panel-A exhibits that if the institution is fragile, the resource extraction maximises (see Chap. 2 for details). The tenet is power, and the rights to access are important variables in shaping the agents' behaviour. Both variables remained dormant in the neo-classical framework. The incorporation of these two variables in the analysis explains why wetlands are disappearing and the stakeholders are getting excluded from the claimant right. The claimant right incorporates a bundle of entitles consisting of access, withdrawal and management (Schlager & Ostrom, 1992). The management of wetlands is not occurring as the agents (both the state body and the cooperative) use to exert power in the distribution of rights and make way to earn an amount of economic rent.

Under the prevailing leasing system in Bangladesh, Panel-B explains that the race of capturing the wetland resources following the Hardin's (1968) law. The additional cost of entry into the market of extraction of resources motivates the users to extract at a maximum rate other than the optimum level of resource extraction. The agents are also interested in issuing more licences with maximum rent as they own the right to extract resources by paying the bid price. Panel-B explains if both the users/agents

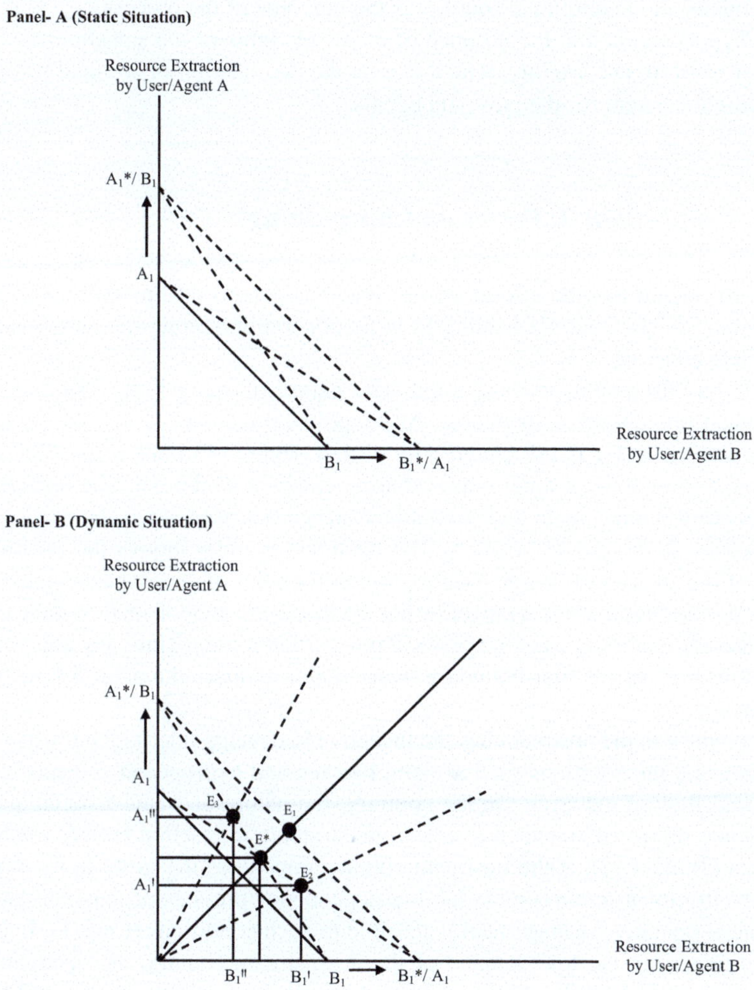

Fig. 4.14 Power and institutional vulnerability. *Source* Prepared by the authors

are powerful under the game theory, they will get equal benefits at points E^* and E_1. The sense of profit-making increases the extraction rate of resources and moves upward. The agents issue more licences and the users extract more resources and thus deplete the overall stock.

The implication in Panel-B for the agents is their participation in bidding. The agents' bid to get the lease follows the game theory. The rational engagement shoots up the price and the bidder seeks political liaison to get the lease. In the graph, agent B will propose a higher price where the agent envisages extracting resources at B_1' point considering the agent A's envision to extract resources at A_1'. Similarly, the agent A will also propose a higher price considering the agent B fixed at B_1'' to extract at A_1''. Thus, heavy political involvement in the distribution of resources downs institutional arrangements and institutional fragility is the outcome of the process.

The hypothesis is that the absence of claimant rights of the stakeholders is the origin of institutional fragility, which causes the degradation of wetland resources and accumulates rent for the powerful agents.

4.5.3 Proposition 3: Power and Bargaining

A powerful agent has the advantage of greater access to information and can set the outcome of the contract leading to non-cooperation among the parties and an incomplete contract.

Water has become a political good. The nature of water is to flow and cross geopolitical boundaries in its course. At the political economy level, the exchange of international water is associated with a high degree of conflict. Conflict arises as the multi-functions of water makes sharing costly at the border. The formulation of cooperative management and equitable sharing does not hold as the parties have asymmetries in the power position. The upstream country enjoys the position of power during such an exchange and determines the rules of the exchange game. The country derives its position of power by having more information on flow data, origin of the international water, and the political power of lobbying. Thus, any intervention by the upstream in the transboundary water region reduces the natural flow to the downstream.

The country in the upstream has the incentive to develop international water. The anthropogenic interventions such as dam, barrage and hydropower projects reduce the natural flow of the river. On the contrary, the nature of water demand and the dependency on international river constitute strategic interaction between both the countries (Table 4.12). If the upstream country intervenes meagrely in the flow of water and the downstream country has low dependency/pressure on that international river, the cooperative management is feasible in such conditions (Case-1). In Case-2, benefiting from the low demand of the downstream country, the upstream has more incentive to launch anthropogenic interventions. Such intentions lead to the unilateral withdrawal of water. In other situations (in Case-3), the high dependency on international water by the downstream and low anthropogenic intervention by the upstream make economic loss for the upstream country. Thus, all these three cases are unstable situations in the negotiation. The dynamic governance thus converges on Case-4 where high conflict and low cooperation are the ultimate outcomes in transboundary water management.

The hypothesis is that the powerful country (upstream country) determines the exchange of international water and any interventions in the natural flow of water reduce the inflow of water to the downstream country.

Table 4.12 Dynamic governance of international water regime

		Upstream country	
		Anthropogenic intervention low	Anthropogenic intervention high
Downstream country	Pressure on water (low)	**(Case-1: low conflict-high cooperation)** • Cooperation on the natural flow of the river • Cooperation across a broad range of issues including further development of river basin • Incentive to move in Case-2 by the upstream	**(Case-2: low conflict-low cooperation)** • Asymmetric information sharing of river flow • Taking projects such as hydropower generation, dam construction and barrage • Ad hoc cooperation • Tactical functional cooperation • Unstable cooperation • Consequences lead to Case-4
	Pressure on water (high)	**(Case-3: low conflict-medium cooperation)** • Narrow and token cooperation • Economic loss for upstream country • Envisaging multi-purpose projects over the international river • Verbal consent to equal water sharing • Unstable cooperation • Consequences lead to Case-4	**(Case-4: high conflict-low cooperation)** • No exchange of information • Launching multi-purpose hydroprojects • Unilateral withdrawal of water • Dominative cooperation • Water scarcity in downstream • Salinity intrusion • Water pollution • Water insecurity

Source Prepared by the authors

4.5.4 Proposition 4: Technology, Scale and Resources

The prevalence of underutilisation of resources may also be subjected to technological limitations, leading to inefficiency in production and suboptimal actualisation of revenue.

Technology has an immense role in the sustainable utilisation of water and aquatic resources. The fourth proposition argues that the lack of appropriate technologies keeps the resources underutilised. The underutilisation curtails the benefit that can be accrued and reduces the economic pie on average. The technological inefficiency and the capital deficiency thus reduce the economic pie and hamper the livelihoods of the dependents.

With the increase in yield, marginal benefit increases and so does the cost of harvesting (Fig. 4.15). While the vertical axis measures the cost and marginal benefit, the horizontal axis the yield. The total cost and the marginal cost are represented by two typical rays. E_1 represents the maximum sustainable yield from where any further harvesting will deplete the stock of resources. The E_2 point represents the economic efficiency as the situation holds the efficiency condition Marginal Cost = Marginal Benefit. If the technology and the capital are inadequate, harvesting occurs below the sustainable as well as below the efficient level at point E_3. Operating at E_3 is costly as the marginal cost is higher at this point than the marginal benefit. The inadequacy of capital (incorporating both investment and technology) limits the economic benefit that can be accrued from the marine resources of Bangladesh.

The hypothesis is that the technological inefficiency limits the potential utilisation of marine fisheries resources.

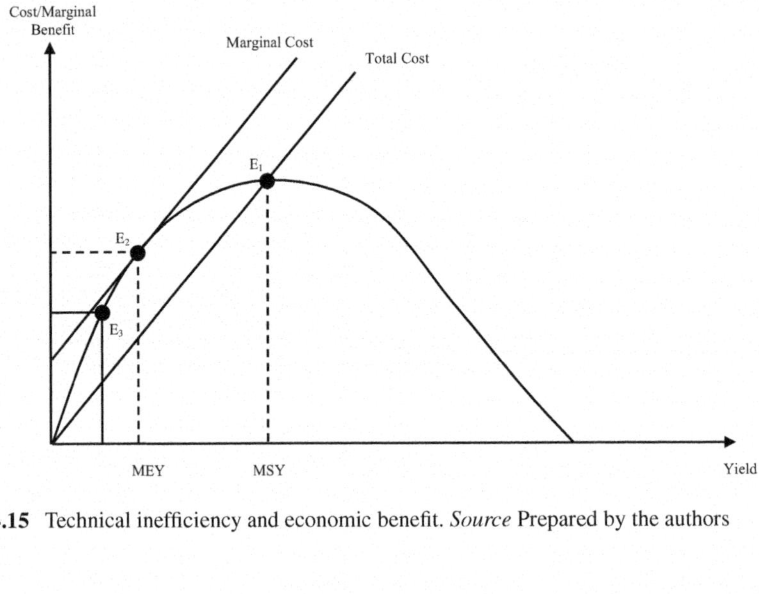

Fig. 4.15 Technical inefficiency and economic benefit. *Source* Prepared by the authors

4.5.5 Proposition 5: Human–Nature Mutuality

The equitable sharing of water, stability of property rights and functioning of the regulatory regime are subject to human sociality.

Historically, preferences and norms make up society and humans as social beings. Human nature is to act accordingly with the formation of formal and informal institutions where the actions are constrained by the social formation of institutions. The human beings thus form an inherent relationship with the nature and the management of resources. The changes in such integral relationships with the expansion of market invasion and the principles of exclusion break the social formation of institutions. Thus, human beings as stakeholders are excluded from nature and institutions become vulnerable. The revitalisation of mutual relations through the strengthening of institutions (formal and informal) and of stakeholder agency over the resources can bring about the sustainable management of water resources.

When the institutions and stakeholders' authority to own the resources are weak, there will be a high rate of extraction and the depletion of water resources (Case-1) (Table 4.13). Secondly, when the stakeholders are powerful enough under the weak institutional set-up the group behaviour will initiate conflicts and unequal benefit sharing (Case-2). Thirdly, where institutions are strong enough and the stakeholders' right to claim is minimal, the resources will be captured, and a heavy entry fee will reduce the economic benefit for the beneficiaries (Case-3). Lastly, Case-4 shows that a strong institutional set-up and stakeholders' rights bring about the social organisation of human sociality and the sustainable governance of water resources.

The hypothesis is that the organisation of socially constructed norms and values as well as formal institutions under a strong authority over the resources by the stakeholders can improve both the quantity and quality of water resources and generate maximum social benefits.

4.6 Provisioning and Access: A Case Study of Groundwater

In terms of ensuring universal access to water, Bangladesh has made significant progress with more than 98% of the population having access to some form of improved water sources in 2021, but the access to safe drinking water is still low at 34.6% (UNICEF, 2021). During the last two decades, the coverage of safely managed drinking water has increased by 3.2% reaching at 58.5%, the basic service drinking water has decreased by 0.3% point reaching at 39.2% and the limited service drinking water coverage has increased by 0.1% point reaching at 1.2%. However, dependency on the unimproved and the surface water has significantly reduced and stood at 0.5% and 0.7%, respectively (Table 4.14).

Moreover, there remains a rural–urban disparity in terms of water coverage. A substantial proportion of rural people are under the coverage of improved drinking water, but the progress over time is very slow. Of the total rural population, 62.1%

Table 4.13 Stakeholders' authority over resource and institutions nexus in the water resources management

		Stakeholders' agency over the resources	
		Weak	Strong
Formal and informal institution	Weak set-up	**(Case-1: water resources depletion)** • Poor management • High extraction rate • Resource depletion • Unstable state when institutions permit human and nature nexus	**(Case-2: group behaviour)** • The differential power will form groups seeking differential benefits • Unequal benefit sharing
	Strong set-up	**(Case-3: accumulation by dispossession)** • Rent seeking and rent dissipation • Impose heavy transaction cost to access resources • Captured by the economic elite • Unstable state when institutions permit stakeholders' to impose sanctions	**(Case-4: cooperative being)** • Human obligation to protect the water resources • Maximum benefit sharing • Sustainability

Source Prepared by the authors

have access to safely managed drinking water, 35.8% get basic service coverage and 0.9% limited service coverage (Table 4.14). All of these improved water sources rely heavily on the water from tube well, which is increasing the risk of arsenicosis. Nevertheless, apart from water for drinking purpose, all other forms of household chores are done mostly using water from ponds, canals and rivers. Therefore, the rural region is not completely under the coverage of safe water for household use and is highly exposed to arsenic contamination and waterborne diseases.

On the other hand, a wide distribution network with water from bore well mostly navigates the provisioning of water in the urban part. However, the distribution is not even. The safely managed drinking water accessible on premises has increased from 41.1% in 2000 to 52.8% in 2020 (Table 4.14). About half of the urban population is still not covered by secured water—44.6% basic service, 1.6% limited service and the rest 1% depend on purely unsafe water sources—who need to spend time, travel a long distance, and pay additional water tariff to get access to the drinking water. Moreover, water supply changes with the seasonal variations and the problem becomes acute in the dry season.

Additionally, the water sources for domestic consumption in the country can be broadly divided into two types such as (a) formal channel and (b) informal channel. The formal source of water is the supply of water distributed through the piped channel connected with the water pump or water treatment plant which is

Table 4.14 State of drinking water coverage

	Drinking water coverage estimates														
	Rural (%)					Urban (%)					Total (%)				
	2000	2005	2010	2015	2020	2000	2005	2010	2015	2020	2000	2005	2010	2015	2020
Safely managed	59.3	59.9	60.5	61.2	62.1	41.1	42.2	42.3	46.9	52.8	55.3	55.2	54.9	56.3	58.5
Basic service	34.3	34.7	35.2	35.5	35.8	56.2	56.2	55.8	50.8	44.6	39.5	40.4	41.5	40.8	39.2
Limited service	1.4	1.4	1.2	1	0.9	0.1	0.2	0.7	1.2	1.6	1.1	1.1	1.1	1.1	1.2
Unimproved	2	1.6	1.1	0.7	0.3	0.7	0.7	0.7	0.7	0.7	1.7	1.3	1	0.7	0.5
Surface	3	2.5	2	1.4	0.9	0.8	0.7	0.5	0.4	0.3	2.5	2	1.5	1.1	0.7

Source WHO/UNICEF (2021)
Safely managed drinking water—an improved water source that is accessible on premises, available when needed and free from faecal and priority chemical contamination
Basic service drinking water—an improved source, provided collection time is not more than 30 min for a round trip including queuing
Limited drinking water—an improved source for which collection time exceeds 30 min for a round trip including queuing
Unimproved drinking water—an unprotected dug well or unprotected spring
Surface water—the water directly from a river, dam, lake, pond, stream, canal or irrigation canal

also regarded as a public water channel. The other improved sources such as deep tube well and swallow tube well are mostly privately owned. The informal channel includes the supply of water from unimproved sources such as unprotected dug well, vendor-provided water and ponds, canals, rivers, etc.

In Bangladesh, a limited number of households are connected through piped water and the progress of expanding the distribution network is extremely slow from 7.4% in 2000 to 15.4% in 2020 (Table 4.15). At the national level, only 15.4% of people are under the piped channel water provisioning coverage by 2020, whereas 83.5% of people have to collect water either from the tube well or by travelling a long distance from the residence to a public water access point. In most cases, the tube wells and the water access points are communally possessed and in some cases, privately owned. Water collection from a privately owned tube well or access point often requires paying rent to the owner. Moreover, nationally at least 1.1% of the total population still fetch water from informal sources.

Moreover, there is a disparity in terms of formal access to water based on urban–rural division. The urban dwellers mostly get access through the piped channel on to premises. About 35.7% of the total urban dwellers are under the coverage of access to water through pipe onto their premises in comparison with only 2.8% of total rural dwellers in 2020 (Table 4.15). The rural households are mostly dependent on the groundwater using tube well, but it severely deteriorates the groundwater table. As the principle of marginal cost does not fit with the arrangement, the source becomes inefficient with a higher rate of extraction.

The provisioning of formal pubic channel water is not only limited but also its distribution is highly skewed to the upper class in terms of socio-economic position (Table 4.16). Nationally, one-fourth (25.61%) or in the urban region more than half (58.17%) of the richest people get water into their dwellings in 2021. Except for a few people in rural areas, the majority of the richest people are connected to the formal public supply of water. Conversely, the poorest group relies mostly on the tube well or other informal sources both in rural and urban settings. A small proportion of middle-income group households (8.46%) get water into the dwelling or other piped connections in urban areas but nationally it is only 0.14%.

Furthermore, the long-term scenario signifies that in the last 25 years, water provisioning through the piped on premises water in the rural part has been limited only to the richest households (4%) while in the urban context also, the richest have more access to the piped water (Fig. 4.16). Apart from the fact that there are disparities in piped on premise public water access among the income groups, the overall coverage of piped water has significantly declined between 1995 and 2020—from 16% to 0.26% for the poorest, 17% to 2% for the poor, 29% to 8% for the middle class, 47% to 38% for the rich and 62% to 58% for the richest.

Therefore, a high proportion of the urban population, especially the poor, depends on water from informal sources. On the contrary, the access to drinking water from other improved sources has increased for all the socio-economic groups. However, the efficiency of water utilisation at the household level can only be achieved if the total population was able to get water through the pipe on premise channel.

Table 4.15 Types of access point of drinking water

Year	Rural (%)			Urban (%)			National (%)		
	Total improved	Piped	Non-piped	Total improved	Piped	Non-piped	Total improved	Piped	Non-piped
2000	95.1	0.1	95.0	98.5	31.3	67.2	95.9	7.4	88.4
2005	96.0	0.7	95.2	98.6	32.4	66.2	96.7	9.2	87.5
2010	96.9	1.4	95.5	98.8	33.5	65.3	97.5	11.2	86.3
2015	97.8	2.1	95.7	98.9	34.6	64.3	98.2	13.3	84.9
2020	98.7	2.8	95.9	99.0	35.7	63.4	98.9	15.4	83.5

Source WHO/UNICEF (2021)

Table 4.16 Access to water based on socio-economic status

	Rural (%)					Urban (%)					National (%)				
	Poorest	Poor	Middle	Rich	Richest	Poorest	Poor	Middle	Rich	Richest	Poorest	Poor	Middle	Rich	Richest
Piped water into dwelling	0.06	0.05	0.06	0.29	4.44	0.26	1.79	8.46	38.17	58.17	0.06	0.06	0.14	1.19	25.61
Piped water to yard/plot	1.58	1.74	1.98	3.45	4.57	4.02	16.59	28.01	11.33	1.13	1.60	2.00	3.02	7.61	9.47
Piped water to neighbour	0.65	0.33	0.15	0.19	0.12	1.05	0.59	0.52	0.46	0.11	0.64	0.31	0.26	0.27	0.24
Public tap/standpipe	0.51	0.42	0.38	0.43	0.61	1.58	3.90	5.00	5.61	3.59	0.55	0.47	0.66	1.16	3.27
Tube well/borehole	90.07	95.53	95.61	93.92	89.05	90.22	74.07	55.88	43.15	34.59	91.01	95.21	94.03	88.06	59.75
Dug well (protected and unprotected)	2.41	0.37	0.43	0.47	0.30	0.55	0.25	0.06	0.03	0.00	2.05	0.36	0.38	0.36	0.16
Spring (protected and unprotected)	0.84	0.03	0.03	0.02	0.03	0.04	0.01	0.01	0.00	0.00	0.66	0.03	0.02	0.03	0.02
Rainwater	0.83	0.43	0.27	0.42	0.21	0.42	0.16	0.10	0.04	0.04	0.73	0.40	0.33	0.24	0.15
Cart with small tank	0.00	0.03	0.03	0.09	0.07	0.19	0.18	0.09	0.14	0.29	0.01	0.04	0.09	0.08	0.15
Water selling plant	0.07	0.07	0.07	0.11	0.19	0.04	0.43	0.14	0.25	0.34	0.05	0.07	0.13	0.17	0.24
Surface water (river, lake, marsh, pond, canal, etc.)	2.92	0.98	0.92	0.56	0.29	1.09	0.27	0.03	0.00	0.00	2.57	0.97	0.77	0.34	0.09
Packaged water	0.04	0.03	0.02	0.03	0.12	0.54	1.75	1.70	0.83	1.72	0.05	0.05	0.17	0.49	0.86

Source WHO/UNICEF (2021)

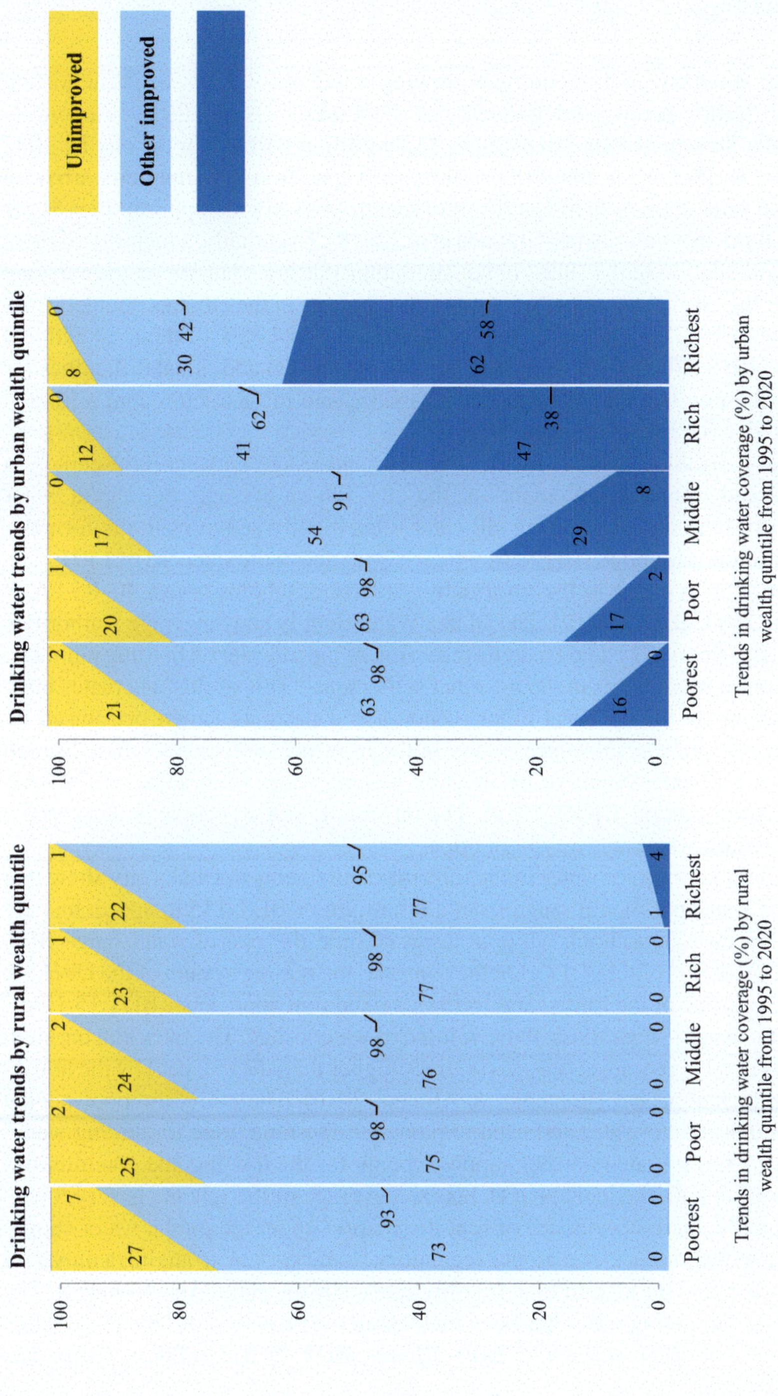

Fig. 4.16 Trends in drinking water coverage by wealth quintile from 1995 to 2020. *Source* WHO/UNICEF (2021)

4.6.1 Financialisation and Rent Dissipation: Case of Dhaka City

Dhaka, the capital city of the country, is growing at the rate of 3.9% each year hosting about 17.4 million people with a density of 47,400/km^2 (Amin, 2018). According to the Dhaka Structural Plan 2016–2035, 22.79 million will live in the city by 2035 (BIGD, 2019). Each year about 0.3 million to 0.4 million new migrants arrive in Dhaka from rural areas as their last resort in search of livelihood opportunities as per the World Bank estimate (Arias-Granada et al., 2018). Eventually, a majority of them seek accommodation in the slums or squatters thus putting a burden on the carrying capacity of the city. The number of slums and squatters in the city has doubled from 1579 slums in 1997 to 3394 in 2014 representing only 24.35% of the total national slum areas (BBS, 2015b). The informal settlements present additional difficulties for Dhaka, as the city has long struggled with inadequate infrastructure and a limited capacity to deliver public services.

There is a sizable disparity in water access and usage between households in the formal and informal settlements of the city. On an average, per capita water usage for drinking and completing all water-related daily activities in the formally settled households is about 310 L/day, which is well above the standard of 150 L/day, whereas per capita usage in the informally settled households is just 40–85 L/day (Onneshan, 2011; BIGD, 2019). The Dhaka Water Supply and Sewerage Authority's (DWASA) distribution system bases its provision of piped water on holding numbers, which excludes households in the informal settlements. This exclusion creates business opportunities for the local elites, vendors and the musclemen patronised by the local political leaders to supply water in the informal settlements, often through privately owned water points or by illegally leaking water from adjacent DWASA pipelines and/or pumps (BIGD, 2019). The informally settled people in this process need to purchase water at a very high price.

The estimated prices of water in the informal settlements in Dhaka city show that the costs of water per month range from a minimum of BDT 155 to approximately BDT 600 for each household, while in terms of unit, the cost of water range BDT 7.33 to BDT 100/m^3 (Table 4.17). On the contrary, the renting system of the DWASA public water is one of the lowest and follows a fixed rent approach—BDT 15.18/m^3 for domestic usage—regardless of the volume of water usage. The rates that the slum households need to pay are at least 7–14 times higher than the rate paid by the households with formal access (Rahaman & Ahmed, 2016). Therefore, the poor people have limited access to water and need to pay a burdensome price for getting water. The DWASA has initiated a water supply scheme for the low-income communities in collaboration with the local NGOs, but the coverage of the scheme is very low.

Additionally, there is evidence of rent dissipation under the public water supply system due to the absence of a proper regulatory framework in arranging a universal supply of water and the institutional fragility caused by the politico environment. The revenue from supplying water has been increasing over a period of time (Fig. 4.17). The revenue collected in the FY 2019–20 was BDT 12,813 million, which has

Table 4.17 Prices of water in the informal settlements

Studies	Water price (in per m^3 or in BDT/month)
Haque (2019)	BDT 7.33–BDT 100/m^3
BIGD (2019)	BDT 209–BDT 600
Arias-Granada et al. (2018)	BDT 155–BDT 308
Cawood (2017)	BDT 300–BDT 400
Onneshan (2011)	BDT 500 for 30 L/day/month

Source Authors' compilation from different sources

augmented more than three times compared to that of the FY 2009–10. Though there has been an overall increase in total revenue, the growth rate of the revenue collection has fluctuated year on year. The dynamics of the growth of revenue can be elucidated in two ways (a) the increment of the non-revenue water supply, which is system loss and (b) the piling up of dues.

The trend of non-revenue water explains the dissipation of revenue (Fig. 4.18). The operational efficiency has improved in the last few years, but a significant amount of revenue is still dissipating each year. Because of the quasi-market structure of water where both the market principles and the government work together, the missing rent is rapidly increasing and causing an overall market failure. Each year, the system loss remains one-fifth of the total production. Besides, the collection ratio remains around 90–95%. Moreover, water connections in the informal settlements are generally diverted from the public water connections illegally. Local politicians and musclemen use their clout to seek rent from this public resource. The condition started improving after FY 2001–02, but the non-revenue water could not be reduced

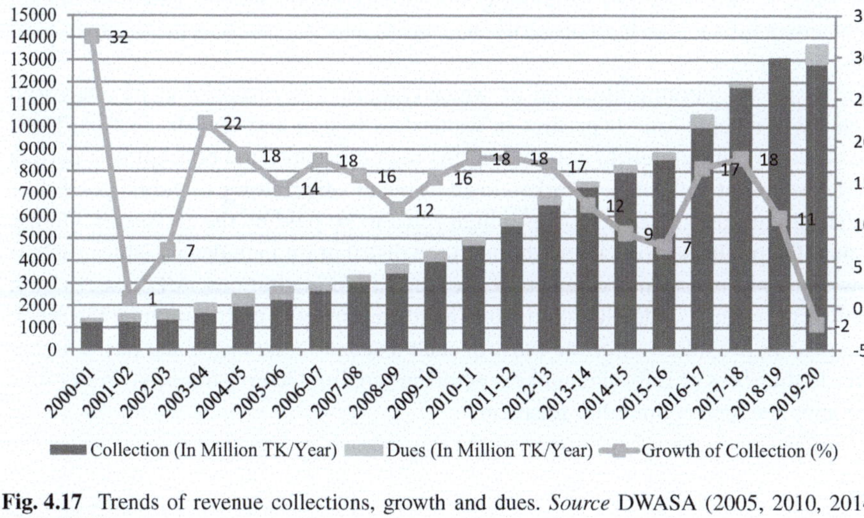

Fig. 4.17 Trends of revenue collections, growth and dues. *Source* DWASA (2005, 2010, 2014, 2017, 2020)

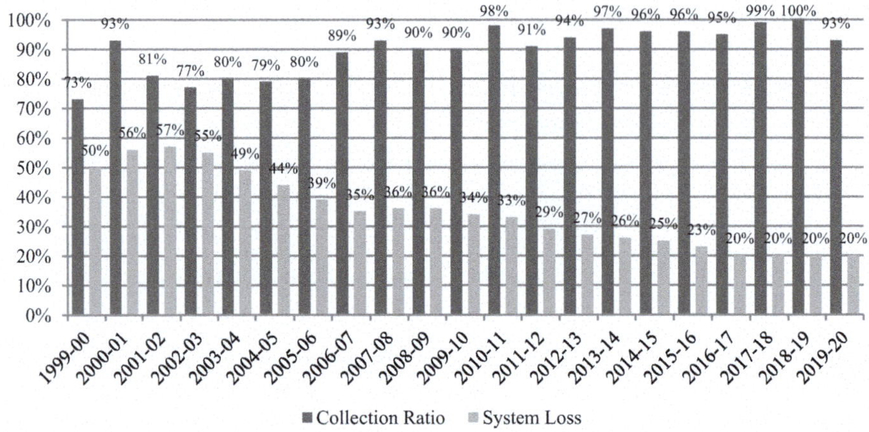

Fig. 4.18 Trends of non-revenue water and revenue collection ratio. *Source* DWASA (2005, 2010, 2014, 2017, 2020)

below 20% in FY 2019–20. Water supply monitoring technologies helped to reduce the non-revenue water. Nevertheless, the amount of dissipated rent highlights the institutional fragility and lax regulatory system.

In addition, a portion of the rent, which becomes non-receivable over time, always remains and the receivable dues have been increasing each year. The amount of dues was the highest at BDT 866 million in the FY 2019–20 (Fig. 4.19). Moreover, the receivable dues have reached BDT 7661 million. Despite all the digitalisation of the water supply system, the responsible authority has failed to collect the total revenue as well as curb the growth of receivable dues. Therefore, the outstanding amount creates the scope to seek rent by the authority and powerful consumers. Thus, the market mechanism fails with a high disparity in terms of access as well as the dissipation of rent.

Overall, the commodification of water resources is responsible for increased groundwater depletion, disparities in water distribution and rent dissipation. First, the efficiency principles do not hold in this regard because the concept of a perfect market is absent in the water market. In addition, the distribution of public goods like water through the market involves a high price to be paid by the consumer. On top of that, the public sector's intervention in the distribution includes variables such as power and politics, which often cause the fragility of institutions. All these result in inefficient revenue collection on the one hand and unequal distribution on the other. The poor households suffer the most in such a scenario, as they must pay a higher price to the elites of informal markets while having limited consumption. Water thus becomes a scarce commodity for them.

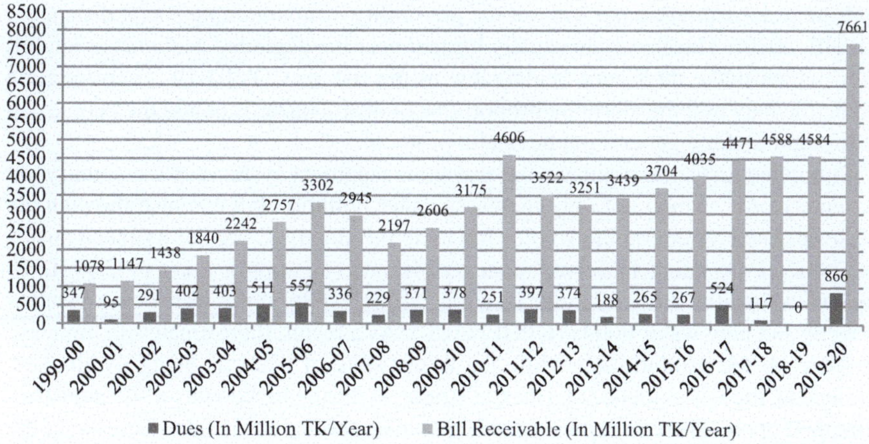

Fig. 4.19 Trends of dues and bills receivable. *Source* DWASA (2005, 2010, 2014, 2017, 2020)

4.6.2 COVID-19, WASH Practice and Groundwater

The COVID-19 pandemic has put extra stress on the usage of groundwater. Because hand hygiene and sanitation practices are found effective, it is recommended that safe water, sanitation and effective waste management be ensured to prevent and protect human health from COVID-19 (WHO & UNICEF, 2020). However, the grim condition of the water supply and the disparity between the rich and the poor to access water heavily affect the WASH practices. Overall, the pandemic has brought three challenges in water management—first, the increased demand for usable water; second, the poor facility and water shortage for the informally settled population; and third, to tackle the growing demand for water, the abrupt policy changes leading to the exploitation of more groundwater.

The provisioning of usable water to all remains critical in Bangladesh. There is already a shortage of water across the country. The pressure had increased on the groundwater level in order to address the higher demand for water in the households for activities like cleaning and sanitising amid the pandemic. A study showed that the usage of per capita water increased at least 13 times which means a person overuses 1.7 L of water per hand wash and at least 14.9 L per day (Sayeed et al., 2021). It further suggests that water overuse while washing hands by keeping the tap on may put additional pressure on the already overstretched groundwater level.

Urban informal settlements, however, are under the greatest threat of exposure to the pandemic due to the lack of WASH facilities. It becomes a burden for the poor to follow hand-washing guidelines when they already have to pay a price to buy drinking water. Predominately, poor quality housing, unhealthy surroundings, unsafe sanitary facility, scarcity of water and congested living conditions have heightened the risk of transmission of the disease (Shermin & Rahaman, 2021). As discussed earlier, to access water from informal sources, more than 20 people are often seen queuing up to draw water from one tap without the possibility of maintaining social distancing (DPHE, 2020). The slums are usually located near the degraded wetlands or landfill sites, which make them more vulnerable to the risk of COVID-19. Therefore, the pandemic may assume alarming proportions of health hazards unless the people can be brought under a general public health protocol.

Simultaneously, the drainage and waste management systems of the country are extremely poor, which not only increases water pollution but can also threaten the freshwater ecosystem on the whole. To tackle the situation, regulatory mechanisms and policies are being frequently changed. In order to supply water to COVID-19 dedicated hospitals and quarantine locations, submersible pumps were being installed in wells and new water points (DPHE, 2020). As a result, there creates an immense pressure on the groundwater while the future state of water is being overlooked. Alongside, there is evidence of a lack of strictness during lockdowns while industries remained open. Therefore, the expected positive gain in terms of a decrease in the usage of water in the industrial sector during lockdowns is certainly not a reality in the context of Bangladesh.

4.7 Unstable Institutions and Power Politics: The Case of Wetlands

This section uses the case of evolving wetland management regimes in Bangladesh to understand the failure of institutional arrangements in protecting the water resources. The degradation of wetland resources is largely associated with fragile institutions. This fragility is primarily derived from the alteration of traditional institutional arrangements, the concentration of power and the exclusion of resource dependents from the management regimes. The property rights that were held by the wetland inhabitants sustained both the livelihoods and the resources of the wetlands. They had the right to stake claim to the resources. However, this claimant rights have changed significantly over time.

With a view to understanding the evolution of property rights, it is important to examine the pattern of settlement that has evolved over the years in the wetland areas. It can be traced back through examining the phases of administration—from the early British period to the present and during these phases, the policies of wetland management have changed several times.

4.7.1 The Pre-British Period

The wetlands were traditionally managed by the locals who lived close to the waterways during the pre-British era. Those indigenous wetland communities imposed several localised tenurial systems for harvesting fisheries resources (Capistrano et al.,

1994; Rahman, 1995). However, the traditional structure of rights varied across different places but that class of people enjoyed exclusive rights over the wetland resources. Initially, the wetland areas of north-eastern Bangladesh were inhabited mostly by Hindus, who had migrated from the Muslim-ruled areas elsewhere in Bengal, along with a few ethnic communities—namely *Garo*, *Hajang*, *Khasi* and *Koch*—who came down from the hills to settle in the area (Soeftestad, 2000). The migration is assumed to have taken place during the thirteenth century, after the extension of Muslim powers in the south and the west (Soeftestad, 2000). Apart from fleeing the Muslim regime, migrants moved to the area because of its fertile land and favourable tenancy terms (Ali, 1990). However, no fishing practice was reported in the early period, only to make the land for agriculture through slash and burn (Islam, 1985). But there might be a link that because of the unique natural settings of the *haor* basin—filling up to the brim during the rainy season and becoming a vast land in the dry season—the fishing system was likely to grow locally among the migrants alongside agriculture. Moreover, the *Kaibarta*—a Hindu caste which was closely associated with the practice of fishing believed that God had entrusted the fishermen with the sacred duty of fishing to serve others with the supply of fish and any form of deviation from their *Jati-Pesha* (caste occupation) would be counted as a sin (Banglapedia, 2021c). Furthermore, the religious belief of propitiating the Goddess (mother Ganges) not only to allow them to make a good catch but also to keep them safe while fishing makes a complex traditional human–nature nexus supplemented by rituals and ceremonies.

The *haor* basin areas had experienced another few rounds of an influx of migrants from elsewhere between the sixteenth and seventeenth centuries. A huge number of Afghans fled landing in the *haor* basin following their defeat by the Mughals in Orissa in 1592 that added a new layer of complexity to the *haor* inhabitants (Soeftestad, 2000). In 1612, Muslims—the Mughals—completely conquered the eastern Bengal and the battles and events caused a rapid growth in population in the *haor* basin areas (Soeftestad, 2000). Therefore, a racial mix caused conflicts but over time, a form of social cohesion had formed. This harmony played a vital role in terms of access and utilisation of the wetland resources. Precisely, in the context of fishing practice, the Muslims had also developed their rituals along with the other native traditional communities and used the wetland resources with care. Therefore, this traditional form of wetland management was marked by extensive terms of coverage regarding resource collection against the intensive in terms of the extraction of natural resources. The Mughals, however, developed a form of taxation in the seventeenth century and started collecting revenue. This was the very first attempt to interpose between the human–nature nexus although to a minimal extent. They divided the land into *Parganas* (a large divided area) which had earlier been settled as *Jalmahals* (wetlands) for revenue purposes and began to collect revenue from the large and profitable fisheries (Islam, 1985).

4.7.2 The British Colonial Period

The Mughals initiated the system of taxing the *Jalmahals,* but the actual change of the wetland management came with the British. The enclosure of land by Regulation XXVII of the Permanent Settlement Act of 1793 in Bengal had brought a swift change in the *Zamindari*[1] system by consolidating the Mughals' land tenure system into a new form of proprietorship (to be a landlord) and selling the possession of the tenure in return for a fixed annual rent (Chaudhuri, 1983). Because of the Act, the customary right of the local people to hold the hereditary land was subject to regular payment of rent, and the right could not be transferred from the *Zamindar* (the landlord) to the people in any form (Khan & Haque, 2010). The British ruler used the *Zamindari* system to extract the rent for natural resources and the proprietorship was deputed to the elites who offered a higher rent to the British Lord and later by the Crown-appointed Governor in 1858 following the transfer of regime from the British East India Company.

The *Zamindari* system of regulating the *Jalmahals* had significantly affected the resource base of the wetlands. *Zamindars* were authorised to define the land tax and its payment system (Gadhwal & Lal, 2008). Notably, the landlords were mostly non-natives who got the proprietorship under the Permanent Settlement. They were nominated either for their loyalty to the British Lord or for their commitment to paying high rent. Therefore, on the one hand, the traditional right of the wetland people on *Jalmahals* was restricted and on the other, the *Zamindars* were only actuated by collecting revenue. However, the landlords could develop and improve the productivity of their lands because the rental fee for their ownership was fixed, in nominal terms. But that did not happen. The rental price was set so high that it was not possible to collect such rent from the local people, who also refused to pay rent. Thus, the question arises as to whether improvements were possible and whether the *Zamindars* had the proper authority to alter the way production was organised at the level of individual water bodies or acted as a proxy intermediary of the British Lord (Khan, 2015). Simultaneously, the failure to pay revenue to the state authority led to the *Zamindari* being auctioned and in reality, over three-fourths of the land was auctioned for not being able to pay a fixed rent to the state.

It was assumed that the *Zamindars* would be able to enlarge profit by residual claim but in most cases, only fulfilling the state demand was not possible. Although historical consensus on Permanent Settlement Act has shifted away from the view that it established 'capitalist' property rights in land (Alavi, 1975; Mukherjee, 1974) to the more plausible view that it tried to improve the efficiency of the pre-existing Mughal revenue structure by giving *Zamindars* strong incentives to pay revenue on time and to increase land productivity (Ray, 1974, 1979; Ray & Ray, 1973, 1975), the fatal flaw in the Act was that the *Zamindars* did not have the requisite right to change the organisation of production. Even in times of economic recession and poor harvesting due to natural calamities, the landlord had to pay the fixed rent. To

[1] Mughal term meaning intermediaries, not the proprietor of the land, between the government and the inferior revenue farmers assigned by the emperor.

facilitate revenue earning, the *Zamindars* leased out the rights over the *Jalmahal*s to some other intermediaries—the *Ijaradars* (Pokrant et al., 1997) and intensified the burden of toll on the fishermen and the wetland resources. Overall, the British regulation caused intensive extraction of resources to fulfil the state's demand.

4.7.3 The Pakistan Period

Management of the wetlands did not change much in the post-colonial (Pakistan) period. In 1947, after the dissolution of the colonial period, the *Zamindari* system was eliminated under the East Bengal State Acquisition and Tenancy Act (EBSATA), 1950 (Khan & Haque, 2010). *Jalmahals* were then *de jure* owned by the state termed as *Khas* land—the state owned-land. At the same time, the local powerful elites filed a petition against the state staking claim to the *Khas* land in order to wrest control of the area surrounding the wetland as their private land. The government attempted to handover temporary property rights over the *Jalmahals* to the fishermen through a leasing process but failed to do so and eventually transferred those to a class of people or outsiders, who did not belong to the fishing community (Toufique, 1998). Thus, the wetland became the private land of the elites when it was dry. However, the contentious issue was when they were filled with water in the rainy season, who got the distribution of the rights to the *Jalmahals* was not clear and aftermath, it was decided to lease the *Jalmahals* in the rainy season.

By then, the wetlands became one of the most important sources of revenue for the state. Therefore, the key beneficiaries of the arrangements were the rural intermediary classes—known as lessees or middlemen or water lords (Rahman, 1987)—who being locally powerful, continued to play a crucial role in shaping the political structures of East Pakistan (Khan, 2015). Following the colonial counterparts, these elites, namely fish merchants (Chowdhury, 1987; Ullah, 1985) began to charge tolls from fishers for access rights under their control. The fishermen cooperatives had started to protest the public auction to secure their right to the water bodies by the mid-1960s (Khan, 2015). The government, however, agreed that cooperative societies should be given priority in leasing if they could bid as high as the current beneficiaries (Huda, 2003). However, as the bids were made public, it gave the local moguls the opportunity to purchase the rights. In addition, the lessees used to trade second-hand rights with the fishing community by imposing a price tag or by employing fishers as wage labour to fish for them.

4.7.4 The Bangladesh Period

After independence in 1971, the Bangladeshi government adopted the *EBSATA*—1950 once again and transferred the property rights of *Jalmahals* under the supervision of the Ministry of Land (MoL). The government tried to find an appropriate approach to manage the wetlands but did not compromise the objective of maximising revenue (Khan & Haque, 2010).

The very first attempt to allocate access rights to the fishermen started with the licencing system in 1973. It was, however, restricted only to the Fisheries Cooperative Societies (FCSs), and they could bid for the lease of the *Jalmahals* only. Then, the society used to issue licences to the registered fishermen. The leasing period of the *Jalmahals* was very short (one year for open fisheries and three years for closed fisheries) which resulted in mounting pressure on the wetlands. The licencing system had its share of problems, too. First, the increased cost—transfer costs—associated with the buying of permits by the fishermen; second, the sense of profitability by the cooperatives increased the pressure on the stock of fisheries; and third, a sizeable portion of the actual poor fishermen was denied the right to access the resources because they remained unregistered at the FCSs.

Another shift in the management of resources happened in 1980. At this point, the authority to manage *Jalmahals* was transferred from the MoL to the Ministry of Fisheries and Livestock (MoFL). The MoFL practised two leasing systems (a) restricted leasing among registered FCSs and (b) direct negotiation with organisations and/or individuals (Khan & Haque, 2010). Both the leasing systems marginalised the fishermen extensively. Previously, all the FCSs were not officially registered, but under this restricted leasing system, the registered cooperatives could participate only. The illiterate poor fishermen failed to take collective action and register as FCSs and eventually were excluded from the leasing system (Khan & Haque, 2010). Nevertheless, leasing through direct negotiation was given to a group of individuals who came from outside the fishing community (Toufique, 2005). In 1984, the authority of allocating property rights was again transferred back to the MoL following the formation of the *Upazila* system—an initiative to decentralise the administrative system of the country under *Upazila Parishads* (subdistrict council) where the chairman and the members of *Upazila Parishad* were elected public representative. Then, the water bodies ranging from 3 to 20 acres were placed under *Upazila Nirbahi Officers* (UNO) of *Upazila Parishads* while water bodies larger than 20 acres were under *Additional Deputy Commissioner (ADC), Revenue* of district council under the jurisdiction of the MoL. This decentralisation of the distribution of property rights did not improve the condition because the council was required to abide by the restricted leasing system.

When the leasing system failed to solve the problem of transferring property rights to the actual fishers, the New Fisheries Management Policy—1985 and the National Fisheries Policy—1998 were other fresh attempts to expedite and ensure benefits to genuine fishers (Huda, 2003). The Department of Fisheries (DoF) was given the charge on a small scale to abolish the dominance of the elites, but it failed due to its institutional incapacity. It attempted to abolish the cooperative societies and issue rights directly to the fishers by paying the lease fee and under this system, in about 270 water bodies, an annual licencing system was introduced for 'genuine fishers' on the condition that they would collectively pay lease at the previous rate plus an additional 10% per year (Valbo-Jorgensen & Thompson, 2007). The imposition of the transfer fee, however, became a huge burden for the poor fishers and identifying the '*genuine fishers*' was difficult. Above all, the endeavour failed because of the rent-seeking practice by the public administrator and the absence of a regulatory regime. The licences issued by the authority were subject to bribes and the local elite and *Mahajans* often had to carry out raids to collect the licence. To curb their controls, the National Fishermen's Committee was entrusted with the task of identifying genuine fishermen, but like most other organisations in Bangladesh, it was dominated by people from the intermediate classes (Khan, 2015). Thus, both the processes of leasing and licencing failed to guarantee sustainable management of wetlands (Khan, 2015).

The inefficiencies in management continuously shifted the policies. There were multiple changes in the authority of leasing and also in the process of distribution of rights after 1996. The government again transferred the authority to lease to the MoL and introduced a decentralised system. The decentralisation process made it easier for different institutions, such as the *Union Parishad* (local government), *Upazila* (subdistrict), District and Divisional Administration, departments and ministries to participate in resource management (Khan, 2011). In 1997, the Ministry of Land transferred smaller water bodies (up to 8.1 ha) to the Ministry of Youth and Sports (MoYS) in order to create employment opportunities for the youths. At that time, leasing was conducted in three distinct ways—(a) the *Upazila Parishad* was responsible for leasing water bodies that were less than 1.21 ha; (b) the Ministry of Youth and Sports was responsible for leasing water bodies that were 1.21–8.1 ha; and (c) the District Administration responsible for water bodies more than 8.1 ha.

After the execution of the Government's *Jalmahal* Management Policy 2005, the Fisheries Cooperative Societies were replaced by the direct bidding system in the method of licencing. The cooperatives had to hold the genuine fishers and the lease would go to the cooperatives. The cooperatives, then, distributed the rights among the fishers. Notably, during the period of 1996–2007, an attempt of community-based wetland fisheries management was made with the support of NGOs and INGOs. It helped reduced the influence of the elites and successfully minimised other vested interests (Sultana & Thompson, 2010), but the projects covered only a small proportion of the vast wetland areas across the country.

The latest management policy, the Government *Jalmahal* Management Policy, 2009, brought some addition to the previous one by dividing the authority of sanctioning property rights into two separate tiers. First, the District *Jalmahal* Management Committee would regulate the leasing of water bodies of more than 8.1 ha and second, the *Upazila Jalmahal* Management Committee would regulate the leasing of water bodies of less than 8.1 ha. Under this policy, the authority of leasing water bodies in the range of 1.21–8.1 ha has been revoked from the supervision of the Ministry of Youth and Sports by the Ministry of Land. The total system of leasing has now become more intricate and dictated by both the bureaucrats and power holders. In the District *Jalmahal* Management Committee, which is responsible for leasing large water bodies, all the top-level bureaucrats have been designated and the political leaders, meaning respective Members of Parliament (MPs), of the wetland areas will be appointed as an adviser. It has, however, kept the provision of including two representatives from registered FCSs but the selection of those representatives will be carried out by the committee and not by the fishers. It therefore clearly indicates that the power has not been decentralised by assigning authority to the fishers rather it has still been controlled by the central committee. Similarly, at the *Upazila* level, the committee members will be represented by the subdistrict government officials and the political members, the MP and the *Upazila* Chairman of the area, and appoint two members from FCPs. The policy further points out that the Upazila *Jalmahal* Management Committee has the authority to identify the genuine fisher of the FCPs and has the decision-making power to cancel the FCP's registration if it does not include the genuine fishers. Ultimately, in reality, the cooperatives held by the elite grabbed the rights by paying bribes while poor fishers were framed by the intermediate class. Therefore, the evolving nature of property rights over the wetlands and open fishery resources of Bangladesh signifies that the allocation of rights was determined by the administrative criteria to collect revenue while the ecological criteria for protecting natural resources were undermined.

Thus, the analysis of the nature of property rights over the wetlands under different regulatory regimes signifies that control over the *Jalmahals* has been predominantly held by the powerful class (Table 4.18). The ever-evolving regulatory frameworks have been distributing the property rights following a market process where the local elite could constitute a form of imperfect market. They are powerful enough to impose heavy transaction costs and also have the ability to enforce the contract acquired by the leasing system.

The intermediate class has been formed since the external intervention into the customarily held property rights. The concentration of power and the rise of this class have evolved since the Mughal period and then fully captured in the colonial period. After the end of the colonial era, the Pakistan regime did not make any effort to introduce any alternative management system, rather proclaimed the *Jalmahals*—the wetlands as *Khas* land (state property). Though the aim of the East Bengal State Acquisition and Tenancy Act (EBSATA)-1950 was to distribute the property rights among the resource dependent people, the distribution of rights did not take place effectively. After independence, following the initiative of the redistribution of land and a series of policy changes, the situation slightly improved but the control over the wetlands remained intact under state control.

The intermediate class, apart from the fishing community, focuses primarily on the profit that can be accrued from the exploitation of resources. These intermediate classes run the fishing cooperatives where the fishers are nothing but a puppet to show. Moreover, the lessees charge tolls on fishers for rights to access the water bodies under their control (Toufique, 1997). On the contrary, the fishing community, in the process, loses its claim to the wetlands and the rights get transferred to the rent-seeking class through appropriation or by enacting legal policies. Historically, the legal arrangements were designed to generate maximum revenue without considering the sustainable management of wetland resources and local livelihood. Therefore, the increased pressure on the resources is the outcome of institutional failure.

Besides, all the associated agents in the whole bidding process, ranging from government officials to local-level moguls, are involved in capturing the wetland resources. The fishermen of these areas barely have money to participate in and

Table 4.18 Different management regimes and the institutional arrangements

Management regime	Management initiatives	Objective	Outcome	Impacts
Pre-colonial period (Before 1757)	No particular management framework	Utilising resources through customary rights	• No report on resource degradation and no shortage of resources	• Through the conventional system of resource management, informal institutions were active
Colonial period (1757-1947)	The Permanent Settlement Act 1793	Governing resources by *Zamindars* for revenue collection	• Rent accumulation from the resources under long-term lease agreement	• No initiatives for resource conservation • Inequitable resource distribution
Pre-independence period (1947-1971)	EBSATA, 1950	Ensuring state ownership of resources through the abolition of the *Zamindari* system	• The MoL became the authorised agency of wetland management in place of *Zamindars* • Revenue generation by the state	• Change in the property rights structure
	The Fish Conservation Act, 1950	Conservation of fisheries	• Ban on usage of harmful fishing gears • Prohibition on catching brood fish and juveniles	• Periodic leasing with increased lease fee • Weak enforcement of the act • No visible positive impact on conservation

(continued)

Table 4.18 (continued)

Management regime	Management initiatives	Objective	Outcome	Impacts
Post-independence period (1971 to the present day)	Continuation of EBSATA, 1950	Continuation of state agency over the wetland resources	• The MoL remained the sole agency to collect revenue	• No major change in wetland resource management
	Licensing system of *Jalmahals* management, 1973	*Jalmahals* are leased to registered FCSs through an annual licence system	• The MoL delegated management to district and sub-divisional administrations	• It appeared initially as a pro-fishermen system, but due to corruption, it became dominated by special interest groups • Poor and genuine fishers remains marginalised in the licence system being excluded from the registration process
	Local Government (*Union Parishad*) Ordinance, 1983	Involvement of locally elected representatives in governance framework	• Transfer of governance of small water bodies (less than 2 acres in size) to *Union Parishad* • Collection of lease fee through *Union Parishad*	• No particular guideline or framework for sustainable management of resources
	Transfer of *Jalmahals* to *Upazila Parishad* (sub-district) in 1984	Decentralization of governance system further	• Transfer of governance of *Jalmahals* (areas between 3 and 20 acres in size) to *Upazila Parishad*	• Collection of lease fee continues • Restricted FCSs • Creation of vested groups among fishers and middlemen
	NFMP, 1986	Management rights of *Jalmahals* distribution between MoL and DoF	• Licence system adoption for genuine fishers to recognise their rights • Involving fishers' association in selection process	• Transfer of insignificant number of *Jalmahals* to DoF • Annual increase in lease fee • Weak institutional capability of the DoF failed to bring success in true sense

(continued)

Table 4.18 (continued)

Management regime	Management initiatives	Objective	Outcome	Impacts
	NFP, 1998	Enhancing fish production to contribute to national economy and alleviating poverty of fishers in particular	• Commercialisation of *Jalmahals* through enhanced production • Overexploitation of resources	• Unsuccessful to establish ownership of *Jalmahals* by the DoF • Low focus on resource conservation
	Government *Jalmahals* Management Policy, 2005	Distribution of governance responsibility among a number of government agencies	• Transfer of *Jalmahals* (up to 20 acres in size) to the MoYS • Transfer of limited number of *Jalmahals* to the DoF and the MoEF	• Community-based wetland management approach practised and lessons learnt • Sign of conservation approach of the government • However, benefits mainly acquired by political agents
	Government *Jalmahals* Management Policy, 2009	To ensure leasing of *Jalmahals* to FCSs of genuine fishers	• Returning of 20 acres size *Jalmahals* from the MOYS under the MoL • Inclusion of two members from registered FCSs in *Jalmahals* Management Committees at the district and sub-district levels	• Indication of willingness of the government to provide benefits to genuine fishers • However, selection of members from genuine fishers for the management committees depends on bureaucratic decisions • Certification of FCSs and its effective operation depends on bureaucratic decisions • Problem of allocation of resource-rich and resource-poor *Jalmahals* • Existing communication methods for bidding process mostly fail to reach targeted stakeholders

Source Khan (2011)

bid for the lease. They have to rely solely on the money (*dadan*) borrowed from the moneylender. *Dadan* is a form of loan consisting of high compound interest by which a fisherman is trapped by the moneylender in a vicious cycle of debt. As a result, the moneylender takes full advantage of the fishermen's inability to pay the debt. The moneylender (usually called *mahajan*) plays a dual role. First, the moneylenders give money for lease to the fishers who return the amount with high compound rates. Second, the moneylender also imposes a share of the catch on the fishers. This uneven imposition of a high return on debt reduces the net income of the fishers. As a result, the fishers inevitably fail to repay the loan from their income. Due to the persistent lending of money to sustain them, the genuine fishers are forced to sell their fishing rights to the *mahajan* (Khan & Haque, 2010). Moreover, heavy liaison and the tradition of bribe-taking play a huge role in the process of distribution. As a result, a form of interlocked and imperfect market prevails in the distribution of rights. Thus, the social power of these intermediate classes is critical in determining their ability to retain property rights over the resource system. In fact, the failure of the state to protect the rights of the appropriators opens up the possibility of the rights being transferred to the most powerful agents but not the direct users of the resource (Toufique, 1997).

Overall, the intermediate class has the power to make an unequal exchange and dysfunction the market for wetlands since the recent practice of distributing rights follows the neo-classical method of internalisation of externalities through price. The market involves transaction costs to transfer the rights. The poor fishers are unable to bear the costs and lose their property rights. In contrast, the intermediate classes, who do not belong to the fishing community, are powerful enough to bear this transaction cost. Hence, the groups capture the property rights and realise the maximum rent generated from the wetlands.

The transaction cost was imposed in order to achieve economic efficiency and internalise the externalities. In addition, the pricing helps the government to earn a significant amount of revenue. Unfortunately, the pricing failed to achieve the economic efficiency as the lion's share of the rent goes to the leaser and to internalise the externality since the leasers are interested not in optimising but in maximising the yield. Nevertheless, under the latest distribution system of property rights which is to lease to the Fishers' Cooperative Society, the society has to pay the incremental fee. Each year, a 5% price increases over the previous lease value. If the cooperative fails to pay the lease amount, a fresh bidding process is thought to be organised by the respective authorities that will be open for all (Khan & Haque, 2010).

The overharvesting in wetlands takes place as the rent claimant has to earn revenue above the cost he bears. The investment cost also increases in order to ensure the enforcement of the contract. There are no such institutional arrangements to enforce property rights by the state for the lease holder. The state only transfers the formal rights of catching fish from the water bodies, which are mostly undefined. The power asymmetry between or among the neighbouring lease holder increases the potential risk of missing rent.

Bangladesh has 34,681 *Jalmahals* in all. Fishing rights in each are leased out by the Ministry of Land for a tenure of one to three years. Notably, the government's earnings from leasing *Jalmahals* have increased by thousand times since independence. The available data on revenue collection from *Jalmahals* shows that in the 2019–20 financial year, the government earned more than BDT 1011.10 million (Table 4.19). The major portion of revenue comes from the Sylhet division. Sylhet is a wetland-rich area where most of the large wetlands in Bangladesh are located, including Tanguar Haor, Hakaluki Haor, etc. 34.49% of the total revenue has been collected from Sylhet division and the rest from different parts of the country.

Overall, the fragility of institutions is the outcome of a market-based solution to manage the wetland resources. The vulnerability stems from insecure property rights, increasing transaction costs in the contract of property rights and the contested exchange in the process of appropriating those rights. Under a market-based solution, property rights belong to the person who can pay for the rights. The exclusion is largely due to the high transaction costs associated with the leasing system. The leasing process comprises different agents and creates a scope for rent-seeking by powerful entities. The regulatory regime thus breaks up and brings inefficiency in the institutional arrangements to manage the wetlands or *haors* sustainably in Bangladesh.

4.8 Power and Unequal Exchange: The Case of Transboundary Water

This section investigates the governance of the transboundary water regime under an asymmetric power relationship in the realm of international political economy. It argues that powerful countries dominate over the less powerful ones in sharing the water of Transboundary Rivers. There are some institutional arrangements in some cases in governing the transboundary water but power is the factor that brings about instability in such arrangements. Alternatively, the success (failure) of a contract depends on the strategic power of negotiating agents and the most powerful agent(s) can set the outcome of the contract.

The arguments would be illustrated with the GBM basin, with a special focus on the Ganges basin, in which case Bangladesh, being a downstream and less powerful country, suffered from the unequal exchange of water, whereas the upstream countries (particularly India) intend to accumulate a larger share of benefit by exploiting the water in various forms. The last part of the discussion provides a comparative analysis between the Ganges and Indus treaties and proposes that a third-party intervention can uplift a less powerful agent to arrange a sustainable governance mechanism. It further contends that the downstream country can retaliate by creating deterrence by restricting floods or by raising the issue at international forums.

The dispute over sharing water of the GBM basin in general and of the Ganges basin in particular between Bangladesh (then East Pakistan) and India came to the forefront in 1951 for the first time although the dispute continues in several phases (Table 4.20). At that time, India planned to divert water from the Ganges through the

Table 4.19 Revenue earnings from *Jalmahals* in different periods based on available data

	Financial year										
Rent in Pakistan period (million BDT)	1962–63	1963–64	1964–65	1965–66	1966–67	1967–68	1968–69	1969–70			
	3.85	4.37	4.42	5.04	4.26	4.58	5.50	6.23			
Rent in Bangladesh pPeriod (million BDT)	2009–10	2010–11	2011–12	2012–13	2013–14	2014–15	2015–16	2016–17	2017–18	2018–19	2019–20
	583.87	515.02	602.65	646.03	728.40	792.61	772.74	920.08	842.42	976.37	1011.10

Khan et al. (2016), MoL (2015, 2017, 2020)

construction of the 2245-m-long Farakka Barrage. The aim was to augment the flow of water for the port of Calcutta (now Kolkata) by reinstating the navigability of the Bhagirathi–Hooghly rivers by diverting 1133 m^3/sec (40,000 ft^3/sec) of water from the Ganges River (Abbas, 1984; Faisal, 2002; Mirza, 1998).

However, Vernon-Harcourt, Stevenson-Moore Committee and Webster were some of the then dominant figures who found no evidence of the deterioration of Hooghly and proposed a series of alternative solutions (Table 4.21). The then Indian regime, however, stood firm on the decision of the Man Singh Committee to construct the Farakka Barrage. The Man Singh Committee report was the last systematic investigation published by the Government of India in 1952, which found that the Hooghly had deteriorated on the stretch between Nabadwip and Kolkata (Parua, 2010), but the signs of deterioration were less pronounced for the lower reaches of the river. Therefore, the decision to construct the barrage was actually a political one.

The position of India was to identify the Ganges as a national river rather than a Transboundary River. The country put the Farakka issue forward by arguing that there would not be any shortage of water for East Pakistan and that the required water would be generated between Farakka and Hardinge Bridge (Abbas, 1984). Despite having a number of technical challenges, the establishment of the Farakka Barrage commenced in 1961 and it was commissioned in 1975. The unilateral diversion of water from the Ganges aggravated the water crisis in Bangladesh. One estimate signifies that in the post-Farakka period, the average inflow of the Ganges during the dry season was reduced to 51% at the Hardinge point in Bangladesh part (Tanzeeema & Faisal, 2001). Such low inflow has resulted in morphological changes, increased salinity in water and soil, lessened agricultural productivity, diversion of fisheries flow and damage to ecosystems of the country. On the contrary, the opening up of the barrage gate during the monsoon causes sudden floods. The population is higher in Bangladesh compared to the other riparian countries. Hence, damages caused by floods have more implications in this country (Khan, 1996). Overall, Bangladesh has

Table 4.20 Major phases of the Ganges water dispute

Period	Main issue
1947–71	Technical investigation to augment the flow of river Hooghly; decision to construct the Farakka Barrage; riparian rights and protests; no comprehensive meeting at the political level during 1961–71
1971–77	Setting up of the Indo-Bangladesh Joint Rivers Commission; sharing of the dry season flow at Farakka point but not shared as per schedule
1977–82	Augmenting the dry season flow by signing Memorandum of Understanding; renewal of the 1977 agreement
1983–87	Unilateral development of the Ganges water
1987–95	Deterioration of relationship and raising the issue at 50th UN General Assembly
1996-onwards	A treaty on sharing water for next 30 years

Source Prepared by the authors based on different sources

Table 4.21 Assessments on the navigability of the Hooghly

Assessment done by	Description	Alternative suggestions
Vernon-Harcourt	Two surveys on the Hooghly River found slow deterioration of it	Slow deterioration could be remedied by river training
Stevenson-Moore Committee	The study found the headwaters of the Nadia rivers—Bhagirathi, Mathabhanga and Jalangi—silting up	Dredging and river training in the Nadia rivers to maintain headwater supply to the Hooghly
Webster	Noted no changes in the river below Kolkata and acknowledged the Hooghly River was dependent on the freshwater supplied by the Ganges through the Nadia rivers; sceptical about the evidence of deterioration of the Hooghly	Recommended construction of a barrage across the Ganges for the diversion of 10,000 cusecs of water as well as dredging and river training in the lower reaches

Source Prepared by the authors based on Crow et al. (1995), Parua (2010)

been experiencing significant ecological, social and economic damages ever since the barrage was built (Baten & Titumir, 2016; Bharati & Jayakody, 2011; Faisal, 2002; Hassan, 2019; Mirza, 2005).

A year after Bangladesh won independence, the Indo-Bangladesh Joint Rivers Commission (JRC) was formed in 1972 as the first attempt to create a niche for advancing institutional arrangements for sharing of transboundary water. Initially, the sharing of Ganges water was kept out of JRC's purview, to be settled at the political level, and after a two-year deadlock on the commissioning of Farakka, the water-sharing negotiation took place in 1977, when it was decided to continue discussions on the augmentation of water flow of the Ganges in the dry period (Faisal, 2002; Haftendorn, 2000; Kawser & Samad, 2016).

Under the 1977 arrangement, a schedule was calculated grounded on 75% availability from the recorded Ganges flows at Farakka during 1948–1973, according to which, Bangladesh would get 60% of the flow at Farakka point and in the context of extremely low flow, Bangladesh would be guaranteed 80% of the scheduled value (Rahaman, 2006). But there was a drastic fall in the quantum of water flow in the dry season contrary to an agreement. The flow stood at 39,000 cusecs against the 59,000 cusecs mentioned in the schedule (Khan, 1976).

In 1982, both parties agreed to a Memorandum of Understanding (MoU) to share Ganges water during the dry seasons of 1983 and 1984, which renewed the already expired first bilateral agreement of 1977 for another five years. The first water-sharing agreement between India and Bangladesh finally expired in 1988 and following that period, there was no effective agreement on the sharing of the Ganges water for the period 1989–96. India started to withdraw the Ganges water unilaterally at Farakka in the dry season and released the overflow in the wet season, which led to several floods in Bangladesh (Kawser & Samad, 2016). Therefore, the release of water to Bangladesh was only at India's mercy (Baten & Titumir, 2016).

It is apparent from the discharge data before the Ganges treaty was signed (Table 4.22) that the inflows of water reduced significantly during different agreement regimes. The pre-Farakka flow of the Ganges ranges between 108,000 cusecs to about 70,000 cusecs depending on seasonal variations. The discharges decreased on an average of 33.11–76.65% with seasonal differences during 1989–95 compared to the average flow during 1934–73. It also manifests that the flow was significantly lower during the months of February–April.

After a series of futile discussions and several rounds of meetings during the 1990–95 period, India and Bangladesh reached a long-term institutional arrangement in 1996, known as the Ganges treaty, which is valid for 30 years (Brichieri-Colombi & Brandnock, 2003; Nishat & Faisal, 2000). The treaty, however, demonstrates the dominance of upstream India. It addressed the dry season sharing of Ganges water but barely made an attempt at sharing information, especially on floods. Moreover, India often breached the principles of dry season cooperation (Kawser & Samad, 2016), which results in unequal water sharing in the actual context. The upstream thus succeeded in exerting power over the negotiation.

Asymmetries in the availability of information also led to the fragility of institutions, which ultimately resulted in the failure of governance of the Ganges basin. Problems related to information asymmetry can be depicted from two different standpoints. First, there is a shortage of water during the dry season due to intensive anthropogenic interventions and hydropower development. Secondly, overflow during the monsoon causes devastating floods downstream. As per the official data, the structures built over the courses of the Ganges have the capacity to divert about 300,000 cusecs before it reaches Bangladesh, though the actual figure is said to be much higher as India keeps the data classified, citing it as 'strategic' (The Dhaka Tribune, n.d). Moreover, The Ganges Treaty (1996) signed for 30 years does not have any resolution regarding the sharing of data on floods. As Bangladesh does not have any data on the flow, the country cannot claim the fair share of this common resource. Bangladesh raised the issue several times at the negotiation table, but India bypassed it by claiming that the flood forecasting information would be shared under the umbrella of SAARC (South Asian Association for Regional Cooperation) (Haftendorn, 2000; Haque, 2008). However, as SAARC is no longer an active regional organisation, such assurances are not a sign of cooperation.

In response to the various levels of non-cooperation, Bangladesh proposed to build another barrage some 98 km downstream known as the 'Ganges barrage' (WARPO, 2001). The objective was to counter the environmental challenges stemming from the Farakka Barrage, particularly to protect the Sundarbans mangrove ecosystem. However, the project was finally cancelled in 2017 by the government based on the argument that it would not be feasible (The Daily Star, 2017). However, it is evident that the underlying reason was India's disagreement with the proposal. Thus, the political contestation and asymmetries of power overwhelmed the Ganges water-sharing procedure. The agreements that have been made on the Ganges so far essentially

Table 4.22 Average flows of the Ganges at Hardinge bridge in Bangladesh before the Ganges treaty (flow figures in cusec)

10 days periods		Pre-Farakka natural flows during 1934–73	Flows after unilateral Farakka withdrawal during 1976	Flows during 1978–1982 *1	Flows during 1983–1984 *2	Flows during 1985 *3	Flows during 1986–1988 *4	Flows during 1989–1995 *5	Comparison of 1989–95 with 1934–73 (reduced by) (%)
Jan	1–10	108,000	77,516	69,397	68,972	59,101	69,011	53,619	50.35
	11–20	101,000	65,721	59,796	61,128	54,377	75,829	47,726	52.75
	21–31	95,000	57,069	52,044	55,492	49,605	62,472	40,781	57.07
Feb	1–10	91,700	55,268	51,388	52,105	46,200	54,299	33,417	63.56
	11–20	89,100	52,866	51,547	46,370	30,049	50,330	26,868	69.85
	21–28/29	85,400	47,074	49,377	44,082	34,587	48,500	24,559	71.24
Mar	1–10	81,700	35,279	46,557	42,492	34,198	49,435	22,868	72.01
	11–20	78,100	28,676	44,054	37,676	31,452	42,761	19,573	74.94
	21–31	75,000	24,508	41,239	32,360	27,333	38,879	17,516	76.65
Apr	1–10	72,300	25,391	42,107	30,574	25,547	36,665	17,177	76.24
	11–20	70,600	25,497	46,736	31,530	27,829	38,702	19,578	72.27
	21–30	69,800	26,239	52,064	33,642	28,822	43,687	21,759	68.83
May	1–10	69,700	35,173	55,117	42,982	31,558	49,699	23,467	66.33
	11–20	73,000	47,781	56,778	52,156	29,914	52,474	32,228	55.85

(continued)

Table 4.22 (continued)

10 days periods	Pre-Farakka natural flows during 1934–73	Flows after unilateral Farakka withdrawal during 1976	Flows during 1978–1982 *1	Flows during 1983–1984 *2	Flows during 1985 *3	Flows during 1986–1988 *4	Flows during 1989–1995 *5	Comparison of 1989–95 with 1934–73 (reduced by) (%)
21–31	78,900	56,256	68,343	58,876	31,833	54,629	50,410	36.11

Source Khan (1996)

*[1]Bangladesh and India shared the flows of the Ganges at Farakka according to the 1977 agreement

*[2]Sharing of flows at Farakka under the Indo-Bangladesh Memorandum of Understanding (MoU) of 1982

*[3]No sharing at Farakka

*[4]Sharing at Farakka under the Indo-Bangladesh (MoU) of 1985

*[5]No sharing at Farakka in the absence of any agreement or understanding

validated India's status quo instead of giving Bangladesh its historic share of water (Haftendorn, 2000).

The excessive unilateral withdrawal of water by the upstream is also evident in the case of the Brahmaputra, particularly, the construction of the barrages on the tributaries of the Teesta River in Sikkim and the diversion in the Gajoldoba. It has fuelled a zero-sum game in this river basin, where one player (Bangladesh) is adversely impacted by the intervention of another player (India) (Arfanuzzaman & Syed, 2018). The average lowest discharge by the Teesta was above 4000 m^3/sec (approximately 141,258 ft^3/sec) before the construction of the two barrages. It has drastically fallen to 529 m^3/sec (approx. 18,681 ft^3/sec) in 2000 and just after five years in 2005, it came down to a mere 8 m^3/sec (282 ft^3/sec) (Rashid, 2014). The intervention has totally destroyed the ecology of Teesta. However, this river is one example in which case Bangladesh also built a barrage for irrigation along with India. The project targeted to divert an amount of 280 m^3/sec (9889 ft^3/sec) of water for irrigation through a canal (Morshed, 2006). The Teesta barrage project could contribute to increasing agricultural production, but the list of environmental consequences was also exhaustive (Baten & Titumir, 2016).

In recent times, the water shortage in Teesta has again become a matter of serious dispute. The withdrawal of water by India in the lean period annihilates the *Aus* production totally, while the Teesta becomes full from downpours during the monsoon. The opening up of the barrage gate at the Gajoldoba point creates flash floods that washes away the *Aman* production and destroys houses, roads, riverbanks and embankments in Bangladesh. Moreover, the most significant development activity on the Teesta is the planning of the construction of dams under approximately 30 hydropower projects in the state of Sikkim by India (Asia Foundation, 2013). So far, India has identified 226 prospective sites for large multi-purpose dams on the rivers of north-east India, most of which are situated in the Brahmaputra basin (IUCN, 2014b).

Finally, there is widespread speculation about China's intention to construct a dam at the great bend on the Yarlung Tsangpo (Brahmaputra) River, which could multiply the reduction in river flow (Chan et al., 2016) and the country is also working on an ambitious South–North water transfer project connecting the Brahmaputra (Barua, 2018). Williams (2022) identified China as the most influential in shaping a new transboundary water resource regime in the South and South East Asia in the upcoming years. In fact, lack of trust, hostile atmosphere, asymmetric information and power-sharing arrangement make transboundary interaction among the Brahmaputra riparian countries intricate and challenging (Barua, 2018; Barua et al., 2017; Yasuda et al., 2018). Such efforts from the upstream side to retain the flow of rivers in the Brahmaputra basin have increased the concern for Bangladesh. Overall, in the case of the Brahmaputra basin, Bangladesh has been facing threats due to the intervention of both India and China.

The power exertion by India over the Meghna basin has emerged as a contentious issue of late. Despite grave dissatisfaction by the Bangladesh side, India envisages construction of the Tipaimukh dam—a 162.8-m-high rock-filled dam with an installed capacity of 1500 MW hydroelectricity generation on the Barak River

500 m downstream on the confluence of the Barak and Tuivai rivers near Tipaimukh village in the Indian state of Manipur (IWM, 2005). The objective of the project is to control floods and produce hydroelectricity. Certainly, Bangladesh as well as Manipur raised concerns against the project (Huda, 2017; Choudhury & Choudhury, 2021). The concern for Bangladesh is that the obstruction in the natural flow of the Barak River will reduce the flow of Shurma and Kushiyara followed by Meghna. When the Tipaimukh dam becomes fully operational, the average annual monsoon inflow from Barak to the Surma–Kushiyara–Meghna River system would be decreased significantly and the highest reduction might be noticeable in the month of July (Table 4.23). Moreover, the water level would also fall across different stations with the largest average fall of 1 m expected at Amalshid station.

The dam will also contribute to climate change by increasing greenhouse gases (GHGs) emissions (Choudhury & Choudhury, 2020). This will specifically have a negative impact on the wetland ecosystem and its diverse biodiversity (e.g., Tanguar Haor, Hakaluki Haor, Hail Haor) in the north-eastern part of Bangladesh.

Finally, the River Linking Project of India can be identified as a prime example of the outcome of an unequal power structure. India had planned the project without considering the low riparian demand for water and the ecological system, particularly of Bangladesh (Chan et al., 2016). The project, as was first conceived in 1982, envisages a large amount of water withdrawing from the Ganges–Brahmaputra basins through networks of channels, reservoirs and dams (Bandyopadhyay & Perveen, 2008). The objective is to enrich its endowment of water for development activities and to transfer water to the southern and western water deficit regions of India. Estimate shows that the project on an overall basis will lead to the addition of 35 million ha for irrigation and the generation of 34,000 MW of power for India (Khera, 2020). Besides, the major benefits will also include navigation, flood, pollution, salinity control and water supply. The progress of the plan, however, was quite slow but got a massive boost under the Modi administration. The first project has already begun in March 2021 to link Ken (Madhya Pradesh) and Betwa (Uttar Pradesh) rivers (Gupta, 2021).

Table 4.23 Reduction in average annual monsoon flow and average fall in water (in July) at different stations under Meghna River system

Month	Reduced average annual amount of flow (%)	Station	Average fall in water level (m)
June	10	Amalshid (Kushiyara)	1.00
July	23	Fenchuganj (Kushiyara)	0.25
August	16	Sherpur (Kushiyara)	0.15
September	15	Markuli (Kushiyara)	0.10
–	–	Kanairghat (Surma)	0.75
–	–	Sylhet (Surma)	0.25

Source Prepared by the authors based on IWM (2005)

Table 4.24 Comparison of hydroelectricity potentiality among the co-riparian countries of GBM basin

Capacities	Bangladesh	Bhutan	China	India	Nepal
Total hydropower potential ('000 MW)	Negligible	50.0	110.0	149.0	83.3
Economically feasible potential ('000 MW)	–	23.8	N/A	84.0	43.0
Actual installed hydropower capacity ('000 MW)	–	1.5	N/A	24.6	0.7
Currently exploited (% of total potential)	–	2.9	N/A	16.5	0.8
Currently exploited (% of economically feasible potential)	–	6.2	N/A	29.3	1.5

Source Rasul (2015)

The project, however, raised several issues regarding social and environmental impact such as displacement of people, climate disorder and transboundary impact on Bangladesh and Nepal (Amarasinghe et al., 2008; Misra et al., 2007). Bangladesh has been showing repeated concerns regarding the probable consequences of the project, but the feasibility reports were prepared without any discussion with Bangladesh (Baten & Titumir, 2016).

Overall, the history of the GBM river basin is characterised by human interventions despite the rivers' significant role to serve the livelihoods of millions of people (Farooque, 2004). The upstream sides, which are comparatively in a more powerful position, however, bring about the interventions. Not exclusively, but frequently, India has been primarily accused of creating the hindrance in the natural flows of the basin through the construction of dams and barrages. In fact, India's craving for developing inland water resources follows a long time of planning. It is likely that such interventions would increase in future more due to the existing hydroelectricity potentiality of the whole basin (Table 4.24). Several studies have suggested regional cooperation on hydroelectricity (Huda & Ali, 2018; Rasul, 2015; Shukla et al., 2017).

However, Bangladesh, being a lower riparian country, having less power and negligible potentiality in hydroelectricity generation, would not get any benefit. Instead, it will be the ultimate sufferer because of rising competition among the countries to govern the rivers for their own interests. Thus, the countries, which are powerful, have the final say regarding the transboundary water and get the maximum benefit possible in multiple ways.

4.8.1 Ganges Treaty and Indus Treaty: A Comparison

Based on the principles of "equity, fairness and no harm to either party" (Article IX, The Ganges Treaty, 1996), the Ganges treaty urged both the Indian and Bangladesh governments to ensure an equitable use of the Ganges water and calls for resolving the water-sharing problems of other Transboundary Rivers (Kliot et al., 2001). According to the treaty, both countries are entitled to share 50% of the discharge available at Farakka when the flow of the river measures less than 70,000 cusecs (Table 4.25). In case the range of flow is 70,000–75,000 cusecs, India is supposed to release 35,000 cusecs to Bangladesh and hold the balance of flow or the maximum of 40,000 cusecs.

If the discharge reaches more than 75,000 cusecs, according to the treaty, India is supposed to receive 40,000 cusecs and the extra flows are to be released to Bangladesh. In this arrangement, first, the upstream country, India, will be holding an additional 50% of the discharge, which is supposed to flow into Bangladesh during the dry season. In case of extra flow, it releases 35,000 cusecs but it is retaining the balance of flow up to 40,000 cusecs; and secondly, the upstream is not sharing the burden of flood if the flow reaches more than 75,000 cusecs. India, in fact, is in a beneficial position under this complicated formula. Another source of concern for Bangladesh has been the lack of a guarantee clause to release the minimum amount of water in the event of abnormally low flows (Baten & Titumir, 2016). Furthermore, the treaty ignores the fact that the Ganges is also a significant contributor to monsoon floods, and thus it focuses only on the lean period to share water. The treaty only divides the water without sharing the value and uses of the river among the participating countries (Hanasz, 2014).

It is evident from the recent discharge data that in most of the cases during the post-treaty period, Bangladesh got less water than the agreed amount of its share according to the indicative schedule of the treaty during the dry season (Table 4.26). On the contrary, the exceeded amount in some years signifies that in those periods, there were floods affecting different parts of the country. Therefore, the treaty could not help increase water availability in Bangladesh in the dry season. One study argued that the treaty performed poorly than the first water-sharing agreement of 1977 during the most critical 10-day periods of March and April (Nishat & Faisal, 2000). It proves that a legally binding agreement does not always bring meaningful cooperation between the parties (Hanasz, 2014). Overall, the treaty could not bring any solution under the cooperative governance mechanisms.

On the contrary, the Indus waters treaty is one of the most successful cases of dissolving power asymmetry between the upstream and downstream parties. The basin consists of a total of six rivers that are partitioned into eastern rivers (Sutlej, Beas and Ravi) and western rivers (Indus, Jhelum and Chenab) (FAO, 2011b). Being a downstream country, Pakistan is supposed to have relatively less power in negotiations with India. However, the negotiating parties under this treaty demarcated the rivers instead of sharing water.

Table 4.25 Resolution under Ganges treaty for January–May

Availability at Farakka	India's share	Bangladesh's share
< 70,000	50%	50%
70,000–75,000	Balance of flow	35,000 cusecs
> 75,000	40,000 cusecs	Balance of flow

Source Annexure 1, The Ganges Treaty (1996)

Table 4.26 Discharge data of the Ganges in the post-treaty period

10 days periods		Bangladesh share as per the indicative schedule of the treaty	Actual release to Bangladesh													
			1997	2000	2005	2010	2011	2012	2013	2014	2015	2016	2017	2018	2019	2020	2021
Jan	1–10	67,516	62,019	91,648	56,582	46,720	56,179	104,410	54,211	70,853	58,528	35,000	57,766	48,783	61,024	60,061	53,957
	11–20	57,673	49,556	70,876	45,112	37,204	48,036	81,133	42,897	67,311	62,330	31,394	45,297	39,555	42,580	78,547	44,508
	21–31	50,154	48,884	57,686	44,257	35,048	38,867	87,522	40,129	77,003	53,141	31,014	43,993	38,480	35,000	79,403	47,136
Feb	1–10	46,323	45,604	52,253	44,455	36,266	33,258	87,603	44,790	76,284	45,071	29,737	38,383	37,035	35,000	73,814	43,607
	11–20	42,859	41,029	47,515	46,341	34,054	32,095	81,493	41,241	70,506	41,017	28,820	35,527	34,331	38,501	62,436	35,000
	21–28/29	39,106	38,387	44,905	36,854	32,412	32,543	67,864	59,624	62,692	35,000	26,865	35,000	31,817	36,832	50,559	33,647
Mar	1–10	35,000	33,489	42,439	33,358	30,371	31,687	48,418	63,879	61,999	33,529	25,419	31,128	29,083	35,000	54,990	31,201
	11–20	35,000	35,028	35,002	35,001	35,000	35,000	35,000	52,492	66,243	39,644	35,000	35,000	35,000	35,000	55,970	35,000
	21–31	29,688	16,528	34,849	25,309	18,769	29,431	32,772	37,757	52,366	44,061	15,606	24,488	22,151	29,926	71,619	23,544
Apr	1–10	35,000	30,137	35,000	35,000	35,000	35,000	35,000	35,000	45,994	60,492	35,000	35,000	35,000	42,349	66,152	35,000
	11–20	27,633	25,613	34,412	25,037	17,191	25,370	40,864	33,302	40,850	54,325	18,282	33,307	21,708	34,997	59,815	21,045
	21–30	35,000	35,065	45,178	35,000	35,000	35,000	35,000	35,000	37,422	55,618	35,000	35,000	35,000	47,825	61,221	35,000
May	1–10	32,351	31,722	48,718	32,015	20,791	22,800	33,059	35,117	35,000	64,165	16,648	35,203	22,676	59,564	67,442	18,839
	11–20	35,000	33,021	50,768	35,001	27,634	32,628	32,327	35,000	35,000	66,203	26,155	35,000	29,571	61,807	72,254	31,957
	21–31	41,854	31,643	98,675	32,951	32,675	43,020	32,958	47,135	41,856	71,193	29,479	37,088	34,392	50,231	89,497	45,658

Source JRC Bangladesh (2021)

The 1947 Partition of India and Pakistan heightened political tension, population displacement and unresolved territorial issues, all inducing to exacerbate hostilities over water (Priscoli & Wolf, 2009). In this setting, in May 1952, Pakistan took the help of the World Bank as a mediator to engage in discussions over the treaty, which was intensively participated and fought based on the understanding of ground reality and the needs of both the parties (Iyer, 2007; Miner et al., 2009; Thatte, 2008). The treaty gave India control over the eastern rivers and Pakistan of three western rivers, providing both countries with a genuine share of water (Kalair et al., 2019; Thatte, 2008). By ensuring the resource as collective goods, both sides are sharing the benefits.

Cooperation in the governance of the Indus basin, therefore, had been achieved by dissolving the imbalance of power through the establishment of sovereign authority over the rights of the utilisation of eastern and western rivers. A non-cooperative game would have been a loss for both countries potentially because it could have increased both internal and external threats. Both the countries agreed to rights-based allocation and joint development under a strong institutional mechanism mediated by the World Bank (Parajuli et al., 2003). The treaty also includes a compensation mechanism for both countries, which has made the non-cooperation costly to the parties.

Provisions of a treaty are the key elements of institutional arrangements. The institutional arrangements under the Indus treaty have clearly identified the scope for mutual utilisation of common waters of the Indus basin (Preamble, Articles I & XI), the substantive rules for water allocation (Articles II, III & IV), the procedural rules for water development and transboundary water management (Articles V, VI, VII & XII), the formation of an authority to periodical monitoring and management (Article VIII) and the mechanism of dispute management and resolution (Article IX) (World Bank, 2018). On the contrary, the Ganges treaty focuses particularly on the water-sharing mechanism at the Farakka point (Table 4.27).

The downstream country in the Indus treaty bolsters interest against the powerful upstream country under third-party mediation. Provisions made under the treaty have ensured a long-term strong resolution by clearly identifying the obligations of the parties. On the contrary, the bilateral settlement between India and Bangladesh ended up with an extractive mechanism by the upstream as a powerful agent. After independence, the five successive treaties have been agreed upon, but these remain unequal in terms of water sharing and fragile in terms of institutional mechanisms for dispute resolution and exchange of information. Bangladesh has tried to involve Nepal, a Ganges basin-sharing country, as a partner and the United Nations to resolve the conflicts and arrange a permanent settlement but failed due to extensive dominance by the upstream in a period of apathy of the then Pakistan regime, followed by weak negotiations of successive regimes in Bangladesh (Brichieri-Colombi & Brandnock, 2003).

Overall, the Ganges treaty is a fragile institutional arrangement under unequal power relations which leads to a suboptimal solution, unequal well-being and non-cooperative governance. In the case of sharing the Ganges water, the upstream country is geographically in an advantageous position which secures enormous potential to

Table 4.27 Comparison of the legal institutional framework of The Ganges Treaty (1996) and the Indus Treaty 1960

Issue areas	Indus treaty	Ganges treaty	Remarks
Scope	Preamble: It addresses the common consent and equal desire to attain the most complete and satisfactory utilisation of the waters of the Indus river system in a cooperative spirit	Preamble: It addresses the common desire and describes it as needed agreement to share water of the Ganges at Farakka for flood management, irrigation and generation of hydro-power	In case of Ganges treaty, the interest of upstream India has been prioritised because it is about water sharing and not management of trans-boundary water. Moreover, only India has the scope to generate hydropower. Furthermore, the design of the treaty ensures availability of irrigation for agricultural development of India during the dry season and keeps the scope for releasing the extra water from India to Bangladesh during the wet season
	Article I: definitions	–	No concrete definitions (e.g., irrigation, hydro-power and flood management) in the Ganges treaty to understand the scope of water sharing
	Article XI: general provisions—ensures the mutual recognition and not to withhold from the provisions mentioned in the treaty. This section has also been supplemented stating the crises and disputes in regard to Indus Waters faced by both the countries *Annexure A: exchange of notes between Government of India and Government of Pakistan*	Article I: agreed sharing the quantum of water at Farakka point	No mutual exchange of notes acknowledging the crises in relation to water extraction at Farakka point and the disputes regarding Ganges water management

(continued)

Table 4.27 (continued)

Issue areas	Indus treaty	Ganges treaty	Remarks
Substantive rules	Article II: provisions Regarding Eastern Rivers *Annexure B: agricultural use by Pakistan from Certain Tributaries of the Ravi* *Annexure H: transitional arrangements* Article III: provisions regarding western rivers *Annexure C: agricultural use by India from the western rivers* *Annexure D: generation of hydro-electric power by India on the western rivers* *Annexure E: storage of waters by India on the western rivers*	Article II: provisions of water sharing by 10-day periods from 1st January to 31st May *Annexure A: water sharing formula* *Annexure B: indicative schedule*	The Indus treaty incorporated the specific provisions regarding the mutual trans-boundary water management and development whereas the Ganges treaty addressed the dry season water sharing between the countries as per the sharing formula and the indicative schedule. In case of Indus waters, partition of the rivers has been done between the countries. The eastern rivers were allocated to India and the western rivers to Pakistan for unrestricted use. The Indus treaty has further articulated the provisions of domestic use, non-consumptive use and agricultural use inside the territories based on the allocated rivers. In addition, it articulated the transition periods to set up the countries' respective water development projects. Although the Ganges is the single stream flowing to Bangladesh, the provisions made under the treaty are more in favour of the upstream. India can secure and withdraw water during the dry season. The treaty, therefore, is designed to address the time period from January to May. Moreover, the sharing formula is extractive in nature, allowing to divert water from the Ganges than to ensure a sustainable transboundary water management

(continued)

Table 4.27 (continued)

Issue areas	Indus treaty	Ganges treaty	Remarks
	Article IV: provisions regarding Eastern Rivers and Western Rivers	Article III: further provision for water diversion	The provisions made under Article IV of Indus treaty ensure each party shall refrain from bringing any material changes to the water flow and material damages to the other party. Whereas in the Ganges treaty, India kept another provision of withdrawing 200 cusecs of water in between the released water of Farakka and the border of Bangladesh
Procedural rules	Article V: financial provisions	–	In order to develop the irrigation system of the western rivers, which was previously dependent on the supply of water from the eastern rivers, inside Pakistan by replacement of the canal projects, India had agreed to a fixed contribution of £62.06 million to Pakistan. The World Bank took the responsibility to disburse the fund by constituting the Indus Basin Development Fund. As per the Ganges treaty, the upstream did not agree to make any financial compensation to Bangladesh in the face of serious damages caused by periodical flash floods and environmental loss

(continued)

Table 4.27 (continued)

Issue areas	Indus treaty	Ganges treaty	Remarks
	Article VI: exchange of data	Articles IV and VI: joint committee to observe daily discharge	There are no specific provisions of data exchange, including no provision for sharing flood data in the Ganges treaty. It only states of keeping the daily discharge record and submission on annual basis to the governments. The governments will arrange a need-based meeting if there remains any discontent. In the Indus treaty, specific provisions and types of data to be shared have been clearly identified for sharing. The data of the Indus waters encompass the mainstream to the link canal and must be shared within a month. Additionally, both the parties are entitled to share data on an urgent request
	Article VII: future co-operation	Articles VIII and IX: consent without any future vision to the trans-boundary water sharing	As per the Indus treaty, both the parties shared a common interest in the optimum development of the rivers to the fullest possible extent. In this regard to ensure mutual trust, each party has been entitled to install hydrological or meteorological stations within the drainage basins of the rivers. The parties agreed to carry out mutual engineering works by sharing the plan and data. On the other hand, the Ganges treaty recognises the need to find out a solution for the long-term problem of augmenting the flows of the Ganges during the dry season. The Ganges treaty is thus too fragile in case of addressing the disputes and the mutual water development and management

(continued)

Table 4.27 (continued)

Issue areas	Indus treaty	Ganges treaty	Remarks
	Article XII: final provisions	Article XII: expiry of the treaty	The final provision of the Indus treaty articulates the stability of the agreement between India and Pakistan as the provisions can be modified time to time through ratification. The Ganges treaty, in contrast, has been signed for a 30-year period without any vision for mutual development of the Ganges water

(continued)

Table 4.27 (continued)

Issue areas	Indus treaty	Ganges treaty	Remarks
Authority	Article VIII: permanent Indus commission	Articles IV, V and VI: duties of joint committee	In the Indus treaty, there is a Permanent Indus Commission headed by two commissioners selected from each country. The commissioners shall ordinarily be a high-ranking engineer and competent in the field of hydrology and water use. The duties and the obligations of the commissioners are articulated in the treaty. Both the commissioners enjoy the same privileges and immunities mentioned in Sections 11, 12 and 13 of Article IV of the Convention on the Privileges and Immunities of the United Nations. The commissioners can also settle the disputes according to Article IX of the Indus treaty As per the Ganges Treaty, a Joint Committee, consisting of equal number of members from each party, has been suggested to keep record of the discharge data and submit an annual report to the governments. The committee is too weak to take part in resolving disputes and usually refers them to the Joint Rivers Commission or to the governments. Because the sharing of water is the central purpose of the agreement, the scope of trans-boundary water management is limited in this treaty

(continued)

Table 4.27 (continued)

Issue areas	Indus treaty	Ganges treaty	Remarks
Dispute avoidance and resolution	Article IX: settlement of differences and disputes	Articles VII, X and XI: dispute management	To resolve the differences and disputes, the Indus treaty has a number of provisions and rules. It encourages the commissioner, government and neutral expert levels of solutions. It promotes renegotiation in presence of mediators if there is any discontent. For the resolution of severe disputes, there is a provision of setting up a Court of Arbitration consisting of seven arbitrators for a permanent solution In contrast, the provisions of dispute resolution are limited in the Ganges treaty, in which it suggests that disputes should be referred to the government for settlement. The sharing arrangement under this treaty can be reviewed at five years interval or earlier but nor earlier than two years to assess the impact and working of the sharing arrangement. In the absence of mutual agreement following the reviews, India will abstract 10% of the water in between the release point at Farakka and Bangladesh border. It indicates the domination of the upstream in the settlement process

(continued)

Table 4.27 (continued)

Issue areas	Indus treaty	Ganges treaty	Remarks
	Annexure F: neutral expert	–	No provision has been kept in the Ganges treaty regarding expert's solution if there arises any differences or disputes
	Annexure G: court of arbitration	–	No provision for setting up a court of arbitration to deal with disputes

Source Prepared by the authors based on different sources
Note Data adapted for Indus treaty from Sarfraz (2013) for making comparison with Ganges treaty

utilise the water. At the same time, both the riparian countries had also undergone a hostile political settlement before the liberation of Bangladesh. That situation has not improved even after the independence of Bangladesh, despite the country's strong interest to build a water-secure region, while the upstream continues to dominate due to the inflows of a number of major rivers culminating in Bangladesh after originating in India.

Alternatively, the Indus treaty is a case of an effective institutional arrangement due to the cooperative manner of the governance of the transboundary water. In the formation of this treaty, a third-party mediation helped to balance the power asymmetry between the contending countries and gave an advantage to the downstream country Pakistan. Moreover, the provisions of the treaty are more comprehensive. This cooperative behaviour ensures the sustainable governance of the transboundary water and allows for maximising the well-being emanating from common resources among the co-riparian countries. Therefore, a stable institutional arrangement plays a key role in ensuring a sustainable governance of transboundary water. A sustainable governance regime is conditioned upon the relative balancing of power among the negotiating agents and thirty-party involvement in the case of a bilateral treaty can bring an effective outcome in this regard.

However, the neutrality in transboundary water sharing can also be guaranteed by ensuring the participation of all riparian countries—not two countries only—in signing a treaty. This means emphasis needs to be put on 'multilateral treaty' instead of 'bilateral treaty' to bring out an effective solution in terms of safeguarding equitable and sustainable transboundary water governance. The underlying implication is that Transboundary Rivers are common resources and therefore, all the riparian countries should collectively make decisions about the fair share of the resources.

4.9 Technology, Institutions and Revenue: The Case of Marine Resources

Bangladesh is yet to utilise its vast reservoir of marine resources. The state of the marine ecosystem clearly indicates that the lack of proper appropriation, conservation and sustainable management are the present trends of marine fisheries resources. The inefficiency in sustainable usage of marine resources involves technological inefficiency, fragile institutions and dominating power relations.

Bangladesh conquered an entitlement to 118,813 km^2 in the Bay of Bengal (BoB) after two verdicts by the United Nations Permanent Court of Arbitration (UNPCA) and the International Tribunal of the Law on the Sea (ITLOS) on the maritime delineation between Bangladesh and India and Bangladesh and Myanmar, respectively (Bhuiyan et al., 2015). The total marine boundary now stands at 121,110 km^2. The tension regarding the maritime boundary was intense because the area, claimed by both parties (India and Myanmar), legally belongs to Bangladesh and is full of resources. According to the verdict, the tension over the property rights problem was dissolved, but security is the next challenge in order to extract resources. Bangladesh owned her maritime boundary, but the *de facto* property rights have not been established. The rival countries are acknowledging the *de jure* transfer of property rights to Bangladesh, but still the external threat is remaining at the same level. Being a less powerful country in the domain of international political economy, Bangladesh's control over the area and its resources could be threatened by the invasion of powerful countries.

Apart from Bangladesh, India and Myanmar, countries such as Thailand, Sri Lanka, Maldives, Malaysia and Indonesia also reside by the BoB. About 450 million people in these countries are directly reliant on the resources, especially fisheries, of the BoB (BOBLME, 2014). Pressure on the resources of the BoB has been intensifying due to the growth of the population in the regions. The increasing pressure in the Bay is likely to arouse a security issue in future regarding the stability of property rights.

Therefore, Bangladesh needs to confirm the stable enforcement of property rights over the resource blocks and the fisheries roaming areas in particular to develop an inclusive and sustainable governance system for its marine wealth. To ensure security, the country can adopt a self-developed patrolling, surveillance and action protocol. Above all, native fishers have to be entitled with property rights, tenure security and protection against foreign threats at the micro-level.

In addition, Bangladesh is a capital-deficient country depending mostly on foreign technology. To explore the potential resources in the maritime region, Bangladesh has to take the help of foreign technology that increases the factor cost and resembles technical inefficiency. Simultaneously, exploration and mining rights are to be leased to foreign firms in exchange for higher payment due to the unavailability or inadequacy of local expertise.

In particular, technological inefficiency is noticeable in the fishery sector. The marine fish species can be categorised based on different fishing grounds in Bangladesh (Table 4.28) and these are being extracted from three tiers (Table 4.29). The existing practice of fishing is mainly congested in the shallow waters of the coast. Local or normal fishing craft (now some of them are motorised) known as artisanal fishery forms the core component of the present-day fishery in the marine water of Bangladesh. The majority of artisanal fishing boats in the country lack adequate safety standards, are old and may not have accessible shelter and sanitation facilities on board (Porras et al., 2017).

Of the total marine landing, artisanal fishing and industrial fishing contribute 83.75% and 16.25%, respectively, at present (DoF, 2019). A long-term trend analysis of marine fisheries production shows that there is a sharp rise in the total production from 1983–84 to 2013–14 and then it increases at a slower pace (Fig. 4.20). The trend is similar in the case of artisanal fishing production. The production of marine fish using artisanal fishing has risen from 150,382 MT in 1983–84 to 552,675 MT in 2018–19. In the most recent case, it increased by 18,075 MT between 2017 and 2019. On the contrary, the rise in industrial (trawler) fishing is significantly low compared to the artisanal one. The production under industrial fishing was 14,500 MT in 1983–84, which reached 120,087 MT in 2017–18. However, the production

Table 4.28 Major fishing grounds of Bangladesh and major commercial fish species found in respective areas

Fishing ground	Area (km^2)	Major commercial species
South patches	6200	Indian salmon (*Eleutheronema tetradactylum*), Hilsa (*Tenualosa ilisha*), Pomfret (*Pampus chinensis*), Ribbonfish (*Trachipteridae lepturus*), Bombay duck (*Harpadon nehereus*), Eel (*Anguilliformes* spp.), Jewfish (*Epinephelus itajara*), Catfish (*Siluriformes*), Sharks (*Selachimorpha*) and Rays (*Batoidea*)
Middle ground	4600	Pomfret, Red snappers (*Lutjanus campechanus*), Ribbonfish, Silver jew, Shrimp (*Penaeid* species), Indian mackerel (*Rastrelliger kanagurta*), Snappers, Groupers, Jewfish
Swatch of no-ground	3800	Shrimps, Hilsa, Pomfret, Ribbonfish, Bombay duck, Jew fish

Source Adapted from Sarker et al. (2018)

Table 4.29 Different tiers of fishing in marine area of Bangladesh

Range of tier depth	Operating boats/trawlers
Up to 40 m from the coastline	Normal fishing boats
From 40 to 200 m	Mid-water trawlers
From 200 m to the end of EEZ	Long-liner trawlers

Source Prepared by the authors based on Islam et al. (2017)

decreased in 2018–19 by 12,851 MT, and it stands at 107,236 MT. Therefore, the annual growth rate is (−) 10.70% for industrial fishing and (+) 3.38% for artisanal fishing.

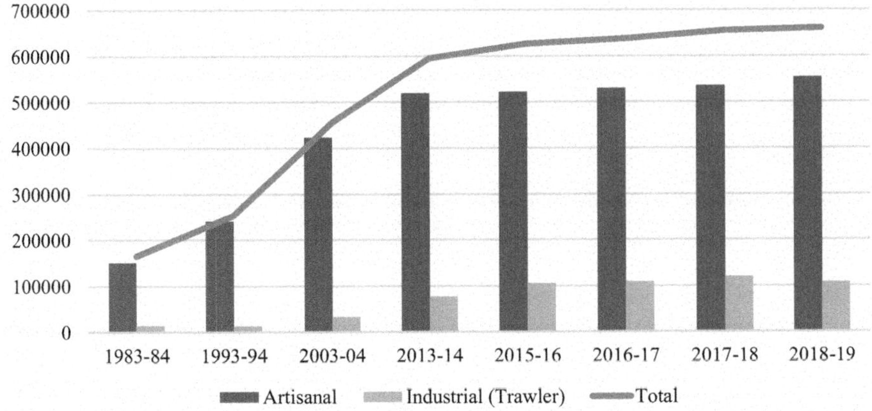

Fig. 4.20 Marine fisheries production trend from 1983–84 to 2018–19 (in MT). *Source* DoF (2019)

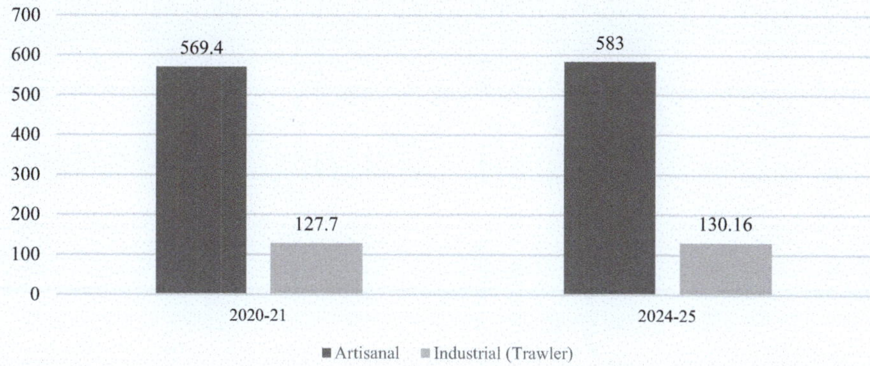

Fig. 4.21 Projected marine fisheries production in 2020–21 and 2024–25 (in '000 MT). *Source* GED (2020)

Projection scenario also indicates that artisanal fishing would dominate marine production in the near future (Fig. 4.21). The heavy dependency on the shallow water fishing under the artisanal fishing is causing the extinction of species, albeit the opportunity to increase production is remaining.

The limited number of fishing in the open sea under the industrial fishing is the result of technological incapacity. A comparative analysis of fishing technologies between two different fiscal years (2013–14 and 2018–19) exhibits that there is no massive improvement in technological capacity during this five-year period and in some cases, the capacity deteriorates (Table 4.30).

The total number of trawlers operating in the deep sea has increased slightly over the two different periods, but the number of shrimp trawlers has reduced. On the contrary, the number of boats that usually operate in shallow water has significantly reduced during the same time, but the number is still higher than that of the trawler. As many as 20,359 mechanised boats and 16,831 non-mechanised boats are currently operating at the near coast. In addition, the total number of gear or net used under artisanal fishing has also increased while the number of gear has decreased in the case of industrial fishing. Specifically, the numbers of gill net and set bag net have increased and in other cases, the numbers have remained constant under artisanal fishing. In the case of industrial fishing, the fish trawling gear number has increased while the number of shrimp trawling gear has decreased.

Apart from the artisanal (small-scale, commercial) and industrial (large-scale, commercial) fishing, there is another type called subsistence (small-scale, non-commercial) fishing (Ullah et al., 2014). Subsistence fishing also puts pressure on the shallow water and sometimes the anglers' disregard for fishing rules and regulations, use of destructive fishing techniques, intention to catch whatever amount of fish is available, including larvae and juveniles all contribute to increased pressure on fisheries (Murshed-e-Jahan et al., 2014). Overall, the coastal people, who are mostly engaged in fishing, have boats and other fishing equipment, but those are not enough for the proper appropriation of fish from the sea for meeting the national demand.

Table 4.30 Total number of boats and gears operating for marine fishery capture

Number of trawlers			Number of boats			Number of gear/net (artisanal)			Number of gear/net (industrial)		
Type	2013–14	2018–19	Type	2013–14	2018–19	Type	2013–14	2018–19	Type	2013–14	2018–19
Shrimp trawler	38	37	Mechanised boat	30,164	20,359	Gill net	114,353	118,353	Shrimp trawling	114	87
Fish trawler	187	218	Non-mechanised boat	27,699	16,831	Set bag net	40,824	42,429	Fish trawling	561	564
						Long line	11,863	11,863			
						Trammel net	422	422			
						Other net	15,640	15,640			
Total	225	255		57,863	37,190		183,102	188,707		675	651

Source DoF (2014, 2019)

It is therefore necessary to establish ancillary fishing industries, especially trawler manufacture. The fishermen would go fishing in the deep sea if they have mechanised boats equipped with modern technology.

Industrial trawling is conducive to capturing fish in the open sea, but there is evidence also that a large quantity of by-catch as caught by the shrimp trawler is generally thrown overboard (Ullah et al., 2014). The fishers usually target a particular species and discard the by-catch. As the governmental authority is responsible in ensuring through strict surveillance that the anglers should comply with the fishing rules and regulations to ensure sustainable production (Islam et al., 2017; Shamsuzzaman et al., 2017), any discrepancy in this case, therefore, signifies the fragility of the existing formal institutional arrangement.

Finally, fishing in shallow water also reduces the enforcement of ownership over the deep sea and creates scope for illegal entry by the fishers of other countries despite patrolling. Bangladeshi coastal and marine areas have always been one of the lucrative places for fishing by the adjacent neighbouring countries. The illegal, unreported and unregulated encroachment and capture by trawlers of neighbouring countries in the maritime boundary of Bangladesh is a regular phenomenon (Shamsuzzaman et al., 2017). The intruders use upgraded technology to capture fish in this case. Moreover, the lack of power to ensure security also paves the way for external invasion. Evidence also suggests that Bangladeshi anglers only catch a very small portion of the total fish caught in the Bay of Bengal (Sanaullah, 2021). Deep-sea fishing is practically non-existent in Bangladesh, but it is practised in neighbouring nations such as India, Malaysia and Thailand. More than 80% of the total water is unexplored by Bangladeshi fishermen, who can only travel up to 70 km of the 660 km freely available for fishing, and fish available beyond 100 m are not being harvested (Sanaullah, 2021; Shehab, 2019). Therefore, Bangladesh still has a lot of untapped sea fishing potential.

Moreover, the conflict between diverse potential users of the same resource is a serious issue in marine water governance, particularly where industrial trawlers operating in inshore water are in competition with artisanal fishers. This has often emerged as a specific problem between Bangladesh, India and Myanmar (Hoq et al., 2013; Murshed-e-Jahan et al., 2014), which signifies the conflicts with outside stakeholders. However, conflicts also arise in the internal context—among the fishers themselves or between small-scale traditional fishers and powerful local individuals (Murshed-e-Jahan et al., 2014).

Conflicts often emerge because of corruption, lack of coordination and overlapping activities of government agencies (Murshed-e-Jahan et al., 2014). Such user conflicts can be recognised as a failure of the regulatory regime of fisheries governance. The conflicts cause disturbances in sustainable resource utilisation by the resource users. For instance, the traditional fishermen who most often get exploited financially (e.g., paying money in the form of bribes to get licences for fishing, paying high interest to get loans from local moneylenders) in the conflicting situation due to their lack of power are forced to extract more resources in the latter period to compensate their monetary loss. The ultimate outcomes of technological incapacity and institutional vulnerability are therefore the degradation of fisheries and the loss of a significant amount of revenue that could be earned by sustainably managing the resources.

Overall, the marine resources of Bangladesh remain underutilised due to a lack of technology, capital deficiency, precarious property rights caused by illegal, unreported and unregulated extraction by the intruders and the use of resources at an unsustainable rate. Consequently, the absence of these preconditions while extracting marine resources causes inefficiency, loss of rent and the overall depletion of the stock of resources.

4.10 Social Norms, Cooperation and Human Sociality in Water Governance

From classical economics to the dominant mainstream neo-classical economics, it has been argued that the individual's rationality derives from self-interest behaviour to maximise their utility. In response, this book argues that human beings are cooperative social beings who behave reciprocally. Because nature provides numerous benefits to humans, humans also have an innate desire to serve the nature by protecting it. Therefore, a mutual relations exists between human beings and nature. With the advent of the market, however, this relationship has become distorted and accordingly, natural resources are being extracted. It is necessary to revitalise the relations to move towards the sustainable resource governance, and this understanding has implications in the case of water resource governance as well.

Water is a vital component of nature that provides resources in a variety of forms for the survival and well-being of human beings and therefore, it is actually synonymous with 'life'. There is a long-standing embeddedness between human beings and water. Humans have had to depend on water for their existence ever since they left their first footprints on the earth. Therefore, they aim to preserve and utilise water sustainably depending on their inherited accumulated knowledge over the course of their lives. Since the initiation of identifying water as a scarce 'commodity', human beings have begun to compete over the ownership of water through non-cooperative behaviour, which ultimately leads to its destruction. Moreover, powerful entities are the ones who benefit the most in the process, creating an inequitable distribution of water resources. This situation needs to be altered through the recognition of social norms and values in the form of informal institutions, by strengthening the formal institutions and by revitalising the sense of cooperativeness among human beings. This would result in the diffusion of power, which would eventually bring justice and equal rights for each individual or entity in the realm of water governance.

As water resources are diverse in nature but connected through a stock and flow process, alternative understanding as stated above would bring sustainability in each domain of water governance ranging from transboundary water flowing across different nation states at the global level to groundwater used in households and commercial activities specifically in the context of a country—in this case,

Bangladesh. When an individual country would recognise that water can be shared equally and sufficiently among the co-riparian countries with an appropriate institutional arrangement under a cooperative environment, it would lead to a sustainable governance of water in the global context. Simultaneously, it would also create a natural flow of water to the individual country and accordingly, ensure the sustainability of freshwater ecosystems such as rivers, wetlands and groundwater in the context of that particular country.

In such a human sociality scenario, there will be no excessive pressure on the water resources at the country level and the concomitant implication is the equal distribution of water among all. The powerful agents would not be capable of accruing wealth by commodifying water and its resources (e.g., fisheries) by exploiting the poor group of people anymore. Furthermore, there is a specific implication of informal and formal institution nexus in the case of wetlands governance, particularly in the context of a transitional economy. The findings on the *haor* management in Bangladesh show that traditional fishers do not get market entry because of the high transaction costs. As the local elites regulate the Fisheries Cooperative Societies, the rent is automatically amassed by them. The power relationship and imperfect market thus result in the dissipation of rent as well as unequal distribution of well-being. The community-based management regimes are relatively efficient because they happen to dissolve power relations between the fishers and the local elites by providing loans and arranging livelihood programmes. This results in the well-being of members of the community, which in turn ensures the conservation of wetland resources. The community people know how to exercise their rights over the resources. Therefore, they tend to use and manage them sustainably based on their traditional norms and beliefs.

Lastly, the natural stock and flow relation as depicted above can also contribute to the marine ecosystem to ensure its sustainability as the rivers ultimately flow into the oceans. However, the marine water covers a major part of the earth, and therefore, the governance of the marine ecosystem is not concerned about the scarcity issue like the freshwater ecosystem. Rather, the concerns are (a) the utilisation of the vast amount of its resources for the well-being of human beings without hampering their sustainability and (b) the enforcement of institutional arrangements, predominantly the formal institutions, to safeguard the open water resources while adhering to demarcation and international law. Findings suggest that the resources are being degraded because of unequal power and thus, violating the human sociality.

This is also envisioned under the concept of 'blue economy'. The human sociality-based understanding can be related here by arguing that sustainable production and creative conservation of marine resources are possible under a cooperative scenario. If human beings were to acknowledge that the oceans are serving them with immense benefits and that they must be protected, then they would cooperate both at national and international levels to prevent overuse and illegal exploitation of the available resources and would innovate and adopt advanced technologies to utilise them sustainably and reasonably. Overall, the alternative understanding, which revolves around human sociality, could help to move towards sustainable water resource governance at various levels.

4.11 Concluding Remarks

The complexities of water governance include the contestation over the control of Transboundary Rivers, sovereignty of marine resources, excessive withdrawal of groundwater for various purposes and the suboptimal production of water resources such as fisheries. Nevertheless, there is a lack of understanding of such complexities in the context of transitional economies as few studies are available on water resources, and they are only concerned about a specific type of problem in the realm of water governance. In such a scenario, this chapter has scrutinised the challenges of water governance in Bangladesh based on a variety of theoretical positions and empirical evidences. Moreover, it offers an alternative understanding that helps in theorising the sustainability of water resources under a comprehensive framework.

By analysing the state of water resources and the underlying factors that explain the emerging facts, the chapter underscores several points regarding specific types of water resources as classified in this chapter as groundwater, transboundary water, wetlands, marine and fisheries resources. The underlying problems in water governance, therefore, cannot be included in one list as they are far too diverse in nature. However, the factors that cause the problems are similar across the diverse resources as defined under the alternative framework. Broadly, the factors are the commodification of water, power, political settlement and fragile institutions. At the policy level, both nationally and internationally, the issue of water has been placed ambiguously and described imperfectly as a scarcity problem, which is a significant flaw.

Considering the state of groundwater, the analysis exhibits that there is a problem of provisioning and access, which in turn puts pressure on the groundwater level. The poor households' condition is the worst in accessing water, and it is acute in the urban context. They have to rely almost entirely on the informal market and pay a high price while consuming little. Furthermore, in the informal market, the powerful intermediaries make a profit at the expense of the loss of national revenue and consumer surplus. The chapter further argues that the whole process is the outcome of the commodification of water. In addition, the outstanding revenues are the outcome of institutional fragility. In the end, the discussion also illustrates that the COVID-19 pandemic has exerted an extra burden on the country's groundwater table, the outcome of which could become a major challenge in the future.

Wetlands are the unique types of aquatic resources of Bangladesh, which are particularly important for the sustainability of the livelihoods of the poor community in the north-eastern region of the country. The historical analysis of the nature of property rights of the wetlands has signified that the community people have the traditional resource right over the wetlands, and they used to conserve the wetlands by following their traditional and cultural beliefs and rituals. However, such institutional arrangement gets distorted due to the inclusion of political interest to capture the resources. The powerful outside stakeholders, both at local and national levels, marginalises the local community in different ways and intend to utilise the aquatic resources of the wetlands for commercial benefits. Thus, the mutual relation between the community people and the nature breaks down, which ultimately causes the degradation of wetland resources.

Transboundary Rivers are vital components for Bangladesh's ecological integrity and the livelihood of the country's millions of people. However, Bangladesh, being a less powerful country among the riparian countries, cannot enjoy equal access to the transboundary water. The existing institutional arrangements in ensuring equality within the governance regime have not been working efficiently as powerful countries, especially India, are not willing to cooperate. The chapter portrays these issues by focusing on the GBM basin in general and the Ganges basin in particular. Moreover, it argues that ensuring third-party involvement in the case of bilateral agreement can be a viable solution to establish a stable and efficient institutional arrangement that can ensure a fair share of Bangladesh in the future. However, the neutrality in transboundary water sharing can also be guaranteed by 'multilateral treaty' instead of 'bilateral treaty' to bring out an effective solution in terms of safeguarding equitable and sustainable transboundary water governance. The human sociality concept has a particular implication in this regard too. The focus is given to multilateral treaty by arguing that all countries need to recognise first that Transboundary Rivers are common resources and then, they would collectively make decisions about the fair share of the resources.

The marine resources of Bangladesh have a great potential to contribute to its economy, but there are several drawbacks to the current governance regime regarding the coastal and marine ecosystem. Most of the resources remain underutilised so far mainly because of technological incapacity. However, the existing nature of the utilisation of the minimum resources gives a negative picture as they are being extracted unsustainably under a weak property rights regime and an unequal power-sharing arrangement both within the country and beyond.

Finally, the chapter argues for integrating the factors such as social norms, cooperation and human sociality into the governance framework of water. The formal and informal institutions can act together under the condition of strong regulatory regimes resulting from human cooperation. Human beings have a profound relationship with water, and this long-standing nexus is a part of evolution. Therefore, the human beings are the ones who can only contribute to conserving the water resources, and this is where the revitalisation of the inherent relation is required first to move towards a sustainable water governance regime. A recent declaration by the Bangladeshi Supreme Court to give rivers of the country their own legal rights can be regarded as a milestone in this regard. The declaration states that each river now has the right to live (Samuel, 2019). It may help create a sense of belongingness between human beings and water within the same ecosystem.

References

Abbas, B. M. (1984). *The Ganges water dispute* (2nd ed.). The University Press Limited.

Ahmed, T. (2013). Transboundary flow and water management in Bangladesh. In *Water Resources in South Asia: Conflict to Cooperation Conference Paper*.

Akter, M. (2016). Conceptualizing environmental governance on the GBM basin. *Bandung Journal of the Global South, 3*(25), 1–17. https://doi.org/10.1186/s40728-015-0025-7

Akther, H., Ahmed, M., & Rasheed, K. (2009). Spatial and temporal analysis of groundwater level fluctuation in Dhaka city, Bangladesh. *Asian Journal of Earth Sciences, 2*(2), 49–57. https://doi.org/10.3923/ajes.2009.49.57

Alam, M. F. (2011). *A background paper for Bangladesh fisheries value chain study*. Institute of Agriculture and Food Policy Studies, Universiti Putra Malaysia.

Alam, S. (2021). *Blue bond—An inventive financing solution for blue economy*. The Financial Express. https://thefinancialexpress.com.bd/views/views/blue-bond-an-inventive-financing-solution-for-blue-economy

Alavi, H. (1975). India and the colonial mode of production. In R. Miliband & J. Saville (Eds.), *The socialist register* (pp. 160–197).

Alcock, F. (2002). Bargaining, uncertainty, and property rights in fisheries. *World Politics, 54*(4), 437–461.

Ali, S. I. (1990). Haor basin eco-system. In A. A. Rahman, S. Huq, & G. R. Conway (Eds.), *Environmental aspects of surface water systems of Bangladesh* (pp. 79–92). University Press Ltd.

Allouche, J. (2016). The birth and spread of IWRM: A case study of global policy discussion and translation. *Water Alternatives, 9*(3), 412–433.

Allouche, J., & Finger, M. (2001). *Water privatisation: Trans-national corporations and the re-regulation of the water industry*. CRC Press. https://doi.org/10.4324/9780203302484

Alom, Z. (2016). *Once in a blue moon: How long can we ignore the crying needs of our enchanting haor areas?* The Daily Star. https://www.thedailystar.net/in-focus/once-blue-moon-1209553

Amarasinghe, U., Shah, T., & Malik, R. (2008). Strategic analyses of the national river linking project (NRLP) of India (series 1). In *India's water future: Scenarios and issues*. International Water Management Institute.

Amarasinghe, U. A., Muthuwatta, L., Surinaidu, L., Anand, S., & Jain, S. K. (2016). Reviving the Ganges water machine: Potential. *Hydrology and Earth System Sciences, 20*(3), 1085–1101. https://doi.org/10.5194/hess-20-1085-2016

Amin, M. A. (2018). Dhaka remains the world's most densely populated city. *The Dhaka Tribune*. https://www.dhakatribune.com/bangladesh/dhaka/2018/10/14/dhaka-remains-the-world-s-most-densely-populated-city

Anderson, T. L., & Leal, D. R. (1988). *Going with the flow: Expanding the water markets*. Cato Institute Policy Analysis No. 104.

Andreucci, D., Garcia-Lamarca, M., Wedekind, J., & Swyngedouw, E. (2017). "Value grabbing": A political ecology of rent. *Capitalism Nature Socialism, 28*(3), 28–47. https://doi.org/10.1080/10455752.2016.1278027

Anokye, N. A., & Gupta, J. (2012). Reconciling IWRM and water delivery in Ghana—The potential and the challenges. *Physics and Chemistry of the Earth, 47–48*, 33–45. https://doi.org/10.1016/j.pce.2011.06.010

Arfanuzzaman, M., & Syed, M. A. (2018). Water demand and ecosystem nexus in the transboundary river basin: A zero-sum game. *Environment, Development and Sustainability, 20*, 963–974. https://doi.org/10.1007/s10668-017-9915-y

Arias-Granada, Y., Haque, S. S., Joseph, G., & Yanez-Pagans, M. (2018). *Water and sanitation in Dhaka Slums: Access, quality, and informality in service provision*. Policy Research Working Paper 8552. The World Bank Group.

Asia Foundation. (2013). *Political economy analysis of the Teesta river basin*. The Asia Foundation.

Ayers, J. C., Goodbred, S., George, G., Fry, D., Benneyworth, L., Hornberger, G., Roy, K., Karim, M. R., & Akter, F. (2016). Sources of salinity and arsenic in groundwater in southwest Bangladesh. *Geochemical Transactions, 17*(4), 1–22. https://doi.org/10.1186/s12932-016-0036-6

BADC. (2020). *Minor irrigation survey report 2018–2019*. Bangladesh Agricultural Development Corporation, Ministry of Agriculture, Government of the People's Republic of Bangladesh, Dhaka.

Bakker, K. (2004). *An uncooperative commodity: Privatizing water in England and Wales*. Oxford University Press.

Bakker, K. (2005). Neo-liberalizing nature? Market environmentalism in water supply in England and Wales. *Annals of the Association of American Geographer, 95*(3), 542–565. https://doi.org/10.1111/j.1467-8306.2005.00474.x

Bakker, K. (2007). The "commons" versus the "commodity": Alter-globalization, anti-privation and the human rights to water in the global south. *Antipode, 39*(3), 430–455. https://doi.org/10.1111/j.1467-8330.2007.00534.x

Bakker, K. (2012). Water: Political, biopolitical, material. *Social Studies of Science, 42*(4), 616–623. https://doi.org/10.1177/0306312712441396

Balamurugan, M., Kasiviswanathanm, K. S., Ilampooranan, I., & Soundharajan, B. S. (2021). COVID-19 lockdown disruptions on water resources, wastewater, and agriculture in India. *Frontiers in Water, 3*(603531). https://doi.org/10.3389/frwa.2021.603531

Bandyopadhyay, J., & Perveen, S. (2008). The interlinking of Indian rivers: Questions on the scientific, economic and environmental dimension of the project. In M. M. Q. Mirza, A. U. Ahmed, & Q. K. Ahmad (Eds.), *Interlinking rivers in India: Issues and concern* (pp. 53–76). Taylor and Francis.

Banglapedia. (2021a). *Groundwater*. National Encyclopedia of Bangladesh.

Banglapedia. (2021b). *Trans-boundary Rivers*. National Encyclopedia of Bangladesh.

Banglapedia. (2021c). *Fisherman*. National Encyclopedia of Bangladesh.

Barlow, M., & Clarke, T. (2003). *Blue gold: The fight to stop the corporate theft of the world's water*. Stoddart.

Barua, A. (2018). Water diplomacy as an approach to regional cooperation in South Asia: A case from the Brahmaputra Basin. *Journal of Hydrology, 567*, 60–70. https://doi.org/10.1016/j.jhydrol.2018.09.056

Barua, A., Vij, S., & Rahman, M. Z. (2017). Powering or sharing water in the Brahmaputra River basin. *International Journal of Water Resources Development, 34*(5), 829–843. https://doi.org/10.1080/07900627.2017.1403892

Baten, M. A., & Titumir, R. A. M. (2016). Environmental challenges of trans-boundary water resources management: The case of Bangladesh. *Sustainable Water Resources Management, 2*(1), 13–27. https://doi.org/10.1007/s40899-015-0037-0

Bayliss, K. (2016). *The system of provision for water in selected case study countries*. Working Paper Series No. 194. Financialisation, Economy, Society and Sustainable Development (FESSUD), Large Collaborative Project, Social Sciences and Humanities, European Union.

BBS. (2015a). *Population projection of Bangladesh: Dynamics and trends 2011–2061*. Bangladesh Bureau of Statistics, Statistics and Informatics Division, Ministry of Planning, Government of the People's Republic of Bangladesh.

BBS. (2015b). *Census of slum areas and floating population-2014*. Bangladesh Bureau of Statistics, Statistics and Informatics Division, Ministry of Planning, Government of the People's Republic of Bangladesh.

Benson, D., Gain, A. K., & Rouillard, J. J. (2015). Water governance in a comparative perspective: From IWRM to a 'nexus' approach. *Water Alternatives, 8*(1), 756–773.

Berkes, F. (1989). *Common property resources: Ecology and community-based sustainable development*. Belhaven Press.

Bharati, L., & Jayakody, P. (2011). Hydrological impacts of inflow and land-use changes in the Gorai River catchment, Bangladesh. *Water International, 36*(3), 357–369. https://doi.org/10.1080/02508060.2011.586200

Bhuiyan, M. N. U., Rahman, M. M., Ali, M. A., & Selim, A. S. M. (2015). Maritime boundary confirmation of Bangladesh: Potentialities of sea resources and challenges ahead. *The Coast and Management, 43*(5), 18–24.

BIGD. (2019). *State of cities 2018: Water governance in Dhaka*. BRAC Institute of Governance and Development.

BirdLife International. (2004). *Important bird areas in Asia: Key sites for conservation*. BirdLife International (BirdLife Conservation Series No. 13).

BOBLME. (2014). BOBLME News Letter, September 2014. Bay of Bengal Large Marine Ecosystem Project. Global Environmental Facility and Food and Agriculture Organization of the United Nations.

Boelens, R., Hoogesteger, J., Swyngedouw, E., Vos, J., & Wester, P. (2016). Hydrosocial territories: A political ecology perspective. *Water International, 41*(1), 1–14. https://doi.org/10.1080/02508060.2016.1134898

Brichieri-Colombi, S., & Brandnock, R. W. (2003). Geopolitics, water and development in South Asia: Cooperative development in the Ganges-Brahmaputra delta. *The Geography Journal, 169*(1), 43–64.

BWP. (2019). *Rapid assessment of greater Dhaka groundwater sustainability*. Bangladesh Water Partnership, 2030 Water Resources Group.

Byomkesh, T., Nakagoshi, N., & Shahedur, R. M. (2009). State and management of wetlands in Bangladesh. *Landscape and Ecological Engineering, 5*, 81–90. https://doi.org/10.1007/s11355-008-0052-5

Capistrano, D., Ahmed, M., & Hossain, M. (1994). *Ecological economic and common property issues in Bangladesh's open water and floodplain fisheries*. Paper presented at the Third Biennial Meeting of the International Society for Ecological Economics, San Jose, Costarica.

Carlson, S. (2020). Nature's contributions to a food secure future: Summary of findings from the IPBES global assessment of biodiversity and ecosystem services. *Agrilinks*. Retrieved December 30, 2020, from https://agrilinks.org/post/natures-contributions-food-secure-future-summary-findings-ipbes-global-assessment-biodiversity

Cawood, S. (2017). *Collective action, service provision and urban governance: A critical exploration of Community Based Organisations (CBOs) in Dhaka's bustee (slum) settlements, Bangladesh*. Doctoral Thesis, University of Manchester, UK.

CEGIS. (2012). *Master plan of haor area*. Main report, Volume II. Center for Environmental and Geographic Information Services and Bangladesh Haor and Wetland Development Board (BHWDB), Ministry of Water Resources, Government of the People's Republic of Bangladesh.

Chan, N. W., Roy, R., & Chaffin, B. C. (2016). Water governance in Bangladesh: An evaluation of institutional and political context. *Water, 8*(9), 403, 1–18. https://doi.org/10.3390/w8090403

Chaudhuri, B. B. (1983). Eastern India. In K. Dharma & D. Meghnad (Eds.), *The Cambridge economic history of India* (Vol. 2: c.1757–c.1970). Cambridge University Press.

Choudhury, N. B., & Dey Choudhury, S. R. (2020). Evaluating the greenness of hydroelectric projects of Northeast India: A study with special reference to the Tipaimukh project. *Decision, 47*(3), 293–302. https://doi.org/10.1007/s40622-020-00251-9

Choudhury, N. B., & Dey Choudhury, S. R. (2021). Implications for planning of hydroelectric projects in Northeast India: An analysis of the impacts of the Tipaimukh project. *GeoJournal, 86*, 1999–2019. https://doi.org/10.1007/s10708-020-10158-8

Chowdhury, A. (1987). *A review of problems of fishing communities in Bangladesh*. Paper presented at the BARC—CIRDAP Workshop on Socio-Economic Aspect of Fishing Community in Bangladesh and Its Role in Rural Development, Dhaka.

Chowdhury, S. R. (2017). *Maritime provinces of Bangladesh*, 2nd edn. (Map). University of Chittagong.

Clement, F., Suhardiman, D., & Bharati, L. (2017). IWRM discourses, institutional Holy Grail and water justice in Nepal. *Water Alternatives, 10*(3), 870–887.

Conca, K. (2006). *Governing water: Contentious transnational politics and global institution building*. MIT Press.

Crow, B., Lindquist, A., & Wilson, D. (1995). *Sharing the Ganges: The politics and technology of river development*. University Press Limited.

Davidson, N. C., & Finlayson, C. M. (2018). Extent, regional distribution and changes in area of different classes of wetland. *Marine and Freshwater Research, 69*(10), 1525–1533. https://doi.org/10.1071/MF17377

Debnath, R. (2016). A review of the sustainability of recent watershed management programmes in Bangladesh. *Lakes & Reservoirs Research and Management, 21*, 152–161. https://doi.org/10.1111/lre.12129

Deinet, S., et al. (2020). *The living planet index (LPI) for migratory freshwater fish.* Technical Report. World Fish Migration Foundation, The Netherlands.

Dinar, A., & Subramanian, A. (1997). *Water pricing experiences: An international perspective.* World Bank Technical Paper No. 386. IRBD, The World Bank.

Dinar, A., Pochat, V., & Albiac-Murillo, J. (Eds.). (2015). *Water pricing experiences and innovations.* Springer International Publishing.

Dinar, A., Rosegrant, M. W., & Meinzen-Dick, R. S. (1997). *Water allocation mechanisms: Principles and examples.* Policy Research Working Paper Series 1779. The World Bank.

DoE. (2015). *Fifth national report to the convention on biological diversity: Biodiversity national assessment 2015.* Department of Environment, Ministry of Environment and Forests, Government of the People's Republic of Bangladesh.

DoF. (2014). *Yearbook of fisheries statistics of Bangladesh, 2013–14.* Fisheries Resources Survey System (FRSS). Department of Fisheries, Ministry of Fisheries and Livestock. Government of the People's Republic of Bangladesh.

DoF. (2019). *Yearbook of fisheries statistics of Bangladesh, 2018–19.* Fisheries Resources Survey System (FRSS). Dhaka: Department of Fisheries, Ministry of Fisheries and Livestock, Government of the People's Republic of Bangladesh.

DPHE. (2020). *Bangladesh strategic paper to respond to water, sanitation and hygiene (WASH) issues during and after the COVID-19 outbreak.* The Department of Public Health Engineering, Ministry of Local Government, Rural Development & Cooperatives, Government of the People's Republic of Bangladesh.

DWASA. (2005). *Dhaka water supply and sewerage authority annual report 2004–05.* Dhaka Water Supply and Sewerage Authority, Ministry of Local Government, Rural Development and Co-operatives, Government of the People's Republic of Bangladesh.

DWASA. (2010). *Dhaka water supply and sewerage authority annual report 2009–10.* Dhaka Water Supply and Sewerage Authority, Ministry of Local Government, Rural Development and Co-operatives, Government of the People's Republic of Bangladesh.

DWASA. (2014). *Dhaka water supply and sewerage authority annual report 2013–14.* Dhaka Water Supply and Sewerage Authority, Ministry of Local Government, Rural Development and Co-operatives, Government of the People's Republic of Bangladesh.

DWASA. (2017). *Dhaka water supply and sewerage authority annual report 2016–17.* Dhaka Water Supply and Sewerage Authority, Ministry of Local Government, Rural Development and Co-operatives, Government of the People's Republic of Bangladesh.

DWASA. (2020). *Dhaka water supply and sewerage authority annual report 2018–2019.* Dhaka Water Supply and Sewerage Authority, Ministry of Local Government, Rural Development and Co-operatives, Government of the People's Republic of Bangladesh.

DWASA & IWM. (2014). *Water supply master plan for Dhaka city.* Final Report, Vol. 1 of 5. Dhaka Water Supply and Sewerage Authority, Ministry of Local Government, Rural Development and Co-operatives and Institute for Water Modelling.

Faisal, I. M. (2002). Managing common waters in the Ganges-Brahmaputra-Meghna. *SAIS Review, 22*(2), 309–327.

FAO. (2011a). *AQUASTAT transboundary River basin overview: Ganges-Brahmaputra-Meghna.* Food and Agriculture Organization of the United Nations.

FAO. (2011b). *AQUASTAT transboundary River basin overview: Indus basin.* Food and Agriculture Organization of the United Nations.

FAO. (2014a). *AQUASTAT country profile—Bangladesh.* Food and Agriculture Organization of the United Nations.

FAO. (2014b). *The state of world fisheries and aquaculture (opportunities and challenges).* Food and Agricultural Organization of the United Nations.

FAO. (2020). *The state of world fisheries and aquaculture: Sustainability in action.* Food and Agricultural Organization of the United Nations.

Farooque, M. (2004). *Environmental order: The security of survival.* University Press Limited.

Figueres, C., Rockstrom, J., & Tortajada, C. (Eds.). (2003). *Rethinking water management: Innovative approaches to contemporary issues.* Earthscan.

Finger, M., Tamiotti, L., & Allouche, J. (2006). Introduction: Conceptual elements. In M. Finger, L. Tamiotti, & J. Allouche (Eds.), *The multi-governance of water* (pp. 1–42). State University of New York Press.

Fridman, A. (2015). Water pricing reform analysis: Alternative scenarios. *Journal of Economic Policy Reform, 18*(3), 258–266. https://doi.org/10.1080/17487870.2015.1057510

Furlong, K. (2010). Neoliberal water management: Trends, limitations, reformulations. *Environment and Society: Advances in Research, 1*, 46–75. https://doi.org/10.3167/ares.2010.010103

Gadhwal, A., & Lal, P. (2008). Abolition of Zamindari system in Rajasthan. Retrieved August 11, 2020 from https://www.legalserviceindia.com/article/l218-Abolition-of-Zamindari-System-in-Rajasthan.html

Gandy, M. (1997). The making of a regulatory crisis: Restructuring New York city's water supply. *Transactions of the Institute of British Geographers, 22*(3), 338–358. https://doi.org/10.1111/j.0020-2754.1997.00338.x

GED. (2020). *8th five year plan (July 2020–June 2025): Promoting prosperity and fostering inclusiveness.* General Economic Division, Bangladesh Planning Commission, Government of the People's Republic of Bangladesh.

Gilman, T., Abell, R., & Williams, C. E. (2004). How can conservation biology inform the practice of Integrated River Basin Management? *International Journal of River Basin Management, 2*, 135–148. https://doi.org/10.1080/15715124.2004.9635228

Glassman, J. (2006). Primitive accumulation, accumulation by dispossession, accumulation by 'extra-economic' means. *Progress in Human Geography, 30*(5), 608–625. https://doi.org/10.1177/0309132506070172

Gondinez-Madrigal, J., Cauwenbergh, N. V., & van der Zaag, P. (2019). Production of competing water knowledge in the face of water crises: Revisiting the IWRM success story of the Lerma-Chapala Basin, Mexico. *Geoforum, 103*, 3–15. https://doi.org/10.1016/j.geoforum.2019.02.002

Grafton, R. Q., Chu, L., Kompas, T., & Ward, M. (2014). Volumetric water pricing, social surplus and supply augmentation. *Water Resources and Economics, 6*, 74–87. https://doi.org/10.1016/j.wre.2014.07.001

Griffin, R. C. (2001). Effective water pricing. *Journal of the American Water Resources Association, 37*(5), 1335–1347. https://doi.org/10.1111/j.1752-1688.2001.tb03643.x

Griffin, R. C. (2006). *Water resource economics: The analysis of scarcity, policies, and projects.* The MIT Press.

Gupta, M. D. (2021). *After nearly 40 years, India's first river-linking project Ken-Betwa could finally begin.* The Print. https://theprint.in/india/governance/after-nearly-40-years-indias-first-river-linking-project-ken-betwa-could-finally-begin/626393/

Habib, K. A., & Islam, M. J. (2020). An updated checklist of marine fishers of Bangladesh. *Bangladesh Journal of Fisheries, 32*(2), 357–367. https://doi.org/10.52168/bjf.2020.32.40

Haftendorn, H. (2000). Water and international conflict. *Third World Quarterly, 21*(1), 51–68. https://doi.org/10.1080/01436590013224

Hanasz, P. (2014). Sharing waters versus sharing rivers: The 1996 Ganges Treaty. In *Global water forum.* Retrieved July 30, 2021 from https://globalwaterforum.org/2014/07/28/sharing-waters-vs-sharing-rivers-the-1996-ganges-treaty/

Haq, T. (2013). A violent threat to conservation of wetlands and the existing laws of Bangladesh: A critical analysis. *Philosophy and Progress, LIII–LIV*.https://doi.org/10.3329/pp.v53i1-2.21950

Haque, M. (2019). Urban water governance: Pricing of water for the slum dwellers of Dhaka metropolis. In B. Ray & R. Shaw (Eds.), *Urban drought: Emerging water challenges in Asia* (pp. 385–397). Springer.

Haque, M. I. (2008). *Water resources management in Bangladesh.* Anushilan.

Haque, M. I., & Basak, R. (2017). Land cover change detection using GIS and remote sensing techniques: A spatio-temporal study on Tanguar Haor, Sunamganj, Bangladesh. *The Egyptian Journal of Remote Sensing and Space Sciences, 20*, 251–263. https://doi.org/10.1016/j.ejrs.2016.12.003

Hardin, G. (1968). The tragedy of the commons. *Science, 162*(3859), 1243–1248. https://doi.org/10.1126/science.162.3859.1243

Haren, N. V. (Ed.). (2007). *Water market and market mechanisms.* Both Ends Working Paper Series. Both ENDS, Amsterdam.

Harris, D., Kooy, M., & Jalloh, G. (2012). *The political economy of the urban water-pricing regime in Freetown, Sierra Leone.* Working Paper No. 348. Overseas Development Institute, UK.

Harvey, D. (2003). *The new imperialism.* Oxford University Press.

Harvey, D. (2004). The 'new' imperialism: Accumulation by dispossession. *Socialist Register, 40*, 63–87.

Hassan, A. B. M. E. (2019). Indian hegemony on water flow of the Ganges: Sustainability challenges in the southwest part of Bangladesh. *Sustainable Future, 1*(100002). https://doi.org/10.1016/j.sftr.2019.100002

Hassan, M. Q., & Zahid, A. (2017). Management of overexploited Dhaka city aquifer, Bangladesh. *Journal of Nepal Geological Society, 43*, 277–283. https://doi.org/10.3126/jngs.v43i0.25615

Haughton, G. (2002). Market making: Internationalisation and global water markets. *Environment and Planning A: Economy and Space, 3*(5), 791–807. https://doi.org/10.1068/a3426

He, F., Zarfl, C., Bremerich, V., David, J. N. W., Hogan, Z., Kalinkat, G., Tockner, K., & Jähnig, S. C. (2019). The global decline of freshwater megafauna. *Global Change Biology, 25*(11), 3883–3892. https://doi.org/10.1111/gcb.14753

Heynen, N., McCarthy, J., Prudham, S., & Robbins, P. (2007). *Neoliberal environments: False promises and unnatural consequences.* Routledge.

Hoq, M. E., Haroon, A. K. Y., & Chakraborty, S. C. (2013). *Marine fisheries of Bangladesh: Prospect and potentialities.* Support to Sustainable Management of the BOBLME Project, Bangladesh Fisheries Research Institute, Bangladesh, 120p.

Hossain, M. A. R., Nahiduzzaman, M., Sayeed, M. A., Azim, M. E., Wahab, M. A., & Olin, P. G. (2009). The Chalan beel in Bangladesh: Habitat and biodiversity degradation, and implications for future management. *Lakes & Reservoirs: Science, Policy and Management for Sustainable Use, 14*(1), 3–19. https://doi.org/10.1111/j.1440-1770.2009.00387.x

Huda, A. T. M. S. (2003). *Fishing in muddy waters: Policy processes for inland fisheries in Bangladesh.* Community Based Fisheries Management Project (CBFM-2) No. 2003/01. World Fish Center Bangladesh.

Huda, M. S. (2017). Envisioning the future of cooperation on common rivers in South Asia: A cooperative security approach by Bangladesh and India to the Tipaimukh Dam. *Water International, 42*(1), 54–72. https://doi.org/10.1080/02508060.2016.1236232

Huda, M. S., & Ali, S. H. (2018). Environmental peacebuilding in South Asia: Establishing consensus on hydroelectric projects in the Ganges-Brahmaputra-Meghna (GBM) Basin. *Geoforum, 96*, 160–171. https://doi.org/10.1016/j.geoforum.2018.08.015

Hukka, J. J., & Katko, T. S. (2003). Refuting the paradigm of water services privatization. *Natural Resources Forum, 27*(2), 142–155. https://doi.org/10.1111/1477-8947.00049

Huq, N., Pedroso, R., Bruns, A., Ribbe, L., & Huq, S. (2020). Changing dynamics of livelihood dependence on ecosystem services at temporal and spatial scales: An assessment in the southern wetland areas of Bangladesh. *Ecological Indicators, 110*(105855). https://doi.org/10.1016/j.ecolind.2019.105855

Hussain, M. (2019). Hilsa Shad fishery success in Bangladesh. *Agrilinks.* Retrieved June 15, 2021, from https://agrilinks.org/post/hilsa-shad-fishery-success-bangladesh

IBRD/Word Bank. (2004). The institutional economics of water: A cross-country analysis of institutions and performance. In R. M. Saleth & A. Dinar (Eds.), *The International Bank for reconstruction and development/World Bank*. Edward Elgar.

Iftekhar, M. S., & Islam, M. R. (2022). How to transform Dhaka into a water sensitive city? *Urban Water Journal*, 1–13. https://doi.org/10.1080/1573062x.2022.2026983

Immerzeel, W. W., Ludovicus, P. H., & Marc, F. P. B. (2010). Climate change will affect the Asian water towers. *Science, 328*(5984), 1382–1385. https://doi.org/10.1126/science.1183188

India-WRIS. (2012). *River basin atlas of India*. India Water Resources Information System, Ministry of Water Resources, Government of India.

IPBES. (2015). *Wetland extent index*. Intergovernmental Platform on Biodiversity and Ecosystem Services. Retrieved April 10, 2017, from http://www.ipbes.net/sites/default/files/Metadata_UNEPWCMC_Wetland_Extent_Trend_Index.pdf

IPBES. (2019). Summary for policymakers of the global assessment report on biodiversity and ecosystem services. In S. Díaz, et al. (Eds.), *Intergovernmental science-policy platform on biodiversity and ecosystem services*. IPBES Secretariat.

IPCC. (2019). Summary for policymakers: IPCC special report on the ocean and cryosphere in a changing climate. In H.- O. Pörtner, D. C. Roberts, V. Masson-Delmotte, P. Zhai, M. Tignor, E. Poloczanska, K. Mintenbeck, A. Alegría, M. Nicolai, A. Okem, J. Petzold, B. Rama & N.M. Weyer (Eds.). In press.

Islam, M. R., Begum, S. F., Yamaguchi, Y., & Ogawa, K. (1999). The Ganges and Brahmaputra rivers in Bangladesh: Basin denudation and sedimentation. *Hydrological Processes, 13*(17), 2907–2923. https://doi.org/10.1002/(SICI)1099-1085(19991215)13:17%3c2907::AID-HYP906%3e3.0.CO;2-E

Islam, M. M., Shamsuzzaman, M. M., Mozumder, M. M. H., Xiangmin, X., Ming, Y., & Jewel, M. A. S. (2017). Exploitation and conservation of coastal and marine fisheries in Bangladesh: Do the fishery laws matter? *Marine Policy, 76*, 143–151. https://doi.org/10.1016/j.marpol.2016.11.026

Islam, M. S., & Islam, F. F. (2017). *Spatial disparity of groundwater depletion in Dhaka city*. Paper for 15th International Conference on Environmental Science and Technology, 31 August–2 September, Rhodes, Greece.

Islam, S. (1985). *Villages in the haor-basin of Bangladesh*. Studies in Socio-Cultural Change in Rural Villages in Bangladesh, No. 4. Institute for the Study of Languages and Cultures of Asia and Africa. Tokyo University of Foreign Studies.

Islam, S. M. (2021). *Groundwater depletion in Dhaka: Too little too late?* The Daily Star. https://www.thedailystar.net/shout/news/groundwater-depletion-dhaka-too-little-too-late-2064933

Islam, S. N. (2010). Threatened wetlands and ecologically sensitive ecosystems management in Bangladesh. *Frontiers of Earth Science in China, 4*(4), 438–448. https://doi.org/10.1007/s11707-010-0127-0

IUCN. (2014a). *River beyond borders: India Bangladesh trans-boundary river atlas*. International Union for Conservation of Nature and Drik Publication.

IUCN. (2014b). *Physical assessment of the Brahmaputra River ecosystems for life: A Bangladesh-India initiative*. International Union for Conservation of Nature and Jagriti Prokashony.

IUCN. (2021). Meghna knowledge forum (MKF) 2021, 22–24 June, Bangladesh and India. Retrieved August 28, 2021 from https://www.iucn.org/regions/asia/about/meghna-knowledge-forum-mkf-2021

IUCN Bangladesh. (2015). National framework for establishing and managing marine protected areas (MPAs) in Bangladesh. International Union for Conservation of Nature. Bangladesh Country Office, pp. xi+106.

IWM. (2005). *Hydrological impact study of Tipaimukh dam of India on Bangladesh*. Institute of Water Modelling.

IWM. (2008). *Resource assessment and monitoring of water supply sources for Dhaka city*. Institute of Water Modelling.

Iyer, R. R. (2007). *Towards water wisdom: Limits, justice, harmony*. Sage Publication.

Johnston, B. R. (2003). The political ecology of water: An introduction. *Capitalism Nature Sociology, 14*(3), 73–90. https://doi.org/10.1080/10455750308565535

JRC Bangladesh. (2011). *Basin map of the Ganges, the Brahmaputra and the Meghna river*. Joint Rivers Commission Bangladesh.

JRC Bangladesh. (2021). *Information related to Ganges water sharing*. Joint River Commission.

Kahil, M. T., Dinar, A., & Albiac, J. (2016). Cooperative water management and ecosystem protection under scarcity and drought in arid and semiarid regions. *Water Resources and Economics, 13*, 60–74. https://doi.org/10.1016/j.wre.2015.10.001

Kalair, A. R., Abas, N., Hasan, Q. U., Kalair, E., Kalair, A., & Khan, N. (2019). Water, energy and food nexus of Indus water treaty: Water governance. *Water-Energy Nexus, 2*, 10–24. https://doi.org/10.1016/j.wen.2019.04.001

Kawser, M. A., & Samad, M. A. (2016). Political history of Farakka Barrage and its effects on environment in Bangladesh. *Bandung: Journal of the Global South, 3*(1), 1–14. https://doi.org/10.1186/s40728-015-0027-5

Keddy, P. A., Fraser, L. H., Solomeshch, A. I., Junk, W. J., Campbell, D. R., Arroyo, M. T. K., & Alho, C. J. R. (2009). Wet and wonderful: The world's largest wetlands are conservation priorities. *BioScience, 59*(1), 39–51. https://doi.org/10.1525/bio.2009.59.1.8

Kelly, A. B. (2011). Conservation practice as primitive accumulation. *The Journal of Peasant Studies, 38*(4), 683–701. https://doi.org/10.1080/03066150.2011.607695

Khaki, M., Forootan, E., Kuhn, M., Awange, J., Papa, F., & Shum, C. K. (2018). A study of Bangladesh's sub-surface water storages using satellite products and data assimilation scheme. *Science of the Total Environment, 625*, 963–977. https://doi.org/10.1016/j.scitotenv.2017.12.289

Khan, A. K. M. F., Mustafa, M. G., & Naser, M. N. (2016). Effective supervision of inland capture fisheries of Bangladesh and its hurdles in managing the resources. *Bandung: Journal of Global South, 3*, 1–12.

Khan, M. H. (2015). *The political economy of community-based fisheries management in Bangladesh*. SOAS University of London.

Khan, S. M. M. H. (2011). *Participatory wetland resource governance in Bangladesh: An analysis of community-based experiments in Hakaluki Haor*. Doctoral Thesis. University of Manitoba.

Khan, S. M. M. H., & Haque, C. E. (2010). Wetland resource management in Bangladesh: Implications for marginalization and vulnerability of local harvesters. *Environmental Hazards, 9*(1), 54–73.

Khan, T. A. (1996). Management and sharing of the Ganges. *Natural Resources Journal, 36*(3/2), 455–479.

Khan, Z. A. (1976). *Basic documents on Farakka conspiracy*. Khoshroz Kitab Mahal.

Khandu, F. E., Schumacher, M., Awange, J. L., & Schmied, H. M. (2016). Exploring the influence of precipitation extremes and human water use on total water storage (TWS) changes in the Ganges-Brahmaputra-Meghna River Basin. *Water Resources Research, 52*, 2240–2258. https://doi.org/10.1002/2015WR018113

Khera, S. (2020). PM Modi scheme of river-linking projects in India. Retrieved 20 August, 2021, from https://www.mapsofindia.com/my-india/government/pm-modi-scheme-of-river-linking-projects-in-india

Kliot, N., Shmueli, D., & Shamir, U. (2001). Institutions for management of transboundary water resources: Their nature, characteristics and shortcomings. *Water Policy, 3*, 229–255.

Kumar, R., McInnes, R. J., Everard, M., Gardner, R. C., Kulindwa, K. A. A., Wittmer, H., & Infante, M. D. (2017). *Integrating multiple wetland values into decision-making*. Ramsar Policy Brief No. 2. Ramsar Convention Secretariat.

Kummu, M., Guillaume, H. A., de Moel, H., Eisner, S., Flörke, M., Porkka, M., Siebert, S., Veldkamp, T. I. E., & Ward, P. J. (2016). The world's road to water scarcity: Shortage and stress in the 20th century and pathways towards sustainability. *Nature Scientific Reports, 9*(6), 38495. https://doi.org/10.1038/srep38495

Lautze, J., de Silva, S., Giordano, M., & Sanford, L. (2011). Putting the cart before the horse: Water governance and IWRM. *Natural Resources Forum, 35*, 1–8.

Lebel, L., Garden, P., & Imamura, M. (2005). The politics of scale, position, and place in the governance of water resources in the Mekong region. *Ecology and Society, 10*(2), 18.

Lee, T. R. (1999). *Water management in the 21st century: The allocation imperative*. Edward Elgar.

Leithead, A. (2018). *The 'water war' brewing over the new River Nile dam*. BBC News. https://bbc.com/news/world-africa-43170408

Lenton, R. (2011). 1.01—Integrated water resources management. *Treatise on Water Science, 1*, 9–21. https://doi.org/10.1016/B978-0-444-53199-5.00002-6

Liu, Y. (2015). Transboundary water cooperation on the Yarlung Zangbo/Brahmaputra—A legal analysis of riparian state practice. *Water International, 40*(2), 354–374. https://doi.org/10.1080/02508060.2015.1015319

Lodrick, D. O., & Ahmad, N. (2018). *Ganges River*. Encyclopedia Britannica. Retrieved September 27, 2018, from https://www.britannica.com/place/Ganges-River

Maltby, E., & Acreman, M. C. (2011). Ecosystem services of wetlands: Pathfinder for a new paradigm. *Hydrological Sciences Journal, 56*(8), 1341–1359. https://doi.org/10.1080/02626667.2011.631014

Mannan, S., Nilsson, H., Johansson, T., & Schofield, C. (2020). Enabling stakeholder participation in marine spatial planning: The Bangladesh experience. *Journal of the Indian Ocean Region, 16*(3), 268–291. https://doi.org/10.1080/19480881.2020.1825200

Mansfield, B. (2004). Neoliberalism in the oceans: "Rationalization", property rights, and the commons question. *Geoforum, 35*(3), 313–326. https://doi.org/10.1016/j.geoforum.2003.05.002

McCarthy, J. (2005). Commons as counter-hegemonic projects. *Capitalism Nature Socialism, 16*(1), 9–24. https://doi.org/10.1080/1045575052000335348

McCarthy, J., & Prudham, S. (2004). Neoliberal nature and the nature of neoliberalism. *Geoforum, 35*(3), 275–283. https://doi.org/10.1016/j.geoforum.2003.07.003

McDonald, D., & Ruiters, G. (2005). *The age of commodity: Water privatization in Southern Africa*. Earthscan.

MEA. (2005). *Ecosystems and human well-being: Wetlands and water synthesis*. A Report of the Millennium Ecosystem Assessment. World Resources Institute.

Mehta, L. (2001). Water, difference, and power: Unpacking notions of water "users" in Kutch, India. *International Journal of Water, 1*(3–4). https://doi.org/10.1504/IJW.2001.002070

Mehta, L. (2003). Problems of publicness and access rights: Perspectives from the water domain. In I. Kaul (Ed.), *Providing global public goods: Managing globalization* (pp. 556–575). Oxford University Press and United Nations Development Program.

Menard, C., & Saleth, R. M. (2012). The effectiveness of alternative water governance arrangements. In M. Young (Ed.), *Towards a green economy* (pp. 152–174). United Nations Environmental Programme & Routledge.

Miner, M., Patankar, G., Gamkhar, S., & Eaton, D. J. (2009). Water sharing between India and Pakistan: A critical evaluation of the Indus Water Treaty. *Water International, 34*(2), 204–216. https://doi.org/10.1080/02508060902902193

Mirumachi, N., & Allan, J. A. (2007). Revisiting transboundary water governance: Power, conflict, cooperation and the political economy. In *International Conference on Adaptive and Integrated Water Management, 12–15 November, Basel, Switzerland*.

Mirza, M. M. Q. (1998). Diversion of the Ganges water at Farakka and its effect on salinity in Bangladesh. *Environmental Management, 22*(5), 711–722. https://doi.org/10.1007/s002679900141

Mirza, M. M. Q. (Ed.). (2005). *The Ganges water diversion: Environmental effects and implications*. Springer. https://doi.org/10.1007/1-4020-2480-0

Misra, A. K., Saxena, A., Yaduvanshi, M., Mishra, A., Bhadauriya, Y., & Thakur, A. (2007). Proposed river-linking project of India: A boon or bane to nature. *Environ Geology, 51*, 1361–1376. https://doi.org/10.1007/s00254-006-0434-7

MoF. (2020). *Bangladesh economic review 2020*. Finance Division, Ministry of Finance, Government of the People's Republic of Bangladesh.

MoFA. (2014). *Press release: Press statement of the hon'ble Foreign Minister on the verdict of the arbitral tribunal/PCA*. Ministry of Foreign Affairs, Government of the People's Republic of Bangladesh.

Mojid, M. A., Parvez, M. F., Mainuddin, M., & Hodgosn, G. (2019). Water table trend—A sustainability status of groundwater development in north-west Bangladesh. *Water, 11*(6), 1182. https://doi.org/10.3390/w11061182

MoL. (2015). *Annual report 2014–2015*. Ministry of Land, Government of the People's Republic of Bangladesh.

MoL. (2017). *Annual report 2016–2017*. Ministry of Land, Government of the People's Republic of Bangladesh.

MoL. (2020). *Annual report 2019–2020*. Ministry of Land, Government of the People's Republic of Bangladesh.

MoLGRDC. (2011). *Sector development plan (FY2011–2025): Water supply and sanitation sector of Bangladesh*. Ministry of Local Government, Rural Development and Cooperatives, Government of the People's Republic of Bangladesh.

Molle, F., Mollinga, P. P., & Meinzen-Dick, R. (2008). Water, politics and development: Introducing water alternatives. *Water Alternatives, 1*(1), 1–6.

Morgan, B. (2004). The regulatory face of the human right to water. *Journal of Water Law, 15*(5), 179–186.

Morshed, M. S. (2006). *A study on the performance evaluation of Teesta barrage project*. MS Thesis. Department of Irrigation and Water Management, Bangladesh Agricultural University, Mymensingh, Bangladesh.

MoWR. (2016). *Classification of wetlands of Bangladesh: Department of Bangladesh Haor and Wetlands Development*. Ministry of Water Resources, Government of the People's Republic of Bangladesh.

MPO. (1991). *National water plan phase II* (Vol. I & II). Master Plan Organization, Ministry of Water Resources, Government of the People's Republic of Bangladesh.

Mucke, P. (2019). Water worldwide: Scarcity versus abundance. In S. J. Day, T. Foster, J. Himmelsbach, M. L. Korte, P. Mucke, K. Radtke, P. Thielbörger & D. Weller (Eds.), *World risk report 2019* (pp. 9–15). Bündnis Entwicklung Hilft and Ruhr University Bochum—Institute for International Law of Peace and Armed Conflict (IFHV).

Mukherjee, R. (1974). *The rise and fall of the East India Company*. Monthly Review Press.

Murshed-e-Jahan, K., Belton, B., & Viswanathan, K. K. (2014). Communication strategies for managing coastal fisheries conflicts in Bangladesh. *Ocean & Coastal Management, 92*, 65–73. https://doi.org/10.1016/j.ocecoaman.2014.01.003

Mutua, M. (2002). *Human rights: A political and cultural critique*. University of Pennsylvania Press.

Nahar, N., Sultana, N., & Mia, J. (2017). Seasonal land cover changes and its effects on essential services of haor and non-haor areas of Kishoreganj district, Bangladesh. *Asia-Pacific Journal of Regional Science, 2*, 399–429. https://doi.org/10.1007/s41685-017-0067-8

Nishat, A., & Faisal, I. M. (2000). An assessment of the institutional mechanisms for water negotiations in the Ganges-Brahmaputra-Meghna system. *International Negotiation, 5*(2), 289–310. https://doi.org/10.1163/15718060020848776

North, D. C. (1991). Institutions. *The Journal of Economic Perspectives, 5*(1), 97–112. https://doi.org/10.1257/jep.5.1.97

O'Connor, J. (1996). The second contradiction of capitalism. In T. Benton (Ed.), *The greening of Marxism* (pp. 197–221). Guilford.

Olmstead, S. M. (2010). The economics of managing scarce water resources. *Review of Environmental Economics and Policy, 4*(2), 179–198. https://doi.org/10.1093/reep/req004

Olmstead, S. M., & Stavins, R. N. (2007). *Managing water demand: Price versus non-price conservation programs*. White Paper No. 39. Pioneer Institute for Public Policy Research.

Ostrom, E. (1990). *Governing the commons: The evolution of institutions for collective action*. Cambridge University Press.

Ostrom, E. (2005). *Understanding institutional diversity*. Princeton University Press.

Ostrom, E., Gardner, R., & Walker, J. (1994). *Rules, games, and common-pool resources*. The University of Michigan Press.

Parajuli, U., Miah, M., Rahman, K., Hamid, S., Mukherjee, S., & Verghese, G. (2003). *Water sharing conflicts between countries and approaches to resolving them*. Water and Security in South Asia (WASSA) Project Reports, Volume 3. SAIS, John Hopkins University and GEE-21.

Parker, D., & Kirkpatrick, C. (2005). Privatisation in developing countries: A review of the Evidence and the policy lessons. *Journal of Development Studies, 41*(4), 513–541. https://doi.org/10.1080/00220380500092499

Parua, P. K. (2010). *The Ganga: Water use in the Indian subcontinent*. Springer, Dordrecht Heidelberg.

Parvin, M. T., & Akteruzzaman, M. (2012). Factors affecting farm and non-farm income of haor inhabitants of Bangladesh. *Progressive Agriculture, 23*(1–2), 143–150. https://doi.org/10.3329/pa.v23i1-2.16578

Patil, P. G., Virdin, J., Colgan, C. S., Hussain, M. G., Failler, P., & Vegh, T. (2018). *Toward a blue economy: A pathway for sustainable growth in Bangladesh*. The World Bank Group.

Penn, J. W. (1983). *An assessment of potential yield from the offshore demersal shrimp and fish stock in Bangladesh water (including comments on the trawl fishery 1981–1982)*. FE: DP/BGD/81/034, Field Doc. 22 p. Fishery Advisory Service (Phase II) Project, Food and Agriculture Organization of the United Nations, Rome.

Piesse, M. (2020). *Global water supply and demand trends point towards rising water insecurity*. Strategic Analysis Paper. Future Directions International.

Pokrant, B., Reeves, P., & McGuire, J. (1997). Riparian rights, the organization of work and market relations among the inland fishers of colonial Bengal, 1793–1950. In C. Tsai & M. Y. Ali (Eds.), *Openwater fisheries of Bangladesh* (pp. 27–48). University Press Ltd.

Porras, I., Mohammed, E. Y., Ali, L., Ali, M. S., & Hossain, M. B. (2017). Power, profits and payments for ecosystem services in Hilsa fisheries in Bangladesh: A value chain analysis. *Marine Policy, 84*, 60–68. https://doi.org/10.1016/j.marpol.2017.06.031

Pradhan, N. S., Das, P. J., Gupta, N., & Shrestha, A. B. (2021). Sustainable management options for healthy rivers in South Asia: The case of Brahmaputra. *Sustainability, 13*, 1087. https://doi.org/10.3390/su13031087

Pratt, L., & Montgomery, W. (1997). Green imperialism: Pollution, penitence and profit. *Socialism Register, 33*, 75–95.

Priscoli, J. D., & Wolf, A. T. (2009). *Managing and transforming water conflicts*. Cambridge University Press.

Prudham, W. S. (2004). Poisoning the well: Neoliberalism and the contamination of municipal water in Walkerton, Ontario. *Geoforum, 35*(3), 343–359. https://doi.org/10.1016/j.geoforum.2003.08.010

Qureshi, A. S., Ahmed, Z. U., & Krupnik, T. J. (2014). *Groundwater management in Bangladesh: An analysis of problems and opportunities*. Cereal Systems Initiative for South Asia Mechanization and Irrigation (CSISA-MI) Project, Research Report No. 2. CIMMYT.

Rahaman, M. M. (2006). The Ganges water conflict: A comparative analysis of 1977 Agreement and 1996 Treaty. *Asteriskos: Journal of International and Peace Studies, 1*(2), 195–208.

Rahaman, M. M. (2012). Water wars in 21st century: Speculation or reality? *International Journal of Sustainable Society, 4*(1/2), 3–10.

Rahaman, M. M., & Ahmed, T. S. (2016). Affordable water pricing for slums dwellers in Dhaka metropolitan area: The case of three slums. *Journal of Water Resource Engineering and Management, 3*(1), 15–33.

Rahman, A. (1987). *The new fisheries management policy and its effects on fishing Community*. Paper presented at the BARC—CIRDAP Workshop on Socio-Economic Aspect of Fishing Community in Bangladesh and Its Role in Rural Development, Dhaka.

Rahman, K. S., Islam, Z., Navera, U. K., & Ludwig, F. (2019). A critical review of the Ganges Water Sharing arrangement. *Water Policy, 21*(2), 259–276. https://doi.org/10.2166/wp.2019.164

Rahman, M. M. (1995). *Wetlands and biodiversity: A case study of common property resources in Bangladesh*. Paper Presented at the Fifth Annual Common Property Conference of the International Association for the Study of Common Property, 24–28 May, Bodo, Norway.

Ramsar Convention Secretariat. (2016). *An introduction to the Ramsar Convention on Wetlands. Ramsar Handbook* (5th ed.). Ramsar Convention Secretariat.

Rashid, H. (2014). *Teesta water sharing: Some hard facts*. The Daily Star. http://www.thedailystar.net/teesta-water-sharing-some-hard-facts-45037

Rasul, G. (2015). Water for growth and development in the Ganges, Brahmaputra, and Meghna basins: An economic perspective. *International Journal of River Basin Management, 13*(3), 387–400. https://doi.org/10.1080/15715124.2015.1012518

Ray, R. (1974). Land transfer and social change under the permanent settlement: A study of two localities. *Indian Economic & Social History Review, 11*(1), 1–45.

Ray, R. (1979). *Change in Bengal agrarian society 1760–1850*. Manohar.

Ray, R., & Ray, R. (1973). The dynamics of continuity in rural Bengal under the British imperium: A study of quasi-stable equilibrium in underdeveloped societies in a changing world. *Indian Economic & Social History Review, 10*(2), 103–128.

Ray, R., & Ray, R. (1975). Zamindars and Jotedars: A study of rural politics in Bengal. *Modern Asian Studies, 9*(1), 81–102.

Rogers, P., de Silva, R., & Bhatia, R. (2002). Water is an economic good: How to use prices to promote equity, efficiency, and sustainability. *Water Policy, 4*(1), 1–17. https://doi.org/10.1016/S1366-7017(02)00004-1

Roy, D., Barr, J., & Venema, H. D. (2011). *Ecosystem approaches in Integrated Water Resources Management: A review of Transboundary River basins*. International Institute for Sustainable Development and UNEP-DHI Centre for Water and Environment.

Saetre, R. (1981). *Survey of the marine fish resources of Bangladesh Nov–Dec 1979 and May 1980*. Report on surveys in the R.V. Dr. Fridtjof Nansen. Institute of Marine Research.

Saleth, R. M. (2018). The institutional economics of water. *Water Economics and Policy, 4*(3), 1802003. https://doi.org/10.1142/S2382624X18020034

Samuel, S. (2019). This country gave all its rivers their own legal rights. Retrieved July 30, 2021, from https://www.vox.com/future-perfect/2019/8/18/20803956/bangladesh-rivers-legal-personhood-rights-nature

Sanaullah, M. (2021). *Blue economy potentials for Bangladesh*. The Daily Sun. https://www.daily-sun.com/arcprint/details/557404/Blue-Economy-Potentials-for-Bangladesh/2021-06-09

Sarfraz, H. (2013). Revisiting the 1960 Indus waters treaty. *Water International, 38*(2), 204–216. https://doi.org/10.1080/02508060.2013.784494

Sarker, S., Bhuyan, M. A. H., Rahman, M. M., Islam, M. A., Hossain, M. S., Basak, S. C., & Islam, M. M. (2018). From science to action: Exploring the potentials of blue economy for enhancing economic sustainability in Bangladesh. *Ocean and Coastal Management, 157*, 180–192. https://doi.org/10.1016/j.ocecoaman.2018.03.001

Sayeed, A., Rahman, M. H., Bundschuh, J., Herath, I., Ahmed, F., Bhattacharya, P., Tariq, M. R., Rahman, F., Joy, M. T. I., Abid, M. T., Saha, N., & Hasan, M. T. (2021). Handwashing with soap: A concern for overuse of water amidst the COVID-19 pandemic in Bangladesh. *Groundwater for Sustainable Development, 13*.https://doi.org/10.1016/j.gsd.2021.100561

Schlager, E., & Ostrom, E. (1992). Property rights regimes and natural resources: A conceptual analysis. *Land Economics, 68*(3), 249–262. https://doi.org/10.2307/3146375

Shamsuddoha, M., & Islam, M. M. (2017). *Bangladesh national conservation strategy: Coastal and marine resources*. Department of Forest, GoB and International Union for Conservation of Nature (IUCN).

Shamsudduha, M., Chandler, R. E., Taylor, R. G., & Ahmed, K. M. (2009). Recent trends in groundwater levels in a highly seasonal hydrological system: The Ganges-Brahmaputra-Meghna Delta. *Hydrology and Earth System Sciences, 13*(12), 2373–2385. https://doi.org/10.5194/hessd-6-4125-2009

Shamsudduha, M., Taylor, R., Ahmed, K., & Zahid, A. (2011). The impact of intensive groundwater abstraction on recharge to a shallow regional aquifer system: Evidence from Bangladesh. *Hydrogeology Journal, 19*, 901–916. https://doi.org/10.1007/s10040-011-0723-4

Shamsudduha, M., Joseph, G., Haque, S. S., Khan, M. R., Zahid, A., & Ahmed, K. M. U. (2019). *Multi-hazard groundwater risks to the drinking water supply in Bangladesh: Challenges to achieving the sustainable development goals*. Policy Research Working Paper. No. 8922. The World Bank Group.

Shamsuzzaman, M. M., Islam, M. M., Tania, M. J., Al-Mamun, M. A., Barman, P. P., & Xu, X. (2017). Fisheries resources of Bangladesh: Present status and future direction. *Aquaculture and Fisheries*, 1–12.https://doi.org/10.1016/j.aaf.2017.03.006

Shehab, M. (2019). Bangladesh's blue economy: The bay of absent policy. *DATABD.CO*. Retrieved June 30, 2021, from https://databd.co/stories/bangladeshs-blue-economy-the-bay-of-absent-policy-1652

Shermin, N., & Rahaman, S. N. (2021). Assessment of sanitation service gap in urban slums for tackling COVID-19. *Journal of Urban Management, 10*(3), 230–241. https://doi.org/10.1016/j.jum.2021.06.003

Shiva, V. (2002). *Water wars: Privatization, pollution and profit*. Pluto Press.

Shivakumar, B. (2020). COVID-19 and water. *Stochastic Environmental Research and Risk Assessment., 35*, 531–534. https://doi.org/10.1007/s00477-020-01837-6

Shopan, A. A., Islam, A. K. M. S., Dey, N. C., & Bala, S. K. (2013). *Estimation of the changes of wetlands in the northwest region of Bangladesh using Landsat images*. Paper for 4th International Conference on Water and Flood Management (ICWFM), 9–11 March, Dhaka, Bangladesh.

Sinha, V., Glémet, R., & Mustafa, G. (2018). IUCN Bridge GBM. *Benefit sharing opportunities in the Meghna Basin. Profile and preliminary scoping study, Bangladesh and India* (vi+53pp). International Union for Conservation of Nature.

Sjödin, J., Zaeske, A., & Joyce, J. (2016). *Pricing instruments for sustainable water management*. Working Paper No. 28. SIWI.

Shukla, A. K., Sudhakar, K., & Baredar, P. (2017). Renewable energy resources in South Asian countries: Challenges, policy and recommendations. *Resource-Efficient Technologies, 3*(3), 342–346. https://doi.org/10.1016/j.reffit.2016.12.003

Soeftestad, L. T. (2000). *Riparian right and colonial might in the haors basin of Bangladesh*. Paper presented at the Conference of the International Association for the Study of Common Property (IASCP), in the Panel "Constituting the Riparian Commons", pp.1–21, May 31-June 04, Bloomington, Indiana, United States.

Sultana, P., & Thompson, P. (2010). Local institutions for floodplain management in Bangladesh and the influence of the Flood Action Plan. *Environmental Hazards, 9*(1), 26–42. https://doi.org/10.3763/ehaz.2010.SI05

Swyngedouw, E. (2005). Dispossessing H_2O: The contested terrain of water privatization. *Capitalism Nature Socialism, 16*(1), 81–98. https://doi.org/10.1080/1045575052000335384

Swyngedouw, E. (2009). The political economy and political ecology of the hydro-social cycle. *Journal of Contemporary Water Research & Education, 142*(1), 56–60. https://doi.org/10.1111/j.1936-704X.2009.00054.x

Tanzeema, S., & Faisal, I. M. (2001). Sharing the Ganges: A critical anlaysis of the water sharing treaties. *Water Policy, 3*(1), 13–28. https://doi.org/10.1016/S1366-7017(01)00004-6

Thatte, C. D. (2008). Indus waters and the 1960 treaty between India and Pakistan. In O. Varis, C. Tortajada, & A. K. Biswas (Eds.), *Management of transboundary Rivers and lakes* (pp. 165–206). Springer.

The Daily Star. (2017). Govt moves to revise Ganges barrage project. https://www.thedailystar.net/frontpage/govt-moves-revise-ganges-barrage-project-1390528

The Daily Star. (2019). Blue economy worth $6.2bn. https://www.thedailystar.net/business/news/blue-economy-worth-62bn-1704688

The Dhaka Tribune. (n.d). The hungry children of Mother Ganges: Special investigation. Retrieved September 28, 2021, from https://ganges.dhakatribune.com/

The Ganges Treaty. (1996). Treaty between the Government of Bangladesh and the Republic of India on sharing the Ganges/Ganga water at Farakka. Delhi.

Thielbörger, P. (2019). Water supply: 'Water Security' through the lens of international law and politics. In S. J. Day, T. Foster, J. Himmelsbach, M. L. Korte, P. Mucke, K. Radtke, P. Thielbörger & D. Weller (Eds.), *World risk report 2019* (pp. 17–24). Bündnis Entwicklung Hilft and Ruhr University Bochum—Institute for International Law of Peace and Armed Conflict (IFHV).

Tiwary, R. (2006). Conflicts over international waters. *Economic and Political Weekly, 41*(17), 1684–1692.

Toufique, K. A. (1997). Some observations on power and property rights in the inland fisheries of Bangladesh. *World Development, 25*(3), 457–467.

Toufique, K. A. (1998). Institutions and externalities in the inland fisheries of Bangladesh. *Land Economics, 74*(3), 409–421.

Toufique, K. A. (2005). *Community empowerment by NGOs: Experience from the fourth fisheries project in Bangladesh.* Graduate School of International Development, Discussion Paper No. 135. Nagoya University.

Trawick, P. (2003). Against the privatization of water: An indigenous model for improving existing laws and successfully governing the commons. *World Development, 31*(6), 977–996. https://doi.org/10.1016/S0305-750X(03)00049-4

TWAP. (2016). *The global transboundary river basins.* Transboundary Waters Assessment Programme. Retrieved February 27, 2018, from http://twap-rivers.org/

Ullah, H., Gibson, H., Knip, D., Zylich, K., & Zeller, D. (2014). *Reconstruction of total marine fisheries catches for Bangladesh: 1950–2010.* Working Paper 2014–15. Fisheries Centre, The University of British Columbia.

Ullah, M. (1985). Fishing rights, production relations, and profitability: A case study of Jamuna fishermen in Bangladesh. In T. Panayotou (Ed.), *Small-scale fisheries in Asia: socioeconomic analysis and policy.* International Development Research Centre.

UNESCO/UN-Water. (2019). *The United Nations world water development report 2019: Leaving no one behind.* UNESCO World Water Assessment Programme. The United Nations Educational, Scientific and Cultural Organization.

UNESCO/UN Water. (2020). *The United Nations world water development report 2020: Water and climate change.* UNESCO World Water Assessment Programme. United Nations Educational, Scientific and Cultural Organization.

UNESCO. (2022). Groundwater: Making the invisible visible. The United Nations World Water Development Report 2022. https://unesdoc.unesco.org/ark:/48223/pf0000380721

UNICEF. (2021). *Better access to safe drinking water: Quality, equitable access and sustainability.* Retrieved August 10, 2021 from https://www.unicef.org/bangladesh/en/better-access-safe-drinking-water

Unnayan Onneshan. (2011). In A. F. M. A. Uddin & M. A. Baten (Eds.), *Water supply of Dhaka city: Murky future, The issue of access and inequality.*

Valbo-Jørgensen, J. & Thompson, P. M. (2007). *Culture-based fisheries in Bangladesh: A socio-economic perspective.* FAO Fisheries Technical Paper No. 499. Food and Agriculture Organization of the United Nations.

WARPO. (2001). *National Water Management Plan, Volume 2: Main Report.* Water Resources Planning Organization, Ministry of Water Resources, Government of the People's Republic of Bangladesh.

WARPO. (2017). *News Letter, issue No. 20. A bulletin of WARPO.* Water Resources Planning Organization, Ministry of Water Resources, Government of the People's Republic of Bangladesh.

West, W. Q. B. (1973). *Fishery resources of the upper Bay of Bengal.* Indian Ocean Programme, Indian Ocean Fisheries Commission, Rome, FAO, IOFC/DEV/73/28. 28pp.

Wetzel, R. (2001). *Limnology: Lakes and river ecosystems.* Academic Press.

WHO & UNICEF. (2020). *Water, sanitation, hygiene, and waste management for SARS-CoV-2, the virus that causes COVID-19: Interim guidance.* WHO. Reference Number: WHO/2019-nCoV/IPC_WASH/2020.4.

Williamson, O. E. (2002). The theory of the firm as governance structure: From choice to contract. *Journal of Economic Perspectives, 16*(3), 171–195. https://doi.org/10.1257/089533002760278776

Williams, J. M. (2022). Murky waters: the future of transboundary water governance in South and Southeast Asia. *International Journal of River Basin Management*, 1–11. https://doi.org/10.1080/15715124.2022.2119992

Winpenny, J. (1994). *Managing water as an economic resource.* Routledge.

Wirsing, R. G., Jasparro, C., & Stoll, D. C. (2013). *International conflict over water resources in Himalayan Asia.* Palgrave Macmillan UK.

Wolf, A. T. (1999). Criteria for equitable allocations: The heart of international water conflict. *Natural Resources Forum, 23*(1), 3–30. https://doi.org/10.1111/j.1477-8947.1999.tb00235.x

Worland, S. C., Hornberger, G. M., & Goodbred, S. L. (2015). Source, transport, and evolution of saline groundwater in a shallow Holocene aquifer on the tidal deltaplain of southwest Bangladesh. *Water Resources Research, 51*(7), 5791–5805. https://doi.org/10.1002/2014WR016262

World Bank. (2000). Bangladesh: Climate change and sustainable development. Bangladesh Report No. 21104-BD, Rural Development Unit, South Asia Region, The World Bank.

World Bank. (2004). *Water resources sector strategy: Strategic directions for World Bank engagement.* The World Bank.

World Bank. (2017). What is the Blue Economy? Retrieved from February 09, 2021, from http://www.worldbank.org/en/news/infographic/2017/06/06/blue-economy

World Bank. (2018). *Indus water treaty 1960.* The World Bank Resources.

WWF. (2021a). *Water scarcity.* World Wildlife Fund for Nature. Retrieved February 27, 2021a, from https://www.worldwildlife.org/threats/water-scarcity

WWF. (2021b). *The world's forgotten fishes.* World Wildlife Fund for Nature.

Xu, T., Weng, B., Yan, D., Wang, K., Li, X., Bi, W., Li, M., Cheng, X., & Liu, Y. (2019). Wetlands of international importance: Status, threats, and future protection. *International Journal of Environmental Research and Public Health, 16*(10), 1818. https://doi.org/10.3390/ijerph16101818

Yasuda, Y., Hill, D., Aich, D., Huntjens, P., & Swain, A. (2018). Multi-track water diplomacy: Current and potential future cooperation over the Brahmaputra River Basin. *Water International, 43*(5), 642–664. https://doi.org/10.1080/02508060.2018.1503446

Young, G., Dooge, J., & Rodda, J. (1994). *Global water resources issues.* Cambridge University Press.

Zahid, A. (2015). *Groundwater management aspects in Bangladesh.* Technical Report. Center for Water and Environment.

Zetland, D. (2021). The role of prices in managing water scarcity. *Water Security, 12*(100081). https://doi.org/10.1016/j.wasec.2020.100081

Chapter 5
Climate Change: Equity and Sustainability

5.1 Introduction

Climate change is probably the greatest challenge of the twenty-first century and the largest threat to future generations (United Nations, 2021; Warren, 2021). There is a considerable amount of scientific studies and policy prescriptions, but evidence suggests anthropogenic climate change is accelerating dangerously. It indicates that there are still gaps in understanding and tackling the problem at its core. The objective of the book in this chapter, accordingly, is to develop an alternative explanation to the causality of climate change in the backdrop of mainstream theoretical underpinnings and policy regimes on the one hand and examine the implications of change on the ground by making a case study of Bangladesh on the other.

An earlier projection by Intergovernmental Panel on Climate Change (IPCC) indicated that if the current rate of increase continues, global warming is likely to reach 1.5 °C between 2030 and 2052 (IPCC, 2018). The Panel's Sixth Assessment Report (AR6) confirms that temperatures would rise by more than 1.5 °C above pre-industrial levels within the next two decades, breaching the ambition of the Paris climate agreement of 2015 (IPCC, 2021). Incidentally, climate change is not just about temperature (Newsome and Ripple, 2019); it has a far-reaching impact, which is already evident and most deeply understood in the light of the COVID-19 pandemic.

The pandemic has caused a slight reduction in the emission of carbon dioxide but that is a short-term gain. Countries are introducing recovery packages to boost economic growth in the post-pandemic situation (Allan et al., 2020; Hepburn et al., 2020), which can again foster the emission of Greenhouse Gases (GHGs). In this connection, there are several policy debates regarding the interrelationship between climate change and the COVID-19 pandemic that are currently ongoing. In economic terms, both of the issues have now been characterised as 'global public bads' and both entail similar policy prescriptions such as adaptation, mitigation, technological innovation and international cooperation (Fuentes et al., 2020). However, climate change is more challenging to comprehend or confront than the pandemic (Klenert

R. A. M. Titumir et al., *Natural Resource Degradation and Human-Nature Wellbeing*,
https://doi.org/10.1007/978-981-19-8661-1_5

et al., 2020) and there are reasons to worry that "the world will leap from the COVID frying pan to the climate fire" (Hepburn et al., 2020, p. 360). The year 2020 was not just the year of the pandemic, but also the year of record temperatures (WMO, 2020) and of floods, drought, storms, wildfires, locust plagues and melting of glaciers at both poles, which are inherently associated with climate change (UNEP, 2020, 2021). The IPCC's AR6 has already warned that there will be a dramatic increase in these extreme weather events in the near future (IPCC, 2021). Therefore, it is an urgent need to combat climate change first to protect the planet from the most devastating impacts that it might face in future, including newer types of epidemics.

The standard neo-classical (or environmental) economics argues for market-based solutions to the problems associated with climate change (e.g., Foster, 2010; Holland & Moore, 2013; Montgomery, 1972; Stavins, 2001; Zhang, 2013). The approach considers climate change as a straightforward problem attributable to a simple error of not charging GHGs emitters for the full cost of their emissions (Atkinson & Hackler, 2010), and accordingly, the suggestion is to impose a price on carbon (Mattauch & Hepburn, 2016). The argument is that once the price is correct, the market will react appropriately and develop the necessary technologies to address the issue. The prevailing policy regimes under this approach are cap and trade policy, carbon tax, financial or regulatory support for renewable energy, direct emission regulations, encouragement for clean energy innovation, clean development mechanism, etc. The governments, in this connection, across the world, including Bangladesh, are undertaking these policies to deal with climate change as part of the ideological project of neoliberalism.

The tools and techniques under the approach, however, are far from adequate to mitigate the problems of climate change (Farmer et al., 2015, 2019; Fletcher, 2012; Gilbertson & Reyes, 2009; Lohman, 2006). First, complexity, irreversibility, heterogeneity and deep uncertainty—inherent characteristics of the climate problem—call into question fundamental economic presumptions, and conventional economic theory seems ill-prepared to address this pressing problem (Farmer et al., 2015; Maréchal, 2012). Second, the approach considers the atmosphere as a 'commodity' and thereby proposes trading of carbon like many other commodities to minimise the changes in climate by ignoring the 'intrinsic value' of the atmosphere. Third, the approach ignores the distributional implications (Farmer et al., 2015) and social goals. As the approach has the disadvantage of incorporating a relatively restrictive ethical structure (Dietz et al., 2008), it fails to provide a technocratic answer to the ethical questions of intergenerational justice. New institutional economics—another branch of market centrism—emphasises stable institutional arrangement. Stable institutions are crucial in addressing the challenge of governing the global commons too and hence, of climate change. The branch, however, fails to consider the distributional aspect and political economy variables that can shape the nature of institutional arrangements. This chapter, therefore, argues that power and political settlement are important to understand the institutional mismatch among different organisations and countries in tackling climate change.

Moreover, the concept of 'sustainability'—inherently connected to climate change issue—remains undertheorised in neo-classical tradition, which defines 'weak sustainability' by arguing that human capital can substitute natural capital (Solow, 1993). Ecological economics, however, is an alternative theoretical approach that defines 'strong sustainability' and reasons that human capital and natural capital are complementary and not interchangeable. Hence, it emphasises the steady state of the economy (Daly, 2008; Ropke, 2004; Smith, 2010). On the contrary, this chapter argues that the growth process needs to continue, but in a manner in which human beings can complement nature, and thus, it does not agree with either concept of sustainability.

This chapter redefines sustainability by emphasising the concept of 'resilience of nature'—one that is grounded on 'sustainability science' (e.g., Berkes et al., 2003; Bousquet et al., 2016; Cumming & Peterson, 2017; Folke 2010; Lebel et al., 2006). The underlying reason is that it is not possible to define sustainability with the metaphors of growth equilibrium and stability and the analysis of the system of steady-state equilibrium is incompatible (Berkes et al., 2003). In essence, the chapter defines sustainability under the framework of dynamic equilibrium and argues that in the process, the power structure is important to consider. The process of dynamic equilibrium of the human–nature mutual relation depends on the existence of an effective regulatory regime where power is a crucial factor.

The chapter based on the underlying critics of the market-centric approach towards climate change and sustainability, thereafter, takes on the political-economic logic, putting the 'atmosphere business' and the issue of 'sustainability' under critical scrutiny. The rationale behind this is that political economy logic offers a thorough examination of the various power dynamics in a given setting (Acosta & Pettit, 2013). In the case of climate change, it necessitates admitting that there are multiple politically negotiated pathways leading to a green future rather than a single path. The solutions, therefore, must function in specific cultural, social and political settings. They must be tuned into and adapted to the local context and negotiated within particular political cultures; this is perhaps even truer at the global scale because solutions must operate across borders (Leach & Scoones, 2006).

Overall, the political-economic approach offers a comprehensive understanding of how the climate markets work or do not work, the various actors involved and how they interact. The chapter, therefore, tries to formulate a conceptual framework to reframe the understanding of climate change and sustainability by taking coupling insights from neo-classical, new institutional, ecological economics and political economy. The discussion, particularly, focuses on the questions of distribution and well-being of human beings and nature. The overall argument is that it is possible to address the problem of climate change only if there is a cooperative effort and sustainability recast in terms of revitalisation of the human–nature mutuality, which are the functions of diffusion of power.

As climate change is a global phenomenon, it is necessary to justify the reason for taking up the integrated case of Bangladesh as a developing country for empirically testing some of the hypotheses. Bangladesh is one of the most vulnerable countries to climate change (IPCC, 2001). Available studies on climate change in Bangladesh only assessed the impacts of climate change across different sectors but have not scrutinised the problem from any theoretical point of view. Moreover, there is a paucity of comprehensive studies addressing the issue concentrating on the distributional and sustainability premises, which can contribute to the understanding of the resource governance system in a developing country amidst the complex international political economy. It altogether indicates the urgency to examine the evidence of climate change and measure the impact of it both theoretically and empirically.

5.2 State of Climate Change

An increase in global average air and ocean temperature is evident from various scientific assessments. IPCC in its First and Second Assessment Reports warned that the global mean surface air temperature had risen by 0.3–0.6 °C over the last 100 years (IPCC, 1990, 1995). The Fourth Assessment Report (AR4) additionally predicted that global temperatures will increase between 1.8 and 4 °C by 2100 (IPCC, 2007). Then, from 1880 to 2012, the globally averaged combined land and ocean surface temperature increased by 0.85 °C, according to the fifth assessment (AR5) (IPCC, 2014). Finally, the Sixth Assessment Report has cautioned that temperatures in the next two decades are likely to rise by more than 1.5 °C above pre-industrial levels and human activities are the key contributors in this regard (IPCC, 2021) (see Chap. 1 for further details on IPCC estimates).

Apart from IPCC, the National Ocean and Atmospheric Administration (NOAA) also estimated that the global average land and ocean temperature rose significantly from 1880 to 2020, reaching 0.94 °C in 2020 (Fig. 5.1). However, the COVID-19 pandemic has caused a slight drop in temperature, with the 2021 estimate standing at 0.90 °C. The projection indicates that by 2100, the mean annual temperature will rise by 1–5 °C depending on the scenario and location (World Bank, 2020). Overall, the key finding is that the temperature has risen over the past several decades and if this continues, the rise will be faster and more acute in the future.

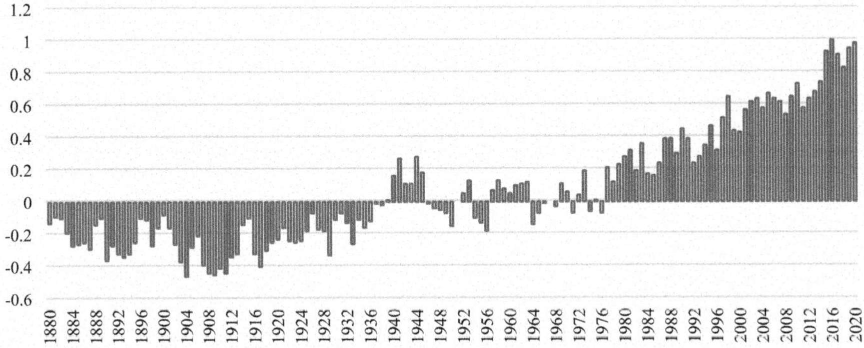

Fig. 5.1 Trend in global (land and ocean) temperature anomalies (C) compared to 1901–2000. *Source* NOAA (2021a)

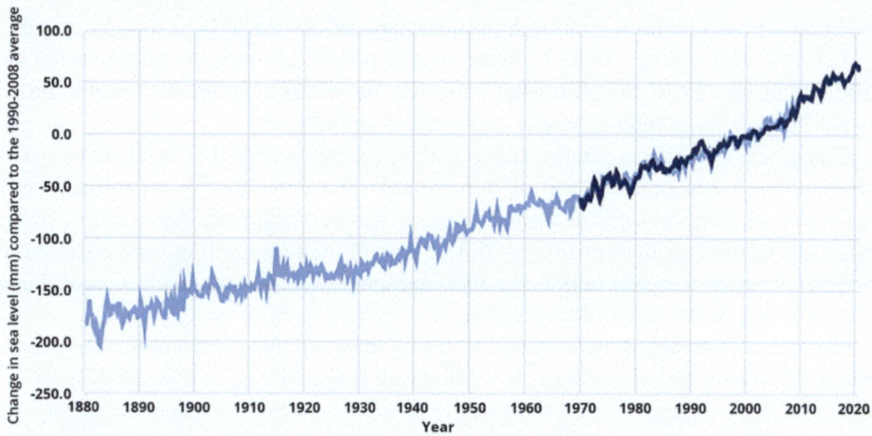

Fig. 5.2 Change in sea level, 1880–2020. *Source* Lindsey (2021)

Temperature rise causes the melting of snow and ice sheets in Greenland and Antarctica, which raises the sea level. IPCC assessment shows that between 1901 and 2010, the average global sea level rose by 19 cm (IPCC, 2014) and between 1901 and 2018 by 20 cm (IPCC, 2021). Therefore, it signifies a steady rise in sea level of 1 cm in less than a decade. The rising trend is also congruent with a recent estimate by NOAA, which reported that the sea level rose from 20.32 to 22.86 cm since 1880. The highest annual average rise recorded was 8.66 cm in 2019, an increase of 0.61 cm since 2018 (Fig. 5.2).

The NOAA estimate also predicted that the global sea level possibly would rise at least 30 cm above 2000 levels by 2100. Projection by the IPCC indicated that by the mid-2090s, the level would reach between 22 and 44 cm above 1990 levels, rising at about 4 mm year^{-1} (IPCC, 2007). Under a very high GHGs emission scenario, the level is estimated to approach 2 m by 2100 and 5 m by 2150 (IPCC, 2021). Many climate scientists, however, view the IPCC prediction as unrealistically optimistic, while asserting that the sea-level rise will accelerate far more rapidly (Luetz, 2008).

Precipitation has augmented at an average rate of 0.08 inches per decade over land areas worldwide since 1901 (IPCC, 2007). Particularly, between 1951 and 2010, both the annual and extreme precipitation increased by 1–2% per decade in dry regions (Donat et al., 2016). This increase in precipitation is also associated with global warming. Every degree increase in global atmospheric temperature causes a 7% increase in extreme rainfall intensity (Westra et al., 2013). The changes in precipitation, however, are not uniform. For instance, wet areas, in particular, exhibit an increase in the size of extreme precipitation but a smaller increase in annual totals (Tollefson, 2016).

The projection also indicates that annual mean precipitation is likely to increase in the high latitudes, equatorial Pacific, and many mid-latitude wet regions, while it is likely to decrease in many mid-latitude and subtropical dry regions and in limited areas of the tropics (IPCC, 2014, 2021). Another prediction signifies that heavy precipitation will be more common in most parts of the world by 2100, increasing by 16–24% (Hausfather, 2018). Climate change is thus driving a fluctuating trend in rainfall patterns across the globe, but there is a general agreement that precipitation would become more intense nearly everywhere in the future.

Changes in climate parameters cause a significant impact on the social, economic, environmental and political life of humankind. The most significant impacts are distortions in the hydrological system, increase in extreme climatic events, ecosystem vulnerability and health hazards (Fig. 5.3). These ultimately result in food insecurity, biodiversity loss, increase in poverty and unemployment, forced displacement, loss in economic growth and violent conflict.

Climate change has already started resulting in significant ramifications for water resources and the hydrological cycle (Chaturvedi et al., 2021; Shamseddin & Chaibi, 2019). Billions of people across the world do not have access to an adequate supply of safe water. Climate change is further aggravating the situation by distorting the hydrological system and causing floods in some places and drought in others (Mengistu et al., 2021; Zhang et al., 2019). Moreover, both groundwater and surface water have been reducing gradually. Such changes in the hydrological system directly affect the agricultural sector and productive activities. In addition, there is an increased probability of negative consequences for different types of ecosystems (e.g., agriculture, forest, wetlands, marine) and society (Malhi et al., 2020). Climate change is accelerating at a higher extent in comparison with the rate at which these ecosystems can adapt and re-establish themselves.

Then, some extreme weather events have changed in frequency and intensity over the last 50 years due to climate change (IPCC, 2007) and might occur more frequently with higher intensity of damage in the near future (IPCC, 2021). It means though disasters also existed in the past, the frequency of such disasters and the severity of damage caused by them have increased manifold over the years. The impact of such extreme weather events is devastating as they affect agriculture, industry, infrastructure, human settlement, human health, biodiversity resources and many other sectors. Study finds that the new generation will suffer more extreme weather events compared to their previous generation (Carrington, 2021). Thus, climate change brings stark intergenerational inequity.

Importantly, global warming and the concomitant changes in rainfall, humidity and solar radiation are, slowly but surely, taking a toll on agricultural production, thereby jeopardising global food security. Climate change is anticipated to have a severe impact on food security by the middle of the twenty-first century, with South Asia accounting for the majority of food-insecure countries (IPCC, 2014). Several studies have already shown that agricultural production in various countries is vulnerable to climate change (e.g., Guoju et al., 2016; Natamba et al., 2018; Rarieya & Fortun, 2010; Sattar et al., 2017). The estimate also indicates that a third of global food production will be in jeopardy by the end of century, if GHGs continue to rise at the current rate (Harvey, 2021).

Climate change also influences human health and diseases in numerous ways. Between 2030 and 2050, climate change is likely to cause approximately 250,000 additional deaths per year resulting from heat stress, malnutrition, diarrhoea and

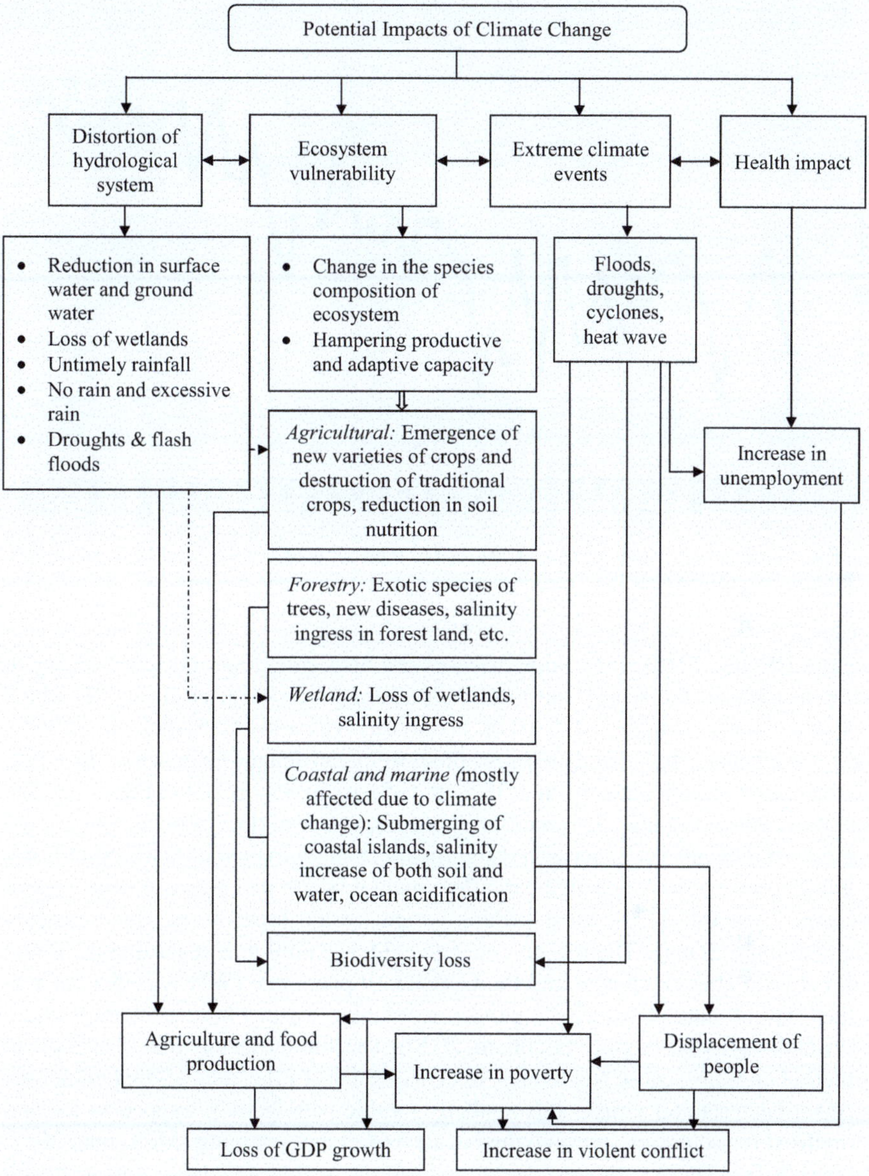

Fig. 5.3 Impacts of climate change: an extended view. *Source* Author's interpretation

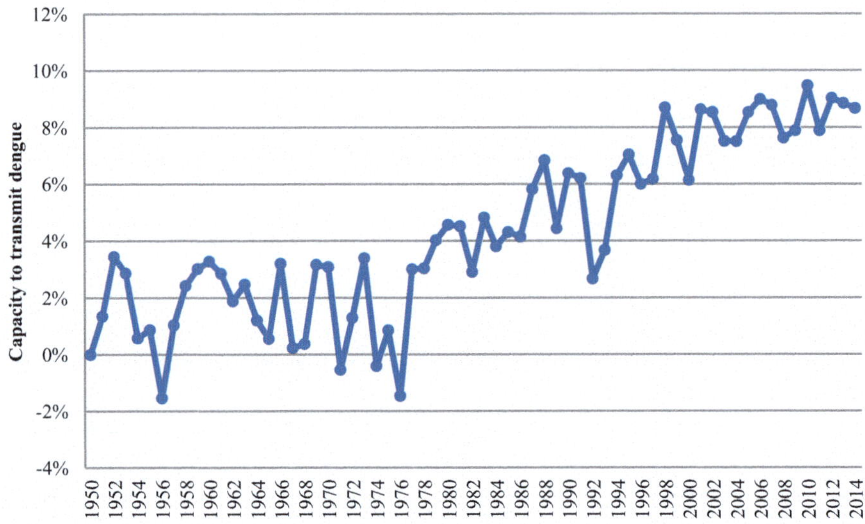

Fig. 5.4 Capacity of mosquitoes (*Aedes aegypti*) to transmit dengue fever worldwide from 1950 to 2014. *Source* Elflein (2017)

malaria (WHO, 2018). Infectious diseases are also emerging and re-emerging in different parts of the world due to climate change (van Wijk et al., 2020) which includes malaria, dengue, HIV/AIDS, hantavirus, Hepatitis C, SARS, etc. (WHO, n.d.). For instance, the capacity of mosquitoes to spread dengue fever across the world has increased significantly from 1.35% in 1951 to 8.68% in 2014 (Fig. 5.4). The countries, where the disease turned into an epidemic resulting from such increased capacity of mosquitoes, experienced significant changes in their climatic conditions.

Furthermore, climate change can affect mental health and increase neurological diseases (Leonard, 2020). Most strikingly, new health threats have been emerging due to climate change. There is no concrete evidence until now regarding the direct link between climate change and the COVID-19 pandemic (WHO, 2020) but it is argued that climate change is "a pandemic enabler, a pandemic accelerator and a multi-pathway crisis engine" (Cadham, 2020). The similar factors that contribute to climate change also cause pandemics. Recent evidence further denotes that numerous dormant bacteria and viruses (also called 'zombie pathogens'), trapped in ice and permafrost for centuries, are reviving because of global warming (McKenna, 2017; Pappas, 2016). These pathogens are a major threat as they can cause serious health hazards in the upcoming decades. Such an impact on the health system, on the other hand, has led to an increase in poverty, unemployment and violent conflict.

Destruction of ecosystems, extreme climate events and, most importantly, sea-level rising are affecting the coastal regions and islands deeply. According to estimates, a 30 cm rise in sea level would intensify floods along the coast by 36–58% (IPCC, 1998). On the other hand, islands such as the Maldives, Marshall Islands,

Federated States of Micronesia, Kiribati, Tuvalu and Arctic islands such as Shishmaref and small islands in Nunavut may become uninhabitable as sea levels rise (Ford et al., 2006; Marino et al., 2009). This imminent danger is forcing local residents to migrate to other places and thereby, hamper the human settlement aspect. IPCC's (2022) recent assessment on the impacts of climate change has confirmed such changes in both ecosystems and human systems on a wider scale.

The multifaceted impacts of climate change vary across different regions of the world, and they are hard to specify as they are interrelated with each other. However, the developing and low-lying countries of the world are the main bearer of the impacts.

5.3 Climate Change in Bangladesh

The temperature in Bangladesh has increased over the years. One of the earlier studies found that since the latter part of the last century, there has been an average increase of 0.5 °C (Warrick et al., 1994), which is congruent with another study that reported an increasing tendency in decadal mean annual temperature especially from 1961 to 70 (Karmakar & Nessa, 1997). Findings regarding monthly, decadal, maximum and minimum temperatures also signify an increasing trend in all the studies (Table 5.1). The temperature, however, has been increasing at a faster rate in recent times as another study reported that it rose predominantly during 1990–2010 (Hasan and Rahman, 2013).

A comprehensive assessment of the historical data (1976–2008) for annual average maximum daily temperatures at 34 meteorological stations of Bangladesh conducted under the *Unnayan Onneshan* revealed that temperature had been increasing at a rate of 0.0186 °C per year (Basak et al., 2013). Some other studies project a future rise in temperature in Bangladesh. One earlier study came up with a scenario of surface warming of 0.30–0.50 °C by 2050 (Mahtab, 1989). Another study predicted that the increase would be 1.4 °C in winter and 0.7 °C in monsoon in 2030 and 2.1 °C and 1.7 °C for winter and monsoon, respectively, in 2075 (Huq et al., 1999). Overall, the temperature has been increasing in Bangladesh over the years and is likely to rise more in the future.

There are several studies on rainfall trends in Bangladesh and some of which examines the impacts of climate change on changing rainfall pattern (e.g., Ahmed & Karmakar, 1993; Ahmed & Kim, 2003; Basher et al., 2019; Islam et al., 2014; Mondol et al., 2018; Murshed et al., 2011; Rahman et al., 1997). Five consecutive decadal annual average rainfall indicates an increasing trend in rainfall intensity in Bangladesh (Islam et al., 2014). Significant changes in the rainfall pattern occurred during the period from 1980 to 2011 (Mondol et al., 2018). The study recorded the highest increasing precipitation trend in the eastern hilly region where the rainfall was increasing at 8.49 mm/year during monsoon and 5.12 mm/year during the pre-monsoon season. The increasing trend in rainfall (at a rate of 8.4 mm/year over

Table 5.1 Trend in temperature in Bangladesh based on various studies

Indicator	(+)	(−)	Increase in temperature (in °C)	Source
Average change in temperature since the latter part of the last century	√		0.5 °C	Warrick et al. (1994)
Monthly maximum and minimum temperatures	√		0.11–0.33 °C	Choudhury et al. (1997)
Annual minimum, maximum and mean temperature	√		0.0066, 0.007 and 0.0034 °C, respectively	Karmakar and Shrestha (2000)
Overall change in annual temperature during 1961–90	√		Overall increase with maximum at Dhaka (0.037 °C)	Quadir et al. (2004)
Decadal mean, mean maximum and mean minimum temperature during 1958–2007	√		0.103, 0.091 and 0.097 °C, respectively	Shahid (2010)
Decadal mean maximum and minimum temperature	√		0.35 and 0.16 °C	Khan et al. (2019)

Source Authors' compilation from different sources

the period 1948–2007) is also observable in the south-west coastal regions of the country—which are more vulnerable to climate change (Hossain et al., 2014).

The comprehensive assessment under *Unnayan Onneshan* shows that 16 (out of 38) stations indicated increasing trends in total rainfall for the winter (December–February) period, 20 for the pre-monsoon (March–May) period, 31 for the monsoon (June–September) period and 30 for the post-monsoon (October–November) period during the 33-year (1976–2008) period under consideration (Basak et al., 2013). On the other hand, a recent study indicates that the monsoonal and annual precipitation have decreased by 87.35 mm/decade and 107 mm/decade, respectively (Khan et al., 2019). Overall, there is a fluctuating trend in the rainfall pattern.

Projection, however, indicates that by 2090, the mean annual rainfall would increase by 7% compared to the 1970–2000 level in Bangladesh although some models project increases up to 24% (MoFA, 2018). A recent study also predicts that rainfall extremes would increase in both pre-monsoon and monsoon seasons in future in the north-eastern part of the country, which is more vulnerable in the face of extreme precipitation events such as flash floods (Basher et al., 2019).

There seem to be an increasing trend in Relative Sea-Level Rise (RSLR) at various locations along the coastline of Bangladesh compared to the global average of 2.0 mm/year or 3.0 mm/year except for the south-eastern part (Table 5.2). Among all the locations, Cox's Bazar recorded the highest RSLR from 1978 to 1998 and the extent was quite similar in Moheshkhali from 1968 to 2002. Overall, there are

Table 5.2 Observed Relative Sea-Level Rise (RSLR) along Bangladesh coastline

Stations	RSLR (mm/year)	SLR (IPCC, 2014)
Hazra et al. (2002) (1985–1998)		2.0 mm/year[a] 3.0 mm/year[b]
Western Sundarbans Coast	3.24	
Khan et al. (1999), SMRC (2003) (1978–1998)		
Hiron Point	4	
Char Changa (Meghna estuary)	6	
Cox's Bazar (East Coast)	7.8	
CEGIS (2011) (1977–2002)		
Hiron Point	5.5	
Sandwip	7	
Cox's Bazar	5.1	
Moheshkhali (1968–2002)	7.5	
MoEF (2012)		
South-east part	1.4	
South-central part	3.9	

Source Authors' compilation from different sources
[a]Valid for the period 1980–2000
[b]Valid for the period 1991–2010

variations in trend across various locations and times, but it is clear that the sea level has been rising in each of the locations. Another study reported that there has been an increase of 0.2 m in the seal level along the Bangladesh coastline (Rawat et al., 2016).

An earlier projection indicated about 10 cm, 25 cm and 1 m rise in sea level along the coastline by 2020, 2050 and 2100 affecting 2%, 4% and 17.5% of the total land mass, respectively (World Bank, 2000). The most recent study, however, has warned that the level could rise twice as much as predicted previously by IPCC (Becker et al., 2020). It claims that in the Ganges–Brahmaputra–Meghna delta, which constitutes about two-thirds of Bangladesh, the sea level could reach 85–140 cm by 2100.

The variability in climate parameters alters the seasonal pattern of Bangladesh. The country now has three seasons instead of six, which can be characterised as a hot summer, a shortened winter and medium to heavy rain during the monsoon (Khan et al., 2019). Bangladesh has always had to deal with floods and cyclones, but now they happen more frequently and especially outside of expected times. There is no doubt that these disorders are caused by climate change. Such changes have negatively affected the economy and the society of Bangladesh in several ways (Table 5.3) (see Fig. 1.8 in Chap. 1 for direct and indirect impacts). The most important consequences are the negative impacts on crop production and thereby, on food security, increased frequency and intensity of natural disasters and submerging of land area as well as associated displacement of people.

Table 5.3 Critical vulnerable areas and most affected sectors due to climate change in Bangladesh

Climate-related elements		Critical vulnerable areas	Most impacted sectors
Temperature rise, increase in heat stress, drought		• North-west	(1) Agriculture (crop, livestock, fisheries) (2) Water (decreased groundwater and surface water) (3) Energy (4) Health
Sea-level rise and salinity intrusion		• Coastal area • Island	(1) Agriculture (crop, fisheries, livestock) (2) Water (waterlogging, deterioration of drinking water) (3) Human settlement (insecure shelter, loss of livelihood and displacement of people) (4) Energy (5) Health
Erratic, excessive and untimely rainfall		• North-east region • Central region	(1) Agriculture (2) Water (3) Health
Natural disasters	Floods	• Central region • North-east region • *Char* land	(1) Agriculture (crop, fisheries, livestock) (2) Water (urban, industry) (3) Infrastructure (4) Human settlement (insecure shelter and displacement of people) (5) Health (6) Energy (7) Degradation of ecosystem and loss of biodiversity
	Cyclone and storm surge	• Coastal zone • Marine zone	(1) Marine fishing (2) Infrastructure (3) Human settlement (insecure shelter and displacement of people) (4) Life and property (5) Degradation of ecosystem and loss of biodiversity

Source Prepared by the authors based on MoEF (2005, 2009)

Agriculture in Bangladesh undergoes a significant transformation to address the associated challenges of achieving food security. With rapid economic growth in recent years, Bangladesh has substantially improved its cropping practices and fulfilled food requirements for its vast population. The success, however, is now under threat because of climate change. The outlook for food security and agriculture in the face of climate change seems to be particularly ominous. Likewise, the country has been experiencing extreme climatic events in recent years claiming thousands of lives and negatively impacting millions of livelihoods.

Every year, Bangladesh faces natural disasters such as drought, flood, cyclone and tidal surge, thunderstorm, river erosion, landslide, among others. The occurrence and intensity of such disasters have increased precipitously due to climate change. Specifically, in the coastal region, there is an increased altitude and intensity of tidal surges, frequent coastal floods and increased salinity. Along with this, in the rainy season, the increased magnitude of floods, flash floods and river erosion becomes a common factor. During summer, susceptibility to increased drought due to lack of rain and erratic rainfall wreaks havoc on agricultural production. In accordance with that, climate change has had an undesirable impact on health, biodiversity resources, urban settlement, safe and secured social structure, energy and many more. All these result in the vulnerability of people's livelihoods especially of the marginalised ones. Overall, Bangladesh faces both 'environmental and biophysical' and 'socio-economic and livelihood' impacts due to climate change (see Ahmed & Khan, 2022 for reviews).

Nevertheless, the country has been trying to develop some capacities for coping with the effects of climate change. However, many regions remain outside the ambit of actions related to climate change. Furthermore, most of the reform measures are neoliberal at their roots. In particular, international development organisations and donor agencies recommend these measures as mitigation and adaptation policies. Mitigation measures focus primarily on the adoption of renewable energy sources for reducing GHGs emissions without jeopardising access to energy. As a vulnerable developing country, however, the adaptation policy is its priority (MoEF, 2009). Most of the adaptation measures are project-oriented, needing a large amount of investment and technology transfer (Table 5.4).

The government receives most of the funds from international organisations and donor agencies in order to finance the projects. When the international organisations administrate climate funds, it means that they approve adaptation projects, be in charge and evaluate them. In this way, they can influence climate policies at the national level.

The country is also incorporating climate change into sectoral plans and national policies (Fig. 5.5), including the national water management plan, coastal zone policy, national agricultural policy, delta plan 2100 and 8th Five Year Plan. The perspective plans such as 'Vision 2021' and 'Vision 2041' also specifically emphasise climate change. Notable among other national efforts is that Bangladesh was the first to complete a National Adaptation Programme of Action (NAPA). Later, Bangladesh adopted a more comprehensive action plan—Bangladesh Climate Change Strategy and Action Plan (BCCSAP)—in 2009.

Table 5.4 Some adaptation measures for climate change in Bangladesh

Adaptation measures	Scope	Nature	
		Traditional/Local	Project-oriented (market-based)
Raising house above normal flood level	Protection from flood	√	
Adjusting crop pattern to take advantage of floodwater	Ensuring food security	√	
Flood management scheme	Protecting low lying rural regions from severe flood		√
Flood protection and drainage scheme	Protecting urban area from rainwater and river flooding		√
Coastal embankments establishment	Protecting coastal region from flooding and salinity intrusion		√
Establishing cyclone shelters	Providing refuge for communities from storm surge		√
Comprehensive disaster management	Tackling the damage caused by natural disasters		√
Coastal green belt project	Plantation of trees along the coastline		√

Source Prepared by the authors

The country is also playing an important role at the international level by signing international conventions and protocols (Fig. 5.5). For instance, Bangladesh is a signatory to the United Nations Framework Convention on Climate Change (UNFCCC) and the Kyoto Protocol—the two most important international documents regarding climate change. Bangladesh also signed and ratified the Paris Agreement in 2016.

Overall, Bangladesh tops the country list of climate change vulnerability according to scientific assessment because of its geographical location and hydrogeological and socio-economic characteristics. Climate change, in fact, has emerged as a major development challenge for Bangladesh. It undercuts the country's hard-earned development successes that have been achieved so far and jeopardises the long-term sustainability of the country's accomplishments in the years to come. Bangladesh, however, is earnestly trying to enhance its resilience to climate change and reduce the risk by implementing adaptation and mitigation measures. Nevertheless, it needs to argue that the marginalised people of the country are still vulnerable and in many cases, the protection measures bring no effective outcome.

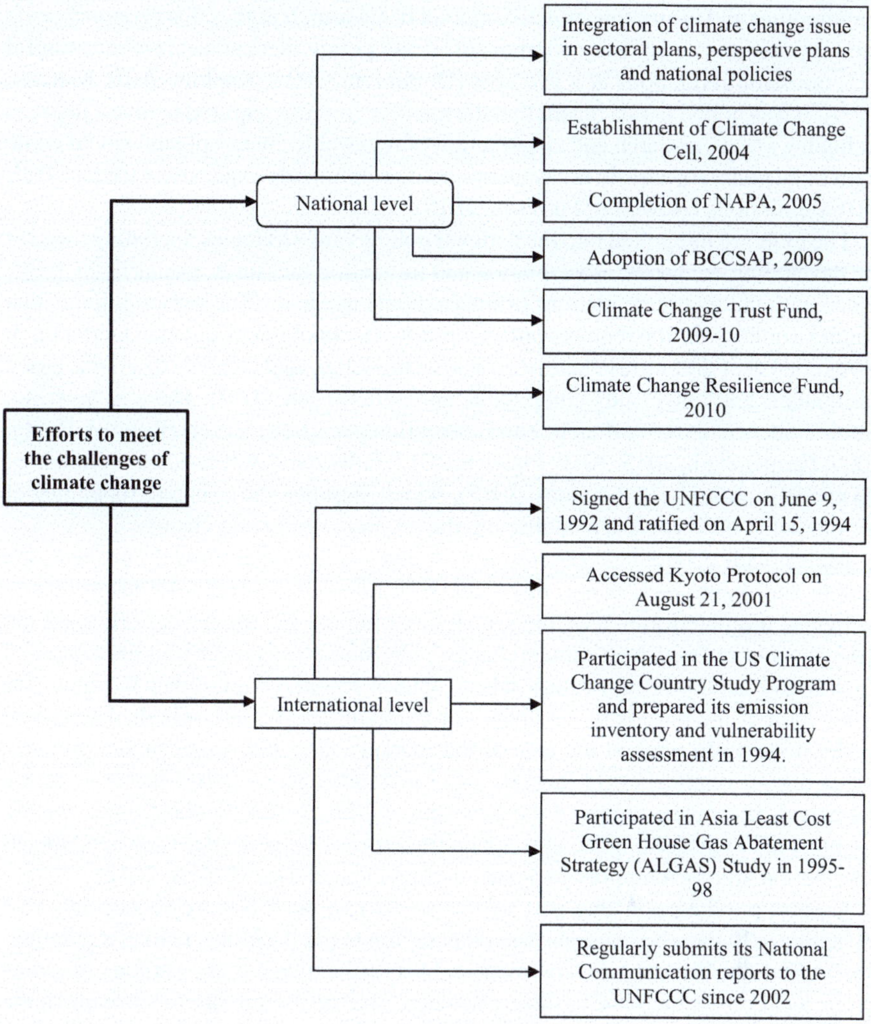

Fig. 5.5 Bangladesh's efforts to meet the challenge of climate change. *Source* Prepared by the authors

5.4 Market Correction of Climate Crisis

The key policy measures for tackling climate change have been drawn from the standard economics at the current time. Nordhaus' (1992, 2001) climate change model is perhaps the most extensively referred one in this regard. Some of the market tools are joint implementation programme, clean development mechanism, emission trading, and climate financing, among others. The Kyoto Protocol primarily established these tools, with the objective of converting GHGS emissions into tradable commodities and promoting effective market behaviour. It argues that economy-wide carbon pricing is a necessary component of any policy that can achieve meaningful CO_2 emission reductions at a low cost (Dominioni, 2022; Kaplow, 2010; Metcalf, 2009). It was hoped in this regard that the massive funding required to invest in green technology (e.g., research and construction of renewable energy plants) would come from large polluters who do not want or find it prohibitively expensive to reduce their GHGs emissions (Newell & Paterson, 2010).

The tools are integrated into the national policies and strategies, including those of the developing countries, as adaptation and mitigation measures, primarily in accordance with the recommendations of international agencies. The presumption is that finding a quick solution requires only a top-down technological and managerial fix. It cedes control of the entire debate to economic modellers and creates a discourse based on monistic and universally comparable numbers (Spash, 2007). There is, however, ample evidence of the inefficiency and corruption of carbon markets (Böhm & Dabhi, 2009; Farmer et al., 2015, 2019; Fletcher, 2012; Gilbertson & Reyes, 2009; Lohman, 2006, 2012; Schmitz & Scoones, 2015), which signifies the concrete loopholes of the market-centric approach addressing the problem of climate change.

Mainstream economics' obsession with endless economic growth, not with the human and nature well-being, has resulted in more than simply the incapacity to recognise the deterioration of the global environment (Clark et al., 2009) and the worst evidence of which is 'climate change'. The market only prioritises the economic growth and capital accumulation where natural resources are either the input for production or the third party receiving the externalities (waste) or both. The failure of the doctrine to explain the role of the market under indistinguishable property rights has an implication in the case of climate change, as it is difficult to define the nature of property rights of the atmosphere. Therefore, complications and risks are involved in discerning the atmosphere in terms of property rights ownership and distribution among a number of agents in the global market (Childs, 2012).

Carbon purchases in one part of the world can be offset by sequestration or switching to low-carbon options in another region as per the global market regulation. Such offset markets rely on 'commensurability' (the idea that the carbon emissions in one location are the same as those elsewhere) and on the smooth functioning of the global market, including a firm guarantee that sequestration or offsetting is actually taking place (Lohman, 2012). In the case of carbon offsets, however, these assumptions have not held up and with a low carbon price and concerns raised about many carbon-offset schemes, the neat solution has not always proven effective (Leach & Scoones, 2015; Newell et al., 2012). For instance, a recent study shows that more than 90% of rainforest carbon offsets are worthless (Greenfield, 2023). Moreover, carbon taxes drive up the price of essential household energy expenditures. In this process, certain socio-economic groups are more negatively affected (Farrell, 2017). Carbon pricing, therefore, cannot ensure equitable distribution either. Above all, the political feasibility of market-based instruments is a crucial concern, partially due to the countries' asymmetric situations (Parson & Zeckhauser, 1995). Climate change thus raises challenging questions of equity and justice, particularly in terms of the distribution of responsibility, burden and benefit between the industrialised and developing nations or between different groups of people within a particular nation (Markandya, 2011; Posner & Sunstein, 2009), which cannot be addressed under the neo-classical model.

Neo-classical economics also introduces the shadow pricing method for dealing with climate change. The method emphasises the necessity of the efficiency gains but ignores talking about the emission that would take place although the system could gain efficiency. In fact, the goal of *Pareto efficiency*—the concrete measure of social welfare—has stifled discussion of distributional issues and social goals other than efficiency in allocation (Gowdy & Erickson, 2005). Moreover, the climate issue, aided by the systematic use of Cost–Benefit Analysis (CBA), had been framed by the optimal growth theory (Ackerman, 2007; Bergh, 2004; Spash, 2007). The CBA is faulty as it moves towards a more stringent emission reduction policy that would inevitably make someone worse off. The intergenerational discounting under the CBA is no longer justified when the intergenerational equity is present (Ackerman, 2007; Spash, 2007). Discounting the future is also inappropriate in calculating the value of a stable climate in the far future (Gowdy & Erickson, 2005; Weitzman, 2009). Above all, the CBA of climate change may have a tendency to overlook some of the most significant ecological consequences of the issue and uncertainty is central to such analyses (Harris et al., 2017).

Along with emphasising the CBA method, the neo-classical economics lays emphasis on technological innovation in tackling climate change. The assumption is the 'perfect substitutability' between human capital and natural capital. Because of how quickly the climate is changing, it would be fallacious to argue that countries should continue to concentrate on economic growth now and in the future, they would be able to address climate change with technological advancement. This is because climate change would threaten their existence by that time. Arguments also prevail that technology could be transferred from developed countries to developing countries. However, technology trajectory can suffer from extreme path dependency or lock-in once set in train by government programmes, commercial investments or development projects (Berkhout, 2002; Unruh, 2000). When infrastructure, incentives and interests combine, they can be difficult to shift (Schmitz & Scoones, 2015). Moreover, technological innovation sometimes makes it worse to the resource base by extracting more of it (Daly, 2008) and support for new green technologies often depends on political mobilisation (Farmer et al., 2019). Overall, the neo-classical economic model of climate change is far from effective in bringing solutions, as four major issues remain inadequately addressed. The issues are "(a) uncertainty, (b) aggregation, heterogeneity and distributional implications, (c) technological change, and (d) realistic damage functions for the economic impact of the physical consequences of climate change" (Farmer et al., 2015, p. 329).

The New Institutional Economics (NIE)—another branch of market centrism—emphasises the stability of institutions for managing resources sustainably (see Chap. 2 for details). This chapter also recognises that stable institutional arrangements are crucial in dealing with climate change. Just as 'fair competition' in commodity markets has to do with institutional arrangements, 'fair competition' in the global discourse is also a matter of institutions (Söderbaum, 1990). However, as the branch excludes the issue of power and political settlement remaining apolitical, it fails to address the issue of climate change too, in which case, factors such as unequal power and political interest are far more obvious. The institutional failure, stemming from the political settlement, the highly asymmetric information among the state parties and the transaction costs involved in the replacement of the production mix are quite apparent in this regard.

The power issue, therefore, is important to understand the institutional mismatch among different organisations and countries in tackling climate change. The underlying implication is that the parties engaged in dealing with mitigating the climate change problem are behaving in a non-cooperative way, as cooperation is costly (Suh, 2016). In fact, developed countries are the ones responsible for such non-cooperative relations, which ultimately result in the failure of international conventions, conferences, treaties and protocols. The developed countries can do that only because they are exercising their power and avoiding the responsibilities.

The market-centric approach is also unable to theorise the concept of sustainability—a concept closely related to climate change—appropriately. The standard economics states that human-made capital (e.g., technology) can replace the natural stock of capital, which is termed as 'weak sustainability' (Cabeza-Gutés, 1996; Solow, 1993). Ecological economics is the branch that stands in stark contrast to this proposition and offers alternatives to the theoretical foundations and policy recommendations of neo-classical economics (Gowdy & Erickson, 2005). In this regard, it particularly concentrates on growth, sustainability and climate change issues (Ackerman et al., 2013; Daily et al., 1991; Daly, 1994; O'Connor, 1994; Magdoff & Foster, 2010). It proposes 'strong sustainability', which is a conservationist approach towards the natural resources. It emphasises the 'limit to growth' concept by conserving all critical resources in spite of their substitution or even by ensuring negative economic growth for the sake of future generations (Daly, 1994). Strong sustainability, at its core, thus accepts that human beings or human-made capital cannot do certain functions that are performed by the environment.

The basis of the strong sustainability argument is a critical scrutiny of the capitalistic production system, which calls for a steady state of the economy (Daly, 2008; Ropke, 2004; Smith, 2010). The issue of climate change is relevant here, as the emissions of GHGs is inextricably interlinked with the expansion of economic growth. Ecological economics itself lays special emphasis on climate change and particularly criticises the pricing the pollution. The argument goes that every product in the marketplace has embodied energy, is associated with GHGs emissions and thus has the 'wrong' price (Kapp, 1978; Spash, 2007).

The steady state of the economy or zero growth, however, could not thwart the exhaustion of resources (Georgescu-Roegen, 1975) while excessive economic growth is also detrimental to the sustainability of nature. This book, therefore, by discarding both strong and weak sustainability concepts, contends that the process of growth necessitates continuing in a manner in which human beings can build a complementary relationship with nature, and accordingly, this chapter defines sustainability in its own way. In this regard, it takes the concept of 'resilience of nature'—a concept under the way after being perturbed determines asustainability science framework—into consideration.

The sustainability science framework consists of understanding the limit to growth, the role of innovation along with the adaptation and resilience of the complex socio-ecological system (Cumming & Peterson, 2017). It proposed that both the anthropocene and growth cause disturbance in the balance of ecology but the latter has the ability to reorganise and can persist with support from innovation. The adaptive capacity of nature in a subtle way after being perturbed determines a newer equilibrium at the end and provides similar ecosystem services defined as resilience. Resilience, therefore, is the capacity to cope with change stemming from the exploitation of resources and continues to develop through transformation, adaptation or regeneration, relates to the social–ecological dynamics in a particular resource governance system (Berkes et al., 2003; Bousquet et al., 2016; Folke et al., 2005, 2010; Lebel et al., 2006; Miller et al., 2010). The ecology thus follows an adaptation cycle of organisation, collapse and renewal to adapt with the changes (see Berkes et al., 2003 for details).

It further extends the cyclical process by connecting the small ecology to the global ecological balances called 'Panarchy', which states that the ecology follows the hierarchical relationship (Gunderson & Holling, 2002; Gunderson et al., 2010). The micro-level ecological change is small and fast which contributes to the intermediary stock of ecology to the global stock of ecology, which is large and slow in the adaptation cycle.

It, however, fails to explain the consumption beyond the capacity to regenerate at the micro-level and the carrying capacity at the global level. It is evident that the climate change is happening, which is an ultimate failure of the adaptive capacity of the global ecological cycles. Moreover, the rates of disturbance among different ecologies by different agents or by spatial differences are not the same that they can be supplemented by the global ecological cycle. Hence, the agents are in a state of overconsumption with inequality resulting in the reduction of resilience.

5.5 Ecological Rift, Ecological Debt and Unequal Exchange

Marx explicitly wrote neither on climate change nor on environmental degradation, but his philosophical contribution still helps understand such problems. Few studies regarding climate change are based on the Marxian framework (e.g., Foster, 2009; Ghotge, 2018; Kovel, 2002). The Marxian politico-economic proponents argued that Marx, in his lifetime, had explained the capitalist invasion of the ecology for enlarging the surplus value. The key argument is that the capitalist mode of production is the root cause of environmental degradation. Marx's concepts of 'social metabolism' and 'metabolic rift' form the central thesis of the political-economic critique of the alienation of nature under capitalism. The notion of social metabolism is the exchange of materials between humans and nature (Foster & Burkett, 2016). The relationship breaks due to accumulation called metabolic rift, defined as the positive relationship between capitalist accumulation and the degradation of the environment (Foster, 1999). This conception of the metabolic rift is the key to understanding the sustainability, according to Marx's explanation.

The relationships between humans and nature or the continued existence of society have changed throughout the history, but with the advent of capitalism, the relationship has taken on an especially unusual shape (Empson, 2015). According to Marx, both nature and labour contribute to the production of wealth, whereas labour can only produce wealth through the exchange of matter between man and nature (Burkett, 1999). Although Marx did not directly incorporate ecological value in the model of capitalist accumulation, he pointed out that the creation of surplus value is inherently linked to the interaction of nature (Marx, 1887). Overall, nature is the third party that actually increases the volume of surplus by receiving the external effect of the production process and in turn, arrested by the expansion of capitalism. Therefore, capitalism, by an automatic process, creates negative external effects. The very fabric of the climate crisis is also woven from the exhaust of intensive and widespread waves of capital accumulation (Bellamy & Diamanti, 2018).

The contemporary Marxian literature further extends the Marxian idea to global imperialism by contrasting the market-centric idea of carbon trading. Carbon trade is an ecological debt that the developed countries have created (Foster & Clark, 2004). It is an indicator of unequal ecological exchange as it premises on the asymmetric power relationship between developed and developing countries. Through carbon trading, the developed countries are now depositing waste in the poor countries. The ecological debts are now being amortised by the poor countries at the expense of the contamination of the atmosphere. Indubitably, ecological imperialism, therefore, has taken place and the unequal exploitation of the global common has reached beyond the capacity to absorb it. The creation and deposition of ecological debt in such a manner has been possible through the exercise of power by the developed countries.

Overall, there are two implications of Marxian analysis of climate capitalism (a) the recognition of the human–nature nexus and (b) power resulting in ecological debt leading to unequal exchange. The political economy tenet, however, does not explicitly analyse the climate change issue, and accordingly, it requires revision. Most importantly, the human–nature relationship aspect needs to be internalised into the framework of climate change governance.

5.6 An Alternative Framework

The chapter, by examining both the market and the non-market politico-economic analyses and by identifying the underlying gaps (Table 5.5) to address the climate change and sustainability issue, develops an alternative framework.

The new framework contrasts the propositions of neo-classical economics strongly as it sets aside the ethical, social and political consideration that is central to the global warming debate. The central argument of the framework is that the degradation of climate is subject to the commodification of nature with the expansion of capitalism

Table 5.5 Underlying gaps of theories regarding climate change

Branches name	Arguments for climate change	Underlying gaps
Neo-classical economics	(a) Market tools can correct negative externality, thereby, solve climate change problem (b) Suggested policy tools are carbon pricing, emission trading, taxation and subsidies as means of mitigation and adaptation (c) Technological progress can solve the climate change problem; emphasis on weak sustainability	(a) Ignores the weak governance and the political transaction cost in developing countries (b) Political feasibility of market-based instruments is questionable (c) Treating atmosphere as a marketable commodity is fallacious (d) Ignored the inherent values of nature and ignores the political and social context in addressing climate change (e) Natural capital is not substitutable with man-made capital
Institutional economics	(a) Financialization institutional stability	(a) Could not incorporate distributional aspects (b) Gap in explaining the political economy factors (power and political settlement) behind unstable nature of institutions (c) Human sociality aspect missing
Ecological economics	(a) Argues for steady state of economy or negative economic growth for future generations (b) Vehemently oppose the neoclassical economics position (c) Criticised pricing of the level of pollution (d) Proposed strong sustainability	(a) The modelling largely leaves out the human economy (b) Zero growth is not a viable solution to protect the atmosphere in the long run as production also has to be continued (c) Gaps in explaining the mutuality of human beings that needs to be emphasised to create a sustainable pathway
Marxism	(a) Capitalism is at root of destruction of nature (theory of metabolic rift) (b) Climate change is an outcome of historical ecological debt between industrialised and developing countries (c) Marxian proponents could identify the nexus between nature and human beings	(a) Could not explicitly internalise the 'human–nature' nexus into the sustainability framework (b) Marxian political economy factors are necessary to be focused more in addressing climate change issue

Source Prepared by the authors

utilising natural capital. The framework particularly emphasises the importance of institutional stability but argues that institutional economics could not capture the key dynamics of the failure of institutions. The power-sharing arrangement is crucial here to analyse such failures and, in this regard, factors of political economy are necessary to consider. Additionally, ecological economics could not provide any viable solution to address the climate change issue. Overall, the existing theoretical positions do not consider the factors such as institutional fragility, power and political settlement and human sociality appropriately or if they do, then inadequately. The new framework by incorporating the elements together shows that these are the key factors to make the market-centric tools fail to address the problem of climate change. Moreover, it theorises the concept of sustainability based on the material balance between humans and nature along with equitable economic growth.

5.6.1 Proposition 1: Externality and Distribution

Market cannot correct negative externality alone due to unachievable property rights over the atmosphere leading towards the unequal rate of pollution and distribution of burden between industrial and non-industrial countries.

The growth of the economy is strongly associated with the increase of negative externality. Simultaneously, the market is inefficient in correcting the negative externality automatically. The inefficiency of the market results from the unworkable property rights over the environment such as the atmosphere. As patterns of the growth of economies across the world are uneven, the negative externalities are shared and distributed unequally. Overall, the level of pollution has increased, and the unequal exchange of pollution is resulting in unequal impacts.

Countries can choose their growth policy in diverse ways; broadly, the trade-offs are two: (a) low economic growth or (b) rapid economic growth (Table 5.6). Based on the available alternatives, it is evident that if both developed and developing countries cooperate, then a rationalised level of pollution is possible (Case-1) which is an ideal scenario. With the advent of capitalism, however, developed countries have little incentive to harmonise economic growth, and therefore, Case 3, signifying unstable cooperation and higher inequality, is the reality. Case 2 is another unstable state where the developing countries are losing the potential economic benefits. In addition, these countries do not have either the physical or the financial capital.

Thus, the developing countries are bearing an uneven share of global emissions and hence the negative consequences of climate change. Case-4, however, signifies the extreme level of pollution caused by rapid economic growth in both by developed and developing countries. Although the developing countries have lower economic growth compared to the developed countries, they are also becoming inclined to increase the rate of growth, which would cause a complete destruction of nature in the upcoming days. Therefore, the lack of cooperation leads to climate change, which has to be sustained for a considerable amount of time. As a result, the developing countries are becoming bearers of the burden of climate change.

Table 5.6 Payoff matrix of the different movements of agents

		Developing countries	
		Low economic growth	Rapid economic growth
Developed countries	Low economic growth	**(Case-1: Rationalised level of pollution)** • The rate of carbon emission will be reduced • Rapid transformation of the composition of production can take place to ensure the growth of economy • The developing countries will get incentives to focus on eco-friendly economic growth • The rate of growth of carbon will decrease but the net stock of carbon can increase or remain the same	**(Case-2: Unstable cooperation)** • The carbon emission will increase slowly • Pressure from developed countries to implement eco-friendly economic growth • External pressure to devise market-based solution • The expansion of economic activity by the developed countries
	Rapid economic growth	**(Case-3: Unstable cooperation and high inequality)** • The income inequality will increase • There will be unequal impact of climate change across the world • The poor countries will become more vulnerable to the effect of climate change	**(Case-4: Extreme level of pollution)** • The rate of pollution will be faster than any level • The effects of climate change will increase • The natural resources will deplete faster

Source Prepared by the authors

The hypothesis accordingly is that there is inequality regarding the level of pollution and sharing the burden of climate change.

5.6.2 *Proposition 2: Capital Deficiency and Non-functioning Market Solutions*

The developing countries do not have adequate technology or financial capital to address climate change. The base of capital to tackle the increasing spatial unequal burden of negative externality by the developing countries is limited and the financial capital to transform the composition of the production process in developing countries is absent.

Technological improvement can reduce social cost and pollution (Fig. 5.6). The marginal damage curve is positively sloped with the cost and pollution where the marginal abatement curve represents that investing in the technology will help reduce pollution. The initial equilibrium presented in the left panel is the *E** where the area symbolised by *DD* is the benefit and the *AA* is the social cost. The positive technological change presented in the right panel, however, shifts the marginal abatement curve_1 to the marginal abatement curve_2. The increased technical efficiency sets the equilibrium at *E*** that is showing the reduction of pollution. The developing countries are lacking such capital in order to transform the composition of production. The deficiency in the total stock of capital reduces the capabilities to implement the adaptation and mitigation measures by these countries. Furthermore, the transformation mechanism of capital from developed to developing countries is fraught in nature due to the unequal power-sharing arrangements. The developing countries, therefore, have been facing capital deficiency and the market-based solutions are ineffective.

The hypothesis, accordingly, is that against the backdrop of a lack of capital in developing countries, the developed countries promise to transfer capital in the name of cooperation; the transfer process, however, is faulty in nature. This results in non-functioning market solutions on the one hand and continuous pollution by developed countries on the other.

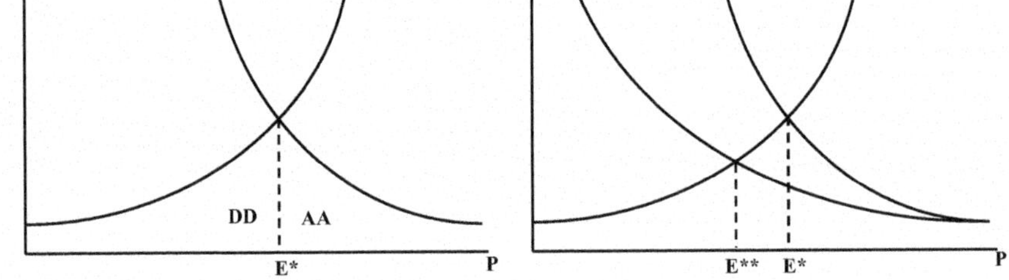

Fig. 5.6 Role of technology in reduction of pollution. *Source* Prepared by the authors

5.6.3 Proposition 3: Institutions and Carrying Capacity

The carrying capacity of developing countries has been decreasing at a faster rate than that of developed countries because of increasing negative externality stemming from increased pollution due to lax institutional arrangements resulting from non-compliance-based international cooperation and unplanned national efforts.

The increasing level of pollution exceeds the carrying capacity of the earth (Fig. 5.7). The carrying capacity of the earth is fixed under the ceteris paribus (keeping the elements of the environment fixed) represented by A. The pollution is positively sloped with the business-as-usual scenario of the current world depicted by P. The saturating point is E because up to this point, the assimilation by the earth is equal to or greater than the pollution ($A \geq P$) but the carrying capacity reduces after point E where the pollution becomes greater ($P > A$) than the assimilation. The environment degrades faster after the equilibrium condition at E depicted by A^* with respect to time. As the developing countries receive an uneven share of global emissions, lack in capital, and urge for rapid economic growth gradually, the carrying capacity of these countries degrades at a faster rate.

The mainstream tools such as taxation, tradable permits, etc. cannot correct this negative externality and can just shift the pattern and the distribution of pollution. However, pollution will increase until pollution justice is ensured. International cooperation is the necessary condition in this context to curb the level of emissions or ensure pollution justice. International cooperation, however, is non-existent due to the fragile nature of institutional arrangements. Moreover, there is institutional fragility stemming from the unplanned decision-making process at the national level.

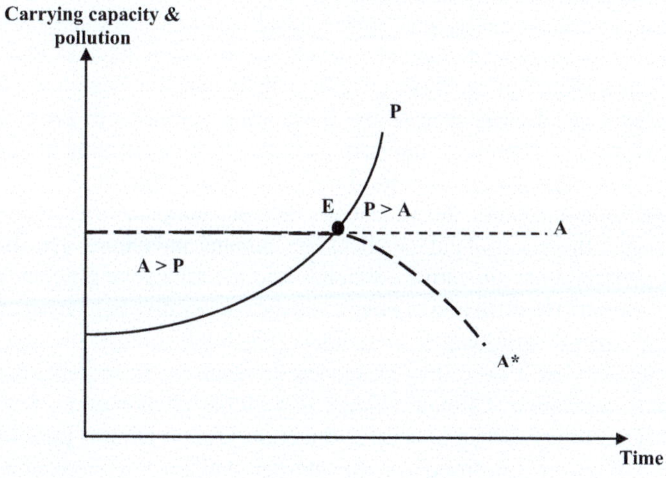

Fig. 5.7 Degradation of carrying capacity. *Source* Prepared by the authors

The hypothesis is that fragile institutions could not bind the developed countries to curb the emission and developing countries to plan appropriately. The ultimate result of this is the decreasing carrying capacity of the developing countries.

5.6.4 Proposition 4: Material Balance and Sustainability

The sustainability of the environment by ensuring equitable economic growth and environmental justice is subject to the material balance between humans and nature, which derives from cooperation among the agents that is subject to the diffusion of power and functioning of regulatory institutions.

One of the prime theses of this chapter is to devise an alternative understanding of sustainability. The previous propositions postulate that the capitalist mode of production is hardly concerned about the negative externality generated from economic activities. This extensive anthropogenic invasion on the environmental quality has reduced the carrying capacity of the earth. The neoliberal market-oriented policy in order to expand production has created a rift in the ecosystem and in the interdependence. Thus, the whole balance of nature and humans ends with a highly unequal distribution of effects and burdens.

Under the sustainability model, the state, the firms and the households interact in the production system. The state allocates the natural resources and conserves nature. Firms produce the commodity and sell it to the consumer. Consumers meet their needs from the produced commodity (Fig. 5.8). In the total process, the environment carries the negative externalities. The book argues that for the environmental crisis, allocation is important both nationally and internationally. The rift takes place due to the disruption of the material balance by the misallocation of resources and uneven growth of the economy worldwide. The developed countries have to bear the cost of transformation of the production system in their land for decarbonisation and to extend technological assistance to the developing countries in order to ensure environmental justice. Therefore, the sustainability of the environment depends on the allocation and management of natural resources and pollution.

The central argument is that the allocative efficiency under an effective regulatory regime can ensure the material balance between humans and nature. The differences between the neo-classical allocative efficiency and the allocative efficiency under a regulatory regime are the diffusion of power and the cooperative behaviour among the agents under perfect institutional arrangements. The book contests the neo-classical allocative efficiency as it describes efficiency as resulting from self-interest with minimum state regulation. Climate change is a global phenomenon, but a global cooperation is yet to crystallise to combat the issue. Cooperation is possible only if the imbalance of power is defused and the internationally acknowledged regulatory framework functions properly.

Overall, the book hypothesises that sustainability should be conceptualised and theorised along with the human sociality aspect rather than only with the development

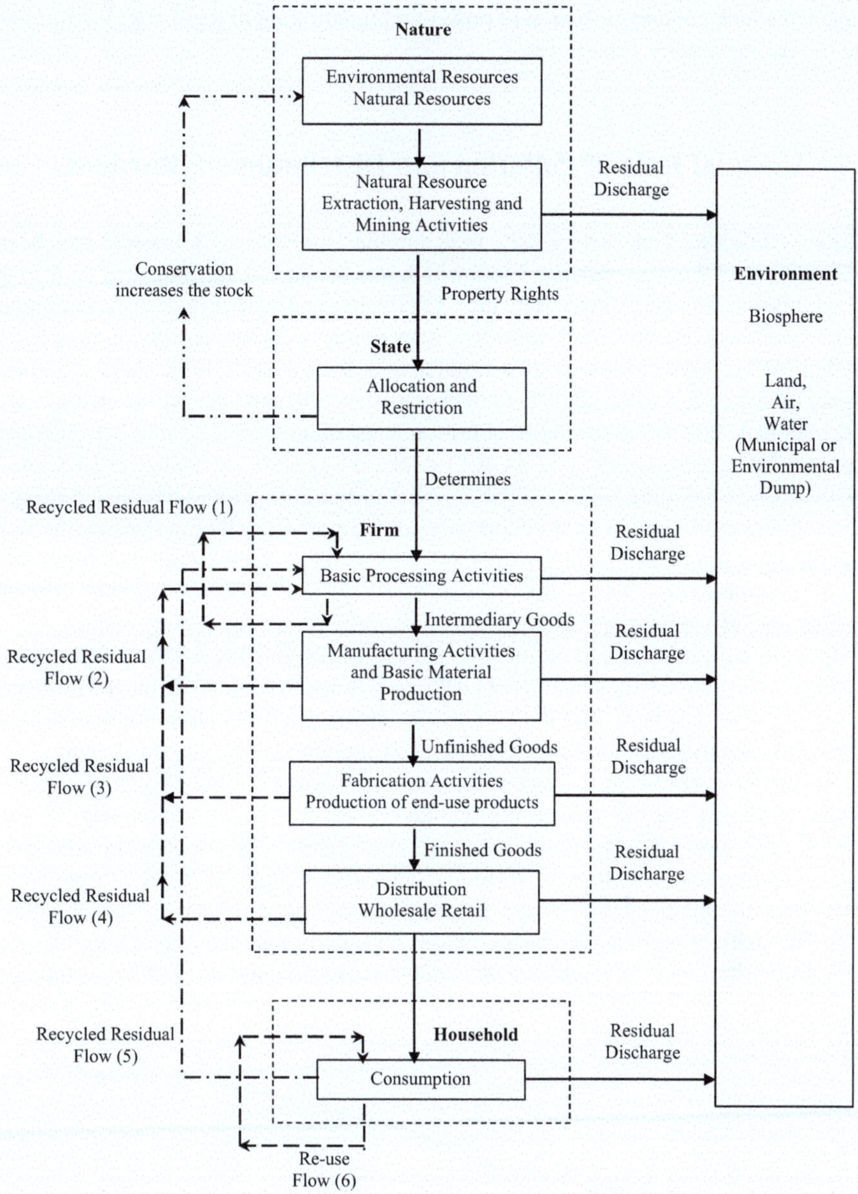

Fig. 5.8 Interactions among states, firms and households in utilising nature. *Source* Prepared by the authors

aspect in which context, diffusion of power and functioning of regulatory institutions are the two necessary conditions.

5.7 Unequal Rate of Pollution and Distribution of Burdens

Global emission scenarios clearly indicate that the rate of emission has been increasing over the years barring 2020, which saw the largest absolute drop (5.8%) in CO_2 emissions in history caused by the COVID-19 pandemic (Fig. 5.9). Other major GHGs such as CH_4 and N_2O, however, kept rising in 2020. Notably, atmospheric methane (CH_4) concentrations are estimated to be at an all-time high, about 2.5 times higher than the pre-industrial level of approximately 750 parts per billion (King et al., 2021; NOAA, 2021b). Accordingly, the total emissions of GHGs continue to rise and have reached a record high of 52.4 $GtCO_2e$ (excluding land-use change or LUC) and 59.1 $GtCO_2e$ (including LUC) (UNEP, 2020). Furthermore, a rapid rebound in energy demand and emissions in many economies in the post-COVID era stresses the risk that the emission of GHGs, including CO_2, will increase significantly to compensate for the economic loss. Therefore, one can surely guess that the fight against climate change is likely to be more difficult and challenging in the upcoming years.

It is difficult, however, to define who should be responsible for the protection of the atmosphere from the emissions of GHGs as it is not possible to define the property rights of the atmosphere. The data on GHGs emissions draws attention to concerns about the effectiveness and fairness in the fight against global climate change.

An estimate based on a 50-year emission scenario (1950–2000) from the last century revealed that the industrialised countries account for approximately 72% of the total CO_2 emissions accumulated in the atmosphere. The contribution to GHGs emissions by these countries in 2000 has decreased compared to the emissions of 1990 while the share of emissions by developing countries has been increasing (Table 5.7). The share of cumulative energy-related CO_2 emissions (country's total historic emissions), however, is larger for the industrialised countries (73.8%) compared

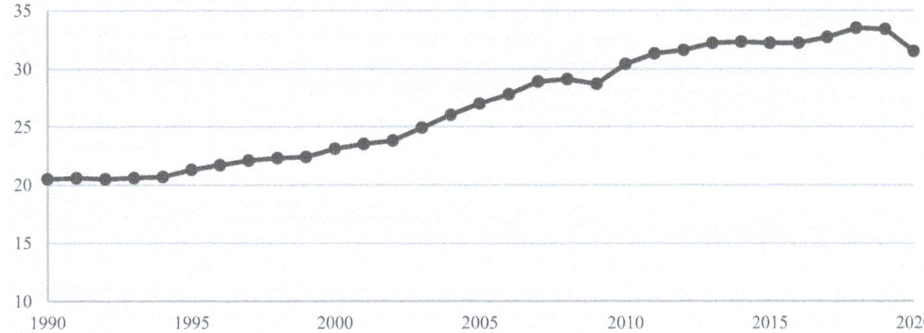

Fig. 5.9 Global energy-related CO_2 emission, 1990–2020 (in Gt CO_2). *Source* IEA (2020)

to the developing ones (26.2%). This is also apparent in the case of emissions of CO_2 with land use change whereas the percentage for industrialised and developing countries is 52.6% and 47.4%, respectively. Moreover, the share of emissions of the top 20 nations indicates that they are accountable for the larger share of emissions compared to the rest of the countries.

During the last decade also, top emitters share the larger proportion of total GHGs emissions in the atmosphere (Fig. 5.10). The top four emitters have contributed to 55% while the top seven contributed to 65% of the total global emission. In addition, the G20 economies, most of which are the industrialised and developed countries, emitted 78% of the total GHGs into the atmosphere.

Individually, China is the largest emitting country in the last few years. In 2019, China, as the top country, alone was responsible for almost 30% of the total global

Table 5.7 Shares of global emissions by the industrialised, developing and top 20 countries

Indicator	Industrialised countries $n = 38$[a] (%)	Developing countries $n = 147$ (%)	Top 20 countries (%)
1990 GHGs emissions (excl. LUC)	53.9	46.1	68.9
2000 GHGs emissions (excl. LUC)	48.4	51.6	70.2
2000 GHGs emissions (incl. LUC)	39.2	60.8	66.3
Cumulative energy-related CO_2 emissions 1950–2000 (excl. LUC)	73.8	26.2	83.0
Cumulative energy-related CO_2 emissions (with LUC)	52.6	47.4	75.4

Source WRI (2008)
[a]Counting the European Union countries individually, excluding the EU as a collective member

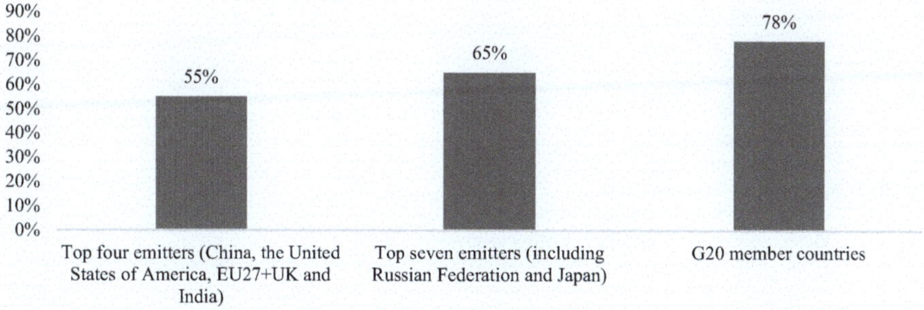

Fig. 5.10 Share of total GHGs emission of the top emitting countries and G20 member countries (excluding LUC), 2011–2020. *Source* UNEP (2020)

CO_2 emission, which was roughly twice the amount emitted by the USA—the second largest emitter (Fig. 5.11). Although China is yet to be officially recognised as a developed country (Boyle, 2021), its performance in the economic sphere during the last couple of years urges that it is high time to accept or admit the fact that it is no longer a developing country (Ahlstrom, 2020). The case of India—as a developing country—is also noticeable. However, most of the largest emitters were the developed countries in this particular year.

Currently, China emits the highest level of CO_2 on an annual basis, but the cumulative emission scenario regarding individual country's contribution shows that China has emitted CO_2 far less than the USA and 28 European Union (EU) countries (Fig. 5.12). The USA has emitted more CO_2 (around 400 billion tonnes since 1751) than any other country to date and is responsible for 25% of the historical emission. EU countries, that jointly set goals, are also significant historical contributors, accounting for 22%. Many of the current top annual emitters—including India and Brazil—are not historically significant contributors. Moreover, the rest of the countries (mainly the developing and underdeveloped) of the world combined were responsible for 22% of the total emission of CO_2 from 1751 to 2017.

The per capita emission scenario also depicts that the developed countries are more responsible for putting pressure on the atmosphere (Fig. 5.13). The USA had the highest rate among major countries, with 17.6 metric tonnes per capita in 2019, which tied with Saudi Arabia for the first time. On the contrary, China is in the top 15 with roughly 6.4 metric tonnes per citizen and most other countries ahead of China are the wealthier countries of Europe and Asia, with France, Poland and the UK all averaging around 7–8 metric tonnes of carbon per citizen. Overall, the noticeable fact is that in terms of per capita emission, most of the high-emitting countries fall in the upper-income category except Russia, Iran and Malaysia.

Despite the decreasing trend of emission by the individual developed countries in recent years, they should be responsible for the historical emission that has caused

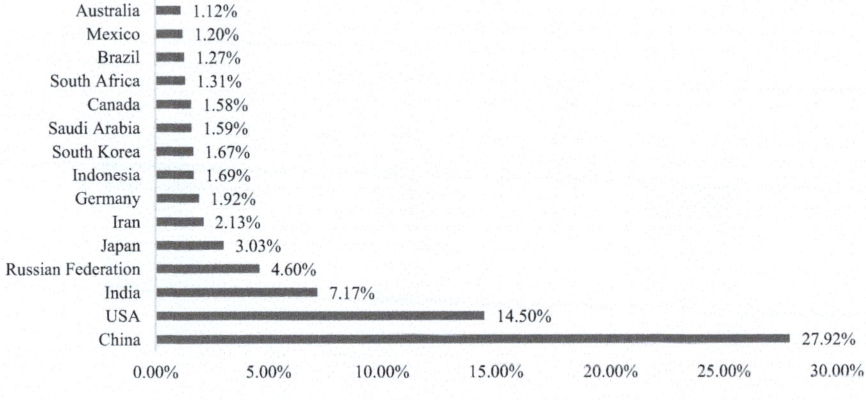

Fig. 5.11 Largest producers of CO_2 emissions worldwide in 2019, by share of emissions. *Source* US Energy Information Administration 2019 (Tiseo, 2021)

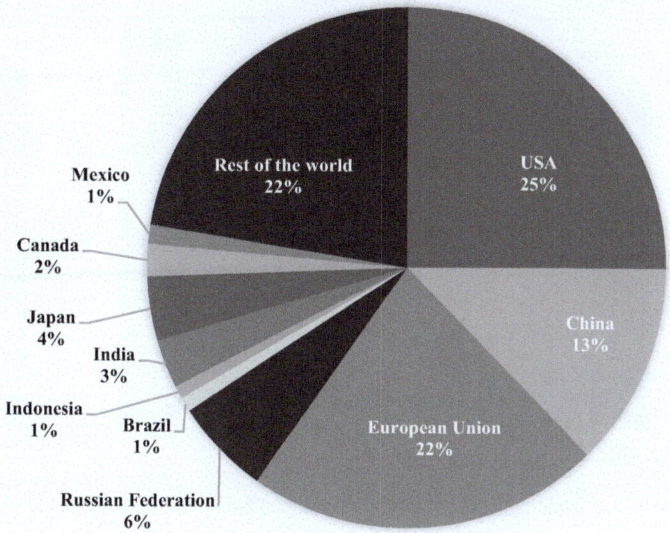

Fig. 5.12 Share of cumulative CO_2 emissions (production-based emission from fossil fuel combustion and cement) over the period 1751 to 2017. *Source* Oxford Martin School (Ritche, 2019)

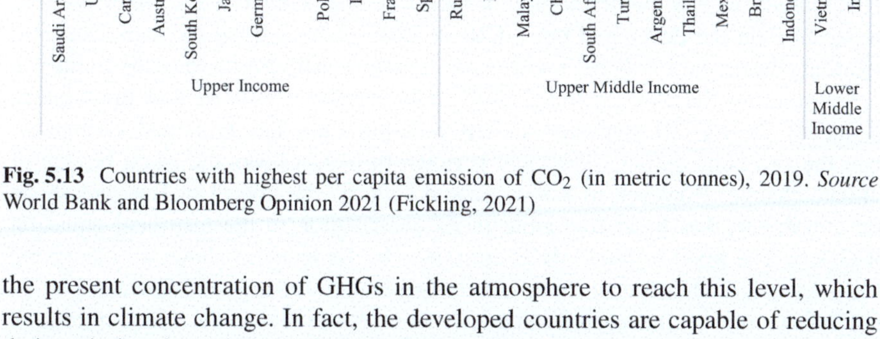

Fig. 5.13 Countries with highest per capita emission of CO_2 (in metric tonnes), 2019. *Source* World Bank and Bloomberg Opinion 2021 (Fickling, 2021)

the present concentration of GHGs in the atmosphere to reach this level, which results in climate change. In fact, the developed countries are capable of reducing their emission through the inclusion of renewable energy sources with technological advancement. The developing countries are still lagging behind in this regard, and accordingly, their emission trend is showing a rise.

Furthermore, it is evident in a recent report that the world's biggest 60 banks, which are of the developed countries, have delivered \$3.8 trillion for fossil fuel companies since 2015 and the financing shows a rising trend over the years, except a

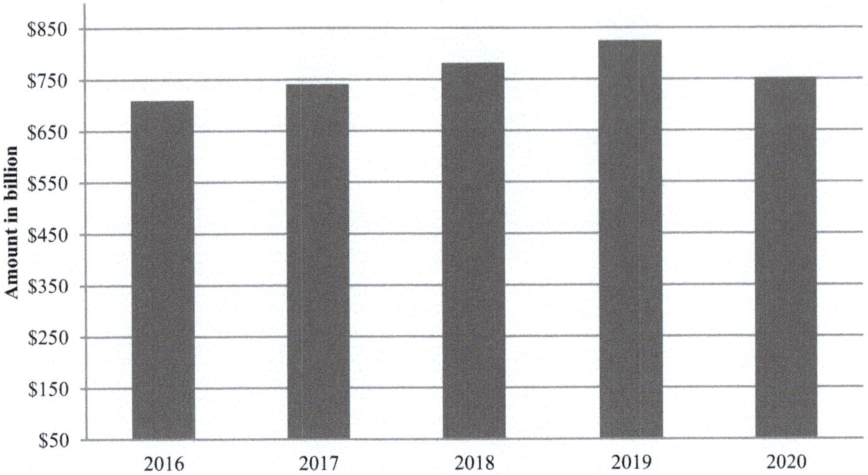

Fig. 5.14 Fossil fuel financing from the world's 60 largest banks was higher in 2020 than in 2016 and 2017. *Source* Rainforest Action Network et al. (2021), The Guardian (2021)

fall in 2020 caused primarily due to the COVID-19 pandemic (Fig. 5.14). It indicates the less inclination of the wealthy nations to fulfil the pledges made at the Paris climate conference to reduce GHGs emissions.

Although the low-income developing countries play a negligible role in rising greenhouse gases in the atmosphere, they are more vulnerable to climate change. According to the Global Climate Risk Index (CRI)-2020, seven of the ten most affected countries and territories between 1999 and 2018 were developing nations with low or lower-middle incomes, two were upper-middle-income countries (Thailand and Dominica), and only one was a high-income country (Puerto Rico) (Table 5.8). It confirms that the developing or less developed countries are typically more impacted than the developed ones. CRI of 2019 also signified the same scenario regarding the period 1998–2017 with some variations in the positions of the countries.

Considering the case of Bangladesh, it is evident from CRI that the country is one of the top 10 countries and its position (7th) has remained unchanged over the two consecutive assessment periods. In fact, it bears the worst burden of climate change. However, as a low-income country, the impact of Bangladesh on the atmosphere has been negligible in its contribution to the historical emission of GHGs. The trend of total GHGs emissions by Bangladesh shows an increasing trend reaching 220.75 metric tonnes (including land use change and forestry-LUCF) and 198.97 metric tonnes (excluding LUCF) in 2018 (Fig. 5.15), which happened primarily due to the country's endeavour to cope with the global trend of economic growth with more industrialisation.

On the contrary, the per capita emission scenario of total GHGs and individual GHG signifies that the contribution of the country is quite non-significant (Fig. 5.16). Per capita emission of GHGs shows an increasing trend with some fluctuations, whereas the trend of per capita CO_2 emission shows an increasing trend. On the

Table 5.8 Long-term Climate Risk Index (CRI) (from 1999 to 2018 annual average)

CRI ranking 1999–2018 (1998–2017)	Country	CRI score	Death toll	Deaths per 100,000 inhabitants	Total losses in million US$ PPP	Losses per unit GDP in %	Number of events
1 (1)	Puerto Rico	6.67	149.90	4.09	4567.06	3.76	25
2 (3)	Myanmar	10.33	7052.40	14.29	1630.06	0.83	55
3 (4)	Haiti	13.83	274.15	2.81	388.93	2.38	78
4 (5)	Philippines	17.67	869.80	0.96	3118.68	0.57	317
5 (8)	Pakistan	28.83	499.45	0.30	3792.52	0.53	152
6 (9)	Vietnam	29.83	285.80	0.33	2018.77	0.47	226
7 (7)	Bangladesh	30.00	577.45	0.39	1686.33	0.41	191
8 (13)	Thailand	31.00	14.00	0.21	7764.06	0.87	147
9 (11)	Nepal	31.50	228.00	0.87	225.86	0.40	180
10 (10)	Dominica	32.33	3.35	4.72	133.02	20.80	8

Source Eckstein et al. (2020)

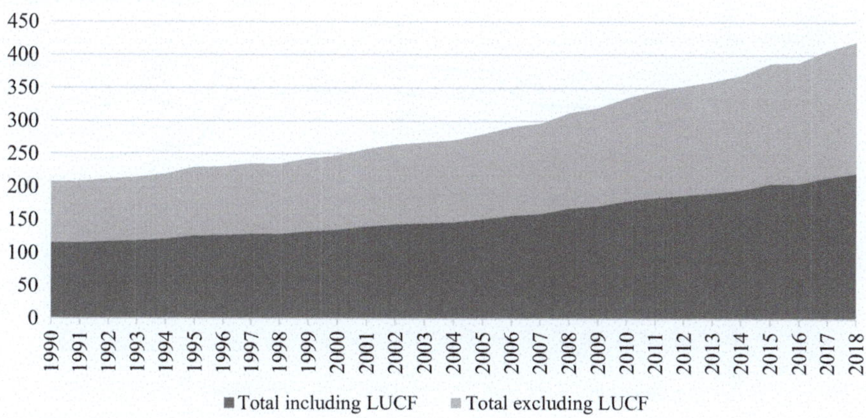

Fig. 5.15 Total GHGs emission by Bangladesh (in Mt). *Source* Climate Watch (2021)

contrary, the trend in the case of NH_4 (methane) and N_2O (nitrous oxide) remains quite stable over the years.

Moreover, the contribution to per capita emission of GHGs by Bangladesh is also non-significant compared to that of the major economies of the world (Fig. 5.17). It is also evident that the per capita emission of GHGs in Bangladesh is lower than the global rate.

Overall, the discussion of this section demonstrates that the developed countries were mainly responsible for GHGs emissions, accordingly for climate change. The developing countries, on the contrary, had a very negligible role in this regard but

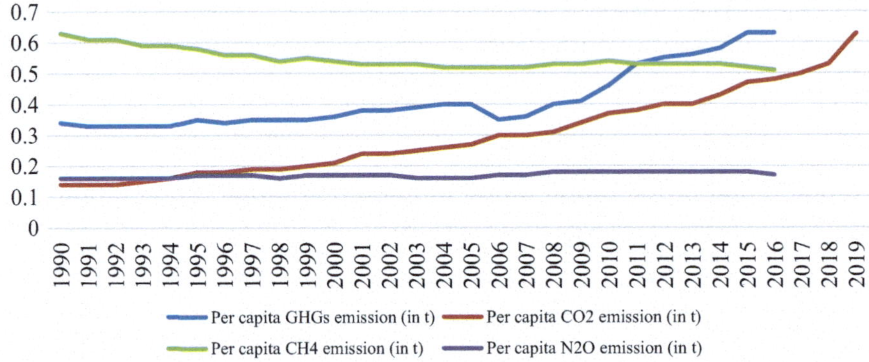

Fig. 5.16 Per capita emissions of GHGs by Bangladesh. *Source* Oxford Martin School (Ritchie & Roser, 2020). *Note* Data available up to 2019 in case of CO_2 only; measured based on data from Global Carbon Project, Gapminder, UN, & Climate Watch

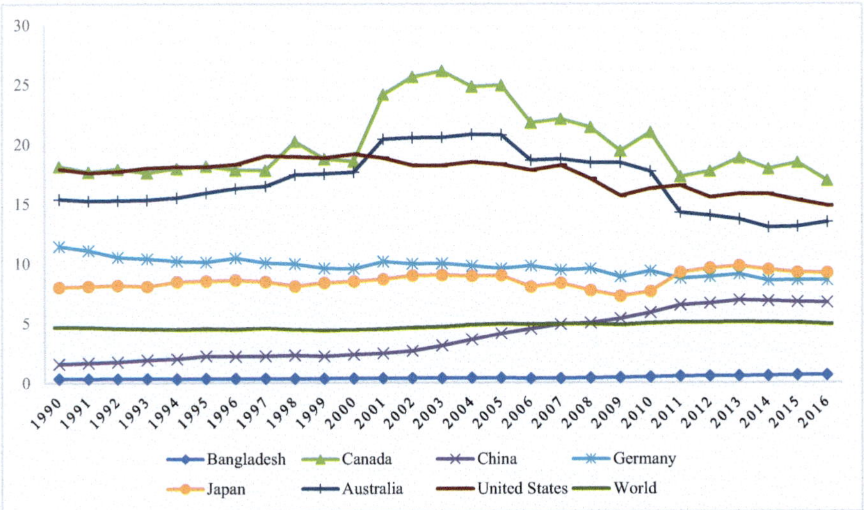

Fig. 5.17 Per capita GHGs* emissions by Bangladesh in contrast to the major economies (in tonnes of CO_2e**). *Source* Oxford Martin School (Ritchie & Roser, 2020). *Note* *Here GHGs include carbon dioxide, methane, nitrous oxide, and F-gases; **CO_2e denotes CO_2 equivalent which means, "…having the same warming effect as CO_2 over a period of 100 years"; measured based on data from Climate Watch

have been forced to bear the brunt. The particular case of Bangladesh in examining the impacts of climate change provides a more concrete understanding in this regard.

5.7.1 *Climate Change, Agriculture and Food Security: Burden for Bangladesh I*

Agriculture is the main source of income and employment for the people of Bangladesh. The deleterious effects of climate change pose severe challenges to food production and livelihood sustenance. This section analyses the impact of climate change on the crop (*boro* rice) production and food security in Bangladesh based on different studies conducted under *Unnayan Onneshan* (Basak, n.d.; Basak, 2009; Basak et al., 2009, 2010; Titumir & Basak, n.d.; Titumir & Basak, 2013). Data for examining the impact on crop production was collected from six locations—Rajshahi, Mymensingh, Satkhira, Barishal, Cumilla and Sylhet—in different Agro-Ecological Zones (AEZs) of Bangladesh, which are the major rice growing areas.

Effects of Maximum Temperature (Tmax) on Boro Rice Production

Temperature increase had varying effects on the yields of different locations. Increase in maximum temperature had the most significant negative impact on Satkhira and the least on Sylhet (Table 5.9). The monthly average maximum temperature in Satkhira in January, February, March, April and May (growing season of *boro* rice) in 2008 was 25.3 °C, 26.7 °C, 32.5 °C, 34.9 °C and 36.1 °C, respectively; increased maximum temperature of 2 °C and 4 °C reduced yields by more than 13% and 28%, respectively, in this region. Yields in other regions of Bangladesh also decreased due to a similar increase in temperature but not at the same rates. The average rates of reduction in yield for all six locations were in excess of 6% and 16% for 2 °C and 4 °C increases in temperature, respectively (Fig. 5.18).

Effects of Minimum Temperature (Tmin) on Boro Rice Production

Increases in minimum temperature, too, had a negative impact on *boro* rice yield; it fell by 0.40–13.1% and 0.11–15.5% due to the increase of 2 °C and 4 °C, respectively. The negative impact of minimum temperature on Satkhira was more pronounced than in the other five regions (Table 5.10). In this region, an increase in monthly average minimum temperatures of 2 °C and 4 °C reduced yield by 13% and 14%, respectively. The results were similar for the other selected regions, but the percentage changes in yields were different. The average percentage reduction of rice yield for all six locations due to the increases in minimum temperature was more than 4% and 8.5% for 2 °C and 4 °C rise, respectively (Fig. 5.19).

Combined Effect of Maximum and Minimum Temperatures on Boro Rice Production

The combined effect of maximum and minimum temperatures on rice yields is more alarming compared to their individual effects. Yields drastically decreased by 3.2–18.7% and 5.33–36.0% for 2 °C and for 4 °C increases, respectively. The most affected area was Cumilla where yield decreased by 18.7% and 36.0% for 2 °C and 4 °C increase, respectively (Table 5.11). The average percentage decreases in rice yield for all six locations were in excess of 10.4% and 22.87% under two different scenarios of temperature rise (Fig. 5.20).

Table 5.9 Percentage change of *boro* rice yield under various maximum temperature scenarios

Location	Climate change phenomena	Rice yield (kg ha^{-1})	Percentage change of rice yield
Rajshahi	Increased maximum temperature 2 °C	3020	– 02.60
	Increased maximum temperature 4 °C	2707	– 12.70
Mymensingh	Increased maximum temperature 2 °C	4167	– 06.74
	Increased maximum temperature 4 °C	4001	– 10.45
Satkhira	Increased maximum temperature 2 °C	3402	– 13.50
	Increased maximum temperature 4 °C	2860	– 28.70
Barishal	Increased maximum temperature 2 °C	4761	– 06.60
	Increased maximum temperature 4 °C	3954	– 22.40
Cumilla	Increased maximum temperature 2 °C	4831	– 11.00
	Increased maximum temperature 4 °C	4271	– 21.30
Sylhet	Increased maximum temperature 2 °C	5569	+ 03.90
	Increased maximum temperature 4 °C	5353	– 00.11

Source Titumir and Basak's calculation based on DSSAT model simulation
DSSAT—decision support system for agro-technology transfer

Effects of Carbon Dioxide on Boro Rice Production

These projections are based on the assumption that the concentration of atmospheric CO_2 progressively increased by 50 ppm, 100 ppm and 200 ppm from the 2005 level of 379 ppm. Had CO_2 levels increased by 50 ppm (i.e. to 429 ppm) in 2005, *boro* rice yields would have increased by 2.1–4.4%; increases in CO_2 levels of 100 ppm and 200 ppm would have resulted in yield increases of 4.0–9.6% and 5.2–18.2%, respectively. For an increase of 200 ppm, the maximum positive effect was found to be in Cumilla (18% increase) and the minimum in Sylhet (only 5.2%) (Table 5.12).

Combined Effects of Maximum, Minimum Temperature and Carbon Dioxide

Scenarios involving different combinations of all three parameters were also examined for two locations: Rajshahi and Cumilla. Tmax 2 °C + Tmin 2 °C + CO_2 200 ppm had a slight positive effect on rice yields but these were not as significant as those of other scenarios. The most significant negative effect was Tmax 4 °C + Tmin

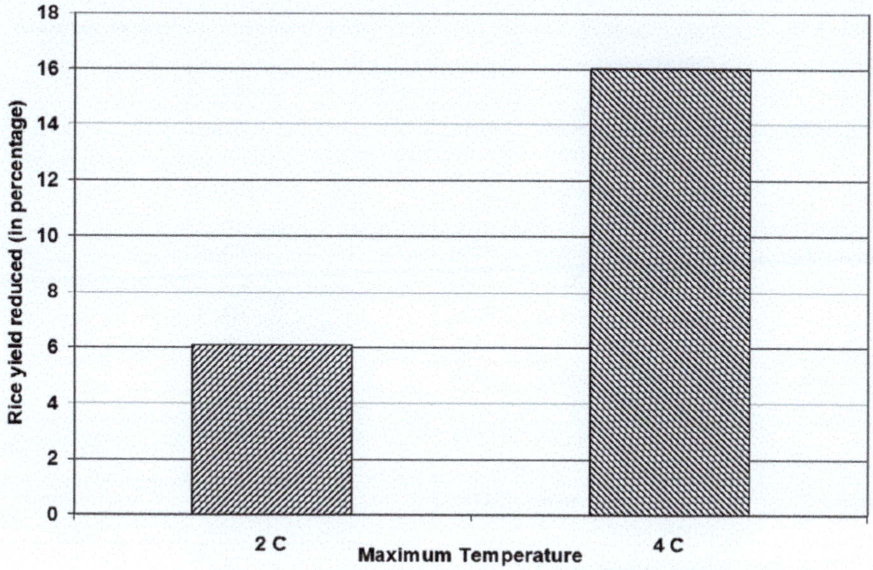

Fig. 5.18 Percentage of average reduction of rice yield due to increases in maximum temperature. *Source* Titumir and Basak's calculation based on DSSAT model simulation

4 °C + CO_2 50 ppm, which resulted in a 33.1% reduction in yield in Cumilla (Table 5.13). From the findings, it is clear that temperature is one of the most dominant climatic factors, which may considerably affect rice yields in future.

In summary, the effect of an increase of both maximum and minimum temperatures on crop yields shows a declining trend though varied across different regions. On the other hand, the level of atmospheric CO_2 concentration is likely to have a positive effect on rice yield. The combined effects of temperature change and CO_2, however, indicate a negative change in the yields of rice.

Future Demand–Production Gap Under Different Climate Scenario (Temperature and CO_2)

Future impacts of climate change on food security can be projected under five different scenarios A, B, C, D and E through simulation (Table 5.14). There is a significant demand production gap of rice in 2050, 2070 and 2100 under the scenarios. Rice yield might be reduced (on average for the selected six regions in Bangladesh) by 8.28% for scenario A, 4.95% for B, 24.66% for C, 21.79% for D and 14% for E.

If the current trend of production persists, on an average, there will be a shortfall of 7.54 million tonnes and 5.49 million tonnes of rice in 2050 under scenarios A and B, respectively. As a result, more than 30 million and 22 million people might face rice shortage, which is equivalent to 11.80% and 8.80% of the projected population in 2050 for respective scenarios (Table 5.14; Fig. 5.21).

Similarly, the amount of rice shortage in 2070 might be 29.38 million tonnes, 27.30 million tonnes and 21.67 million tonnes on an average under scenarios C, D

Table 5.10 Percentage change of *boro* rice yield under various minimum temperature scenarios

Location	Climate change phenomena	Rice yield (kg ha^{-1})	Percentage change of rice yield
Rajshahi	Increased minimum temperature 2 °C	3051	– 01.67
	Increased minimum temperature 4 °C	2927	– 05.60
Mymensingh	Increased minimum temperature 2 °C	4450	– 00.40
	Increased minimum temperature 4 °C	4323	– 03.24
Satkhira	Increased minimum temperature 2 °C	3418	– 13.10
	Increased minimum temperature 4 °C	3384	– 14.00
Barishal	Increased minimum temperature 2 °C	4917	– 03.50
	Increased minimum temperature 4 °C	4334	– 14.90
Cumilla	Increased minimum temperature 2 °C	4936	– 09.04
	Increased minimum temperature 4 °C	4588	– 15.50
Sylhet	Increased minimum temperature 2 °C	5502	+ 02.70
	Increased minimum temperature 4 °C	5353	– 00.11

Source Titumir and Basak's calculation based on DSSAT model simulation

and E, respectively. The number of population might face such rice shortage amounts of more than 119 million, 111 million and 88 million, equivalent of 35.04%, 32.56% and 25.96% of the projected population in 2070 under three scenarios (Table 5.14; Fig. 5.22).

The gap between the demand and production of rice might increase further in the year of 2100. A continuation of the prevailing trend might witness a demand–production gap of rice at 68.07 million tonnes, 55.47 million tonnes and 48.39 million tonnes under scenarios C, D, and E, respectively. Consequently, 45.90% (273.02 million), 43.85% (226.40 million) and 38.25% (197.50 million) of the projected population may face difficulty in accessing rice in 2100 (Table 5.14; Fig. 5.23).

Increased incidence of natural disasters—a direct outcome of climate change—is also seriously threatening the overall food security of the country. Moreover, the decrease in arable land, land fragmentation, depletion of groundwater, decline in soil fertility, waterlogging and salinity intrusion are also emerging as newer problems in the agricultural sector in the face of climate change.

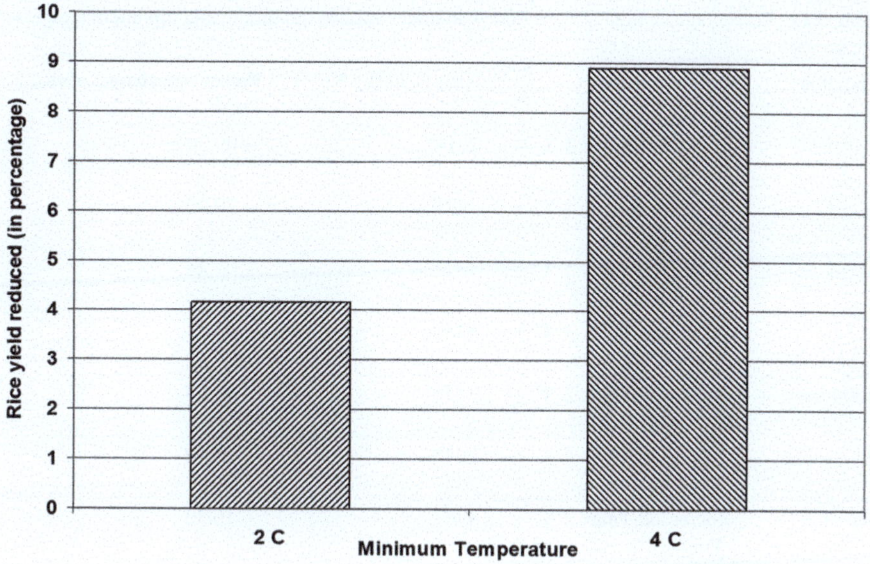

Fig. 5.19 Percentage of average reduction of rice yield for minimum temperature. *Source* Titumir and Basak's calculation based on DSSAT model simulation

Floods have the most detrimental effect on crop production in Bangladesh. An important loss in potential production would result from protracted floods delaying the *Aman* rice plantation. Flash floods have caused the loss of *boro* rice on a regular basis in the *haor* areas in recent years. Throughout the years, cyclones have also wreaked havoc on Bangladesh's agriculture (Table 5.15). Bangladesh is primarily affected by drought during the pre-monsoon and post-monsoon seasons. Additionally, rising sea levels can have a variety of effects on agriculture, including salinity intrusion and flooding, as well as damage caused by increased cyclone frequency. The combined effects of these three factors reduce agricultural output, particularly in the coastal zone of Bangladesh.

5.7.2 *Frequency and Intensity of Natural Disasters: Burden for Bangladesh II*

Bangladesh is one of the most disaster-prone countries in the world due to its geographical location. The intensity and frequency of natural disasters, however, have increased in recent decades because of climate change. As per the World Risk Report 2020, Bangladesh is the thirteenth (13th) most natural disaster-prone country among 181 countries of the world. According to the index values, Bangladesh falls under the category 'very high' in terms of risk, exposure and lack of coping and

Table 5.11 Percentage change of *boro* rice yield under various maximum and minimum temperature scenarios

Location	Climate change phenomena	Rice yield (kg ha^{-1})	Percentage change of rice yield
Rajshahi	Increased Tmax 2 °C + Tmin 2 °C	2816	– 09.20
	Increased Tmax 4 °C + Tmin 4 °C	2481	– 20.00
Mymensingh	Increased Tmax 2 °C + Tmin 2 °C	4323	– 03.24
	Increased Tmax 4 °C + Tmin 4 °C	3230	– 05.33
Satkhira	Increased Tmax 2 °C + Tmin 2 °C	3225	– 18.00
	Increased Tmax 4 °C + Tmin 4 °C	2945	– 25.10
Barishal	Increased Tmax 2 °C + Tmin 2 °C	4267	– 16.30
	Increased Tmax 4 °C + Tmin 4 °C	3446	– 32.30
Cumilla	Increased Tmax 2 °C + Tmin 2 °C	4413	– 18.70
	Increased Tmax 4 °C + Tmin 4 °C	3471	– 36.00
Sylhet	Increased Tmax 2 °C + Tmin 2 °C	5518	+ 03.00
	Increased Tmax 4 °C + Tmin 4 °C	4369	– 18.50

Source Titumir and Basak's calculation based on DSSAT model simulation

adaptation capacities, and under the category 'high' in terms of vulnerability and susceptibility (Table 5.16).

Floods are a common occurrence in Bangladesh as part of the normal hydrological cycle, generally affecting 22–30% of the country (MoEF, 2009). Floods have become more intense and frequent, as the climate has changed over the last three to four decades. Recent notable and catastrophic floods have occurred in 1988, 1998, 2004, 2007, 2017 and 2022. A significant percentage of the total land area was affected by floods across different times (Fig. 5.24). It is also evident that the majority of the severe floods, which covered more than 30% of the country, occurred after 1974.

Floods caused extensive damage by inundating significant land areas and thus, affected the lives and livelihood of thousands of people (Table 5.17). The most recent flood in 2022 has severely impacted a total of 70 upazilas (subdistricts) in the divisions of Sylhet, Rangpur, Mymensingh and Chittagong (The Daily Star, 2022).

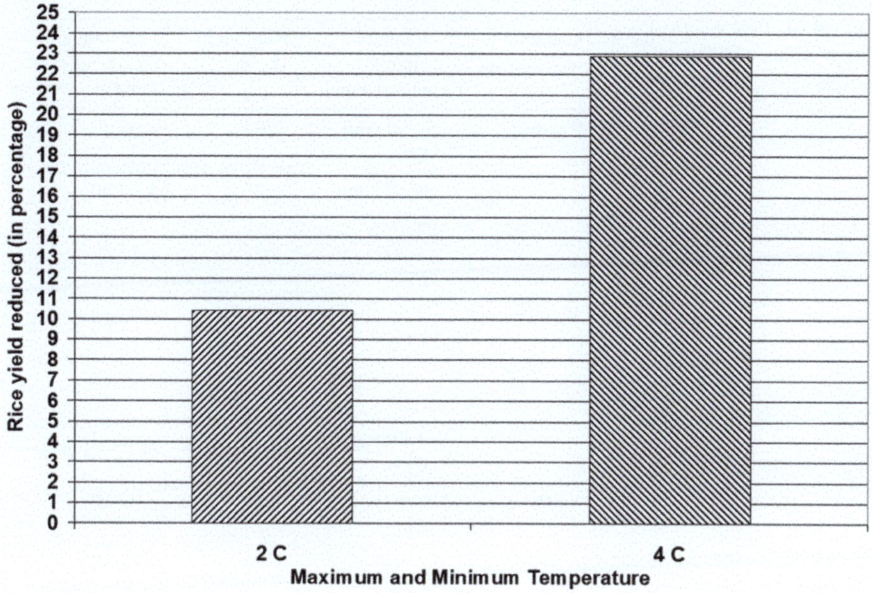

Fig. 5.20 Percentage of yield reduction for maximum and minimum temperature. *Source* Titumir and Basak's calculation based on DSSAT model simulation

Overall, the frequency, intensity and duration of floods have increased over the last few decades.

The impact is also noticeable in terms of the number of normal and extreme floods. Over the years, the number of normal or moderate floods has decreased while the number of extreme floods or extreme low floods has increased (Table 5.18). Severe floods such as those of 1988 and 1998 are expected to occur at intervals of 50–100 years but climate change is accentuating the occurrence more often (Unnnayan Onneshan, 2008).

Cyclones and storm surges hit Bangladesh's coastal area on a quite regular basis (Table 5.19) causing devastating effects. The anomalies are already visible in recent times, with the occurrence of cyclones like *Sidr* in 2007 and the *Aila* in 2009, causing huge economic and social losses.

Then, Bangladesh experienced severe drought in 1951, 1957, 1961, 1972, 1976, 1979, 1986, 1989 and 1997 (CEGIS, 2013). Lack of surface water resources as well as climate change are factors contributing to the drought in Bangladesh (FAO, 2007). Only five devastating droughts occurred in the century between 1800 and 1900, but since 1981, four major droughts have occurred in the last 25 years, mostly in north-western Bangladesh (Unnayan Onneshan, 2008). This is because climate change alters the pattern of rainfall, which increases the frequency of drought. Generally, the western regions of the country are more vulnerable to drought than its eastern side (DoE, 2019). While drought has severe ramifications for the agriculture sector

Table 5.12 Percentage of change in rice yield under various CO_2 concentrations

Location	Climate change phenomena	Rice yield (kg ha^{-1})	Percentage change of rice yield
Rajshahi	+ 50 ppm CO_2	3234	+ 04.30
	+ 100 ppm CO_2	3346	+ 07.90
	+ 200 ppm CO_2	3638	+ 17.30
Mymensingh	+ 50 ppm CO_2	4612	+ 03.22
	+ 100 ppm CO_2	4686	+ 04.88
	+ 200 ppm CO_2	4858	+ 08.72
Satkhira	+ 50 ppm CO_2	4055	+ 03.10
	+ 100 ppm CO_2	4107	+ 04.40
	+ 200 ppm CO_2	4367	+ 11.00
Barishal	+ 50 ppm CO_2	5320	+ 04.40
	+ 100 ppm CO_2	5502	+ 07.80
	+ 200 ppm CO_2	5675	+ 11.40
Cumilla	+ 50 ppm CO_2	5626	+ 03.70
	+ 100 ppm CO_2	5947	+ 09.60
	+ 200 ppm CO_2	6413	+ 18.20
Sylhet	+ 50 ppm CO_2	5470	+ 02.10
	+ 100 ppm CO_2	5574	+ 04.00
	+ 200 ppm CO_2	5638	+ 05.20

Source Titumir and Basak's calculation based on DSSAT model simulation

Table 5.13 Percentage of change in rice yield under various climatic scenarios

Location	Climate change phenomena	Rice yield (kg ha^{-1})	Percentage change of rice yield
Rajshahi	Tmax 2 °C + Tmin 2 °C + 50 ppm CO_2	2907	– 06.30
	Tmax 2 °C + Tmin 2 °C + 100 ppm CO_2	3030	– 02.30
	Tmax 2 °C + Tmin 2 °C + 200 ppm CO_2	3237	+ 04.40
	Tmax 4 °C + Tmin 4 °C + 50 ppm CO_2	2616	– 15.70
	Tmax 4 °C + Tmin 4 °C + 100 ppm CO_2	2712	– 12.60
	Tmax 4 °C + Tmin 4 °C + 200 ppm CO_2	3196	+ 03.00
Mymensingh	Tmax 2 °C + Tmin 2 °C + 50 ppm CO_2	4245	– 05.00
	Tmax 2 °C + Tmin 2 °C + 100 ppm CO_2	4369	– 02.21
	Tmax 2 °C + Tmin 2 °C + 200 ppm CO_2	4604	+ 03.04
	Tmax 4 °C + Tmin 4 °C + 50 ppm CO_2	3162	– 29.23
	Tmax 4 °C + Tmin 4 °C + 100 ppm CO_2	3229	– 27.73
	Tmax 4 °C + Tmin 4 °C + 200 ppm CO_2	3401	– 23.88
Satkhira	Tmax 2 °C + Tmin 2 °C + 50 ppm CO_2	3340	– 15.10
	Tmax 2 °C + Tmin 2 °C + 100 ppm CO_2	3438	– 12.60
	Tmax 2 °C + Tmin 2 °C + 200 ppm CO_2	3631	– 07.70
	Tmax 4 °C + Tmin 4 °C + 50 ppm CO_2	3036	– 22.80
	Tmax 4 °C + Tmin 4 °C + 100 ppm CO_2	3144	– 20.10
	Tmax 4 °C + Tmin 4 °C + 200 ppm CO_2	3327	– 15.40
Barishal	Tmax 2 °C + Tmin 2 °C + 50 ppm CO_2	4419	– 13.30
	Tmax 2 °C + Tmin 2 °C + 100 ppm CO_2	4618	– 09.40

(continued)

Table 5.13 (continued)

Location	Climate change phenomena	Rice yield (kg ha^{-1})	Percentage change of rice yield
	Tmax 2 °C + Tmin 2 °C + 200 ppm CO_2	5016	– 01.60
	Tmax 4 °C + Tmin 4 °C + 50 ppm CO_2	3442	– 32.50
	Tmax 4 °C + Tmin 4 °C + 100 ppm CO_2	3607	– 29.20
	Tmax 4 °C + Tmin 4 °C + 200 ppm CO_2	3924	– 23.00
Cumilla	Tmax 2 °C + Tmin 2 °C + 50 ppm CO_2	4622	– 14.80
	Tmax 2 °C + Tmin 2 °C + 100 ppm CO_2	4809	– 11.40
	Tmax 2 °C + Tmin 2 °C + 200 ppm CO_2	5217	– 03.90
	Tmax 4 °C + Tmin 4 °C + 50 ppm CO_2	3631	– 33.10
	Tmax 4 °C + Tmin 4 °C + 100 ppm CO_2	3820	– 29.60
	Tmax 4 °C + Tmin 4 °C + 200 ppm CO_2	4222	– 20.20
Sylhet	Tmax 2 °C + Tmin 2 °C + 50 ppm CO_2	5619	+ 04.80
	Tmax 2 °C + Tmin 2 °C + 100 ppm CO_2	5800	+ 08.20
	Tmax 2 °C + Tmin 2 °C + 200 ppm CO_2	6095	+ 14.00
	Tmax 4 °C + Tmin 4 °C + 50 ppm CO_2	4577	– 14.60
	Tmax 4 °C + Tmin 4 °C + 100 ppm CO_2	4743	– 11.50
	Tmax 4 °C + Tmin 4 °C + 200 ppm CO_2	5112	– 04.60

Source Titumir and Basak's calculation based on DSSAT model simulation

in particular (Table 5.20), it can have indirect effects on land degradation, livestock population, employment and health as well.

Finally, salinity intrusion is the direct result of climate change. As sea levels rise, the soil and water in coastal areas become contaminated with saline water. An estimate shows that soil salinity increased from 0.833–1.056 million ha between 1973 and 2009; however, only in nine years from 2000 to 2009, it affected 3.5% of the coastal land (Table 5.21). Approximately 0.328, 0.274, 0.189, 0.161 and

Table 5.14 Percentage of deprived population from rice under different climate scenarios

2050			2070				2100			
P	A	B	P	C	D	E	P	C	D	E
3.85	11.80	8.80	13.75	35.04	32.56	25.96	28.20	45.90	43.85	38.25

Source Titumir and Basak's calculation based on DSSAT model, FAOSTAT and World Bank data
P—demand–production gap due to huge population pressure on rice
Scenario A—2 °C increase of Tmax and Tmin with 50 ppm CO_2, from base year 2008 to 2050
Scenario B—2 °C increase of Tmax and Tmin with 100 ppm CO_2, from base year 2008 to 2050
Scenario C—4 °C increase of Tmax and Tmin with 50 ppm CO_2 from base year 2008 to 2070 and 2100
Scenario D—4 °C increase of Tmax and Tmin with 100 ppm CO_2 from base year 2008 to 2070 and 2100
Scenario E—4 °C increase of Tmax and Tmin with 200 ppm CO_2 from base year 2008 to 2070 and 2100

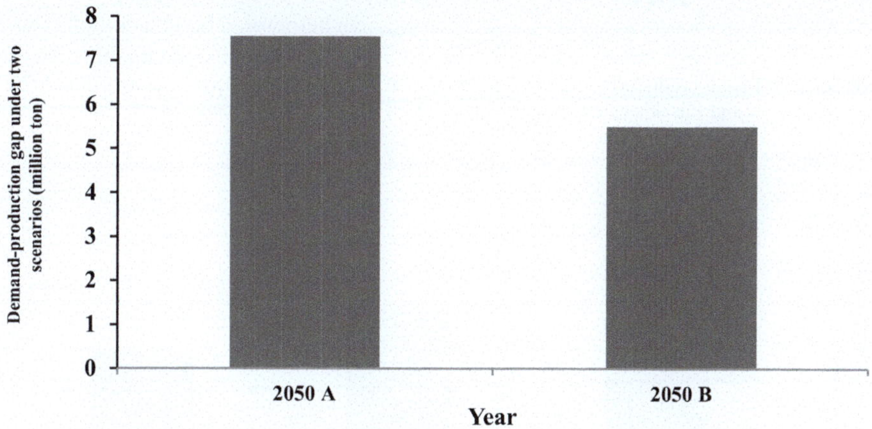

Fig. 5.21 Demand–production gap (million tonnes) of rice under scenarios A and B in 2050. *Source* Titumir and Basak's calculation based on FAOSTAT and World Bank data

0.101 million hectares of land are affected by very slight (S1), slight (S2), moderate (S3), strong (S4) and very strong salinity (S5), respectively. Moreover, salinity has increased by about 26% nationwide over the past 35 years, reaching the non-coastal areas as well (The Daily Star, 2019).

Apart from the increase in intensity and frequency of existing natural disasters, several newer types of disasters have become evident with climate change. For instance, during 1900–1981, the major types of natural disasters in Bangladesh were flood, drought and cyclone (Fig. 5.25a) whereas during 1982–2011 disasters such as landslide, extreme temperature, tsunami and earthquake have emerged (Fig. 5.25b).

It is also noticeable that the total number of disasters increased manifold in 1982–2011 with having 6.7 disasters per year, whereas per year disaster was 0.91 during 1900–81 (Table 5.22). The number of people killed, however, shows a negative trend.

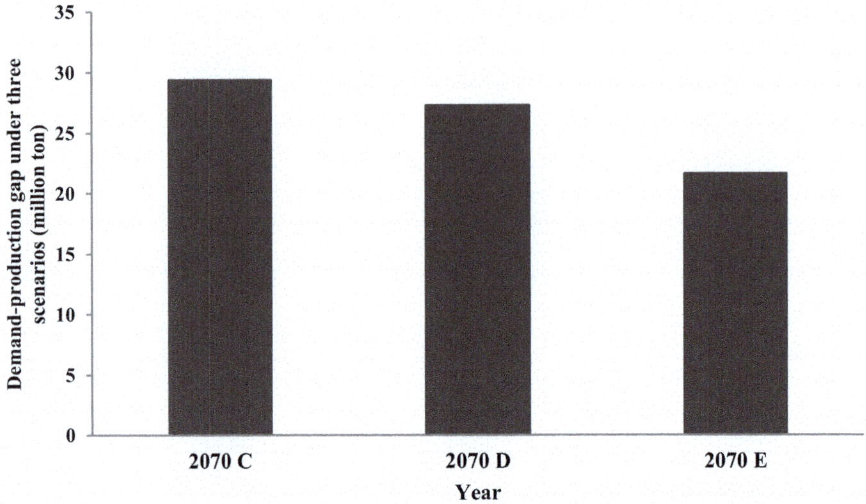

Fig. 5.22 Demand–production gap (million tonnes) of rice under scenarios C, D and E in 2070. *Source* Titumir and Basak's calculation based on FAOSTAT and World Bank data

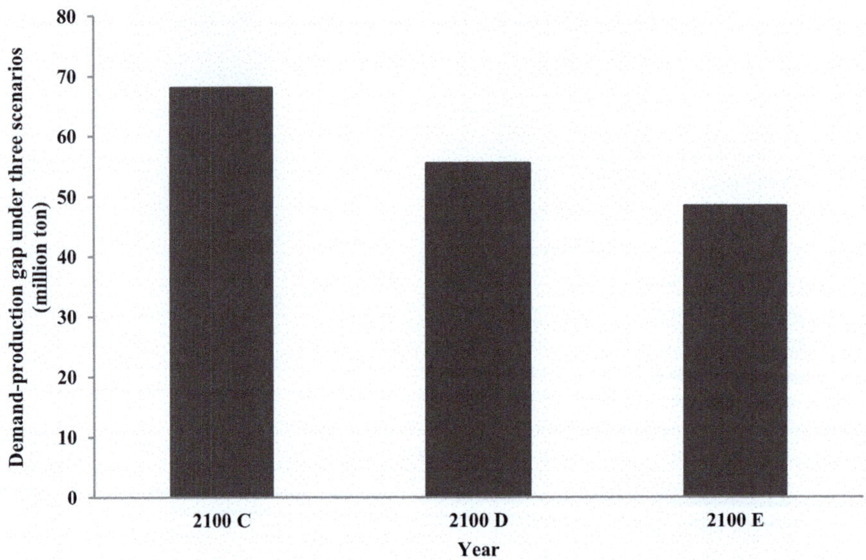

Fig. 5.23 Demand–production gap (million tonnes) of rice under scenarios C, D and E in 2100. *Source* Titumir and Basak's calculation based on FAOSTAT and World Bank data

Table 5.15 Comparison of agricultural losses resulting from recent cyclones

Cyclone	Damaged entities	Damages occurred
12 November 1970	Crop lost	Tk. 4.41 billions
	Loss of cattle	280,000
	Loss of poultry	500,000
26 May 1985	Damage to crops	90,381 ha
	Livestock lost	135,033
	Trees destroyed	1200
29 April 1991	Damage to crops in acreage	133,272 (fully), 791,621 (partly)
	No. of domestic animals killed	1,061,028
	boro rice, *aus* rice and other food crops (e.g., potatoes and vegetables) damaged	210,000, 36,000 and 3500 tonnes, respectively
15 November 2007 (*Sidr*)	Livestock killed	1,778,507
	Crops damaged (fully)	505,660 ha
	Crops damaged (partly)	1,177,086 ha
	Agricultural crop lands damaged	1.51 million ha
	Damage of the betel nut and other trees	40–50%
	Vegetable crops damaged	80%
	Water area for fish culture was affected	17,700 ha
25 May 2009 (*Aila*)	Crops damaged	77,486 acres (fully), 245,968 acres (partly)
	Livestock deaths	150,131
	Damaged of shrimp fields and freshwater fishponds	38,885 ha

Source Authors' compilation from Titumir and Basak (n.d.), MoFDM (2008), Farukh et al. (2019), Subhani and Ahmad (2019)

Table 5.16 Bangladesh's position in World Risk Report 2020

Indicators	Value	Category[a]
World Risk Index	16.40	Very high (10.76–49.74)
Exposure	28.28	Very high (19.70–86.77)
Vulnerability	57.98	High (48.13–61.49)
Susceptibility	33.21	High (27.94–45.13)
Lack of coping capacity	54.91	Very high (85.21–93.80)
Lack of adaptation capacity	85.81	Very high (52.73–69.72)

Source Behlert et al. (2020)
[a]Categories are 'very low', 'low', 'medium', 'high', and 'very high'

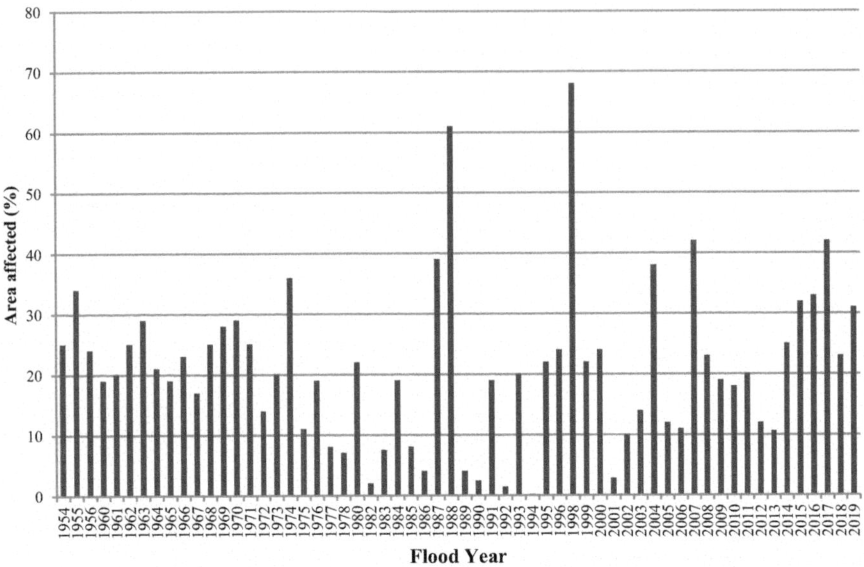

Fig. 5.24 Percentage of total land area affected by floods from 1954 to 2019. *Source* FFWC (2019)

The reason is the improved disaster preparedness of the country. Nevertheless, the number of affected people increased from 1.5 to 10.4 million per year. The total and average economic loss has also increased significantly.

Overall, Bangladesh, being a low-lying country, has suffered immense losses incurred by natural disasters over the years and the loss has multiplied in recent decades as climate change has increased the intensity of those disasters.

5.7.3 Decreasing Carrying Capacity and Displacement: Burden for Bangladesh III

Climate-induced migration is on the rise signifying the decreasing carrying capacities of the countries. The Internal Displacement Monitoring Centre (IDMC) reported that 17.2 million people were internally displaced for the first time in 2018 on a global scale as a result of climate-induced disasters and natural hazards (UNDRR, 2019). According to another estimate by the UNHCR Protection and Return Monitoring Network, approximately 883,000 new internal displacements occurred in 2018, of which 32% were associated with floods and 29% with drought (UNDRR, 2019).

Based on the sea-level projection for 2050, land that currently houses 300 million people will fall below the elevation of an average annual coastal flood, and by 2100, land that is now home to 200 million people may permanently be below the high tide line (Climate Central, 2019). Climate-induced displacement, therefore, occurs

Table 5.17 Impacts and loss resulting from recent catastrophic floods

Item	1974	1987	1988	1998	2004	2007	2017
Affected area of Bangladesh ('000 km^2)	53	57	90	100	56	62	31 district[a]
People affected (million)	30	30	47	31	33	14	6,775,352 (in number)
Fatalities	28,700	1657	2379	918	285	1110	N/A
Houses damaged ('000s)	N/A	989	2880	2647	895	1000	N/A
Roads damaged (km)	N/A	N/A	13,000	16,927	27,970	31,533	4815
Livestock killed (Nos.)	N/A	N/A	172,000	26,564	8318	40,700	9331
Crops damaged (million ha)	N/A	N/A	2.12	1.7	1.3	2.1	N/A
Rice production losses (million tons)	N/A	N/A	1.65	2.06	1.00	1.2	N/A
Asset losses (million $)	936	1167	1424	2128	1860	1100	N/A
Asset losses (as % of GDP)	7.5	4.9	5.5	4.8	3.3	1.6	N/A
Total losses (million $)	N/A	N/A	1.4	2.0	2.3	1.1	N/A

Source Authors' compilation from World Bank (2007, 2010), DDM (2017)
Note N/A means data not available in common sources. Please note that the available flood damage information is not always complete and consistent

Table 5.18 Normal floods and extreme floods during different time spans

Time-span	Average flooding area (%)	Number of normal floods	Number of extreme high floods	Number of extreme low floods
1954–1972	23.6	13	01	02
1973–1990	16.9	04	03	09
1991–2007	20.7	06	03	07

Source Ali (n.d.)

mainly for two reasons: submerging of the land area due to sea-level rise and destructive effects of natural hazards, although there are some other indirect reasons as well (Fig. 5.26). For instance, people from coastal areas, whose livelihood depended on farming or fisheries, decided to shift to urban areas as their livelihood options were impacted by climate change. Moreover, the destruction of natural resources in those

Table 5.19 Types of various major cyclones hitting Bangladesh over the years

Date	Types of cyclone	Landfall area	Maximum wind speed (in km/h)	Tidal surge height (in ft)	Death toll
23.10.1970	Severe Cyclonic Storm of Hurricane intensity	Khulna-Barishal	163	–	–
12.11.1970	Severe Cyclonic Storm with a core of hurricane wind	Chittagong	224	10–13	300,000
28.11.1974	Severe Cyclonic Storm	Cox's Bazar	163	9–17	–
10.02.1981	Cyclonic Storm	Khulna	120	7–15	–
15.10.1983	Cyclonic Storm	Chittagong	93	–	–
09.11.1983	Severe Cyclonic Storm	Cox's Bazar	136	5	–
24.04.1985	Severe Cyclonic Storm	Chittagong	154	15	11,069
29.11.1988	Severe Cyclonic Storm with a core of hurricane wind	Khulna	160	2–14.5	–
18.12.1990	Cyclonic Storm (crossed as a depression)	Cox's Bazar Coast	115	5–7	–
29.04.1991	Severe Cyclonic Storm with a core of hurricane wind	Chittagong	225	12–22	138,882

(continued)

Table 5.19 (continued)

Date	Types of cyclone	Landfall area	Maximum wind speed (in km/h)	Tidal surge height (in ft)	Death toll
02.05.1994	Severe Cyclonic Storm with a core of hurricane wind	Cox's Bazar-Teknaf Coast	220	5–6	–
25.11.1995	Severe Cyclonic Storm	Cox's Bazar	140	10	–
19.05.1997	Severe Cyclonic Storm with a core of hurricane wind	Sitakundu	232	15	155
27.09.1997	Severe Cyclonic Storm with a core of hurricane wind	Sitakundu	150	10–15	–
20.05.1998	Severe Cyclonic Storm with core of hurricane winds	Chittagong Coast near Sitakunda	173	3	–
28.10.2000	Cyclonic Storm	Sundarban Coast near Mongla	83	–	–
12.11.2002	Cyclonic Storm	Sundarban Coast near Raimangal River	65–85	5–7	–
19.05.2004	Cyclonic Storm	Teknaf-Akyab Coast	65–90	2–4	–
15.11.2007	Severe Cyclonic Storm with core of hurricane winds (*Sidr*)	Khulna-Barisal Coast near Baleshwar River	223	15–20	3363
25.05.2009	Cyclonic Storm (*Aila*)	West Bengal-Khulna Coast near Sagar Island	70–90	4–6	190

(continued)

Table 5.19 (continued)

Date	Types of cyclone	Landfall area	Maximum wind speed (in km/h)	Tidal surge height (in ft)	Death toll
16.05.2013	Cyclonic Storm (*Mahasen*)	Noakhali-Chittagong Coast	100	–	–
30.07.2015	Cyclonic Storm (*Komen*)	Chittagong-Cox's Bazar Coast	65	5–7	–
21.05.2016	Cyclonic Storm (*Roanu*)	Barisal-Chittagong Coast near Patenga	128	4–5	–
30.05.2017	Severe Cyclonic Storm (*Mora*)	Chittagong-Cox's Bazar Coast near Kutubdia	146	–	–

Source BMD (2017)

Table 5.20 Chronology of major drought in Bangladesh and their impacts

Year	Description
1791	Drought-affected Jessore District, prices doubled or tripled
1865, 1866, 1872, 1874	Reported to occur in Dhaka, Bogra and the Sundarbans. Crop suffered greatly in most cases; drought preceded Dhaka famine in 1865
1951	Severe drought in north-west Bangladesh substantially reduced rice production
1973	Drought responsible for the 1974 famine in northern Bangladesh
1975	Drought affected 47% of the country and more than half of the total population
1978–1979	Widespread damage to crops reducing rice production by about 2 million tonnes, directly affecting about 42% of the cultivated land and 44% of the population
1981	Severe drought adversely affected crop production
1982	Drought caused a loss of rice production of about 53,000 tonnes
1989	Drought dried up most of the rivers in north-west Bangladesh in several districts, including Naogaon, Nawabganj, Nilpahamari and Thakurgaon
1994–1996	Immense crop damage, especially to rice, jute and bamboo clumps
	No comprehensive study has been done on the droughts that occurred after 1995–96

Source Adapted from FAO (2007), CEGIS (2013)

Table 5.21 Soil salinity during last four decades (1973–2009) in coastal areas of Bangladesh. *Source* SRDI (2010)

Salt affected area (000' ha)			Salinity class and area (000' ha)											
			S1 2.0–4.0 dS/m			S2 4.1–8.0 dS/m			S3 8.1–16.0 dS/m			S4 > 16.0 dS/m		
1973	2000	2009	1973	2000	2009	1973	2000	2009	1973	2000	2009	1973	2000	2009
833.45	1020.75	1056.26	287.37	289.76	328.43	426.43	307.20	274.22	79.75	336.58	351.69	39.90	87.14	101.92

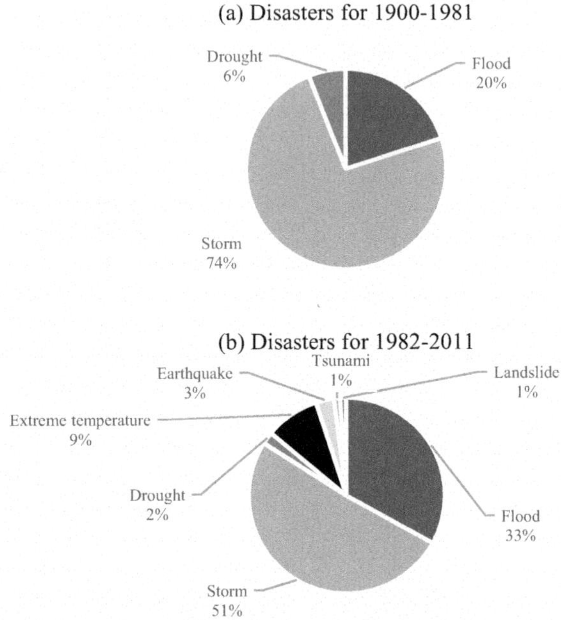

Fig. 5.25 Types of disasters in two different periods. *Source* Ali (n.d.)

Table 5.22 Intensity of loss occurred by disasters

Criteria	Disaster year	
	1900–1981	1982–2011
Number of disasters (total)	74	202
Number of disasters (average)	0.91 per year	6.7 per year
Number of people killed (total)	2.4 million	0.18 million
Number of people killed (average)	29,600 per year	6000 per year
Number of affected people (total)	119 million	308 million
Number of affected people (average)	1.5 million per year	10.4 million per year
Economic loss (total in USD)	1.25 billion	16.9 billion
Economic loss (average in USD)	15 million	563 million

Source Ali (n.d.)

localities created scope for competition and conflict among the inhabitants, which also forced many to move to other places.

In Bangladesh, the displacement of people, especially of the marginalised, has been increasing over the years. Natural disasters have displaced nearly 700,000 Bangladeshis each year over the last decade (CEDMHA, 2020). Data shows that a potential number of people are newly displaced in different years because of natural disasters and conflicts (Fig. 5.27a, b). The notable fact is that displacement by natural disasters shows an increasing trend whereas displacement by conflict exhibits a fluctuating trend. It signifies that although conflicts are a potential threat for displacement, natural disasters, induced by climate change, are likely to impact more on the displacement trend over the years. For instance, in the year 2019, the number of newly displaced people drastically rose to 4,086,000 whereas the number of internally displaced people was 88,000. Additionally, projections by IDMCs indicated earlier that one out of every seven Bangladeshis would experience displacement by 2050 (Siddiqui, 2019).

Indubitably, climate-related migration in Bangladesh is significant, but very few empirical studies have been conducted to date. As a result, Bangladesh's current policies and plans fall short of adequately addressing this growing issue. As per the government report, approximately 8.7 million people lived in cyclone high-risk areas (HRAs) in 2012, and that figure could rise to 33.67 million (without climate change) and 38.33 million (with climate change) by the decade of 2050s (MoEF, 2012). The increased number of natural calamities such as cyclones and floods with their extreme intensity have forced many people of the coastal area of Bangladesh to migrate to other places, particularly urban areas. Cyclone *Sidr* alone displaced 650,000 people, while nearly 20,000 people were displaced by cyclone *Bijli* and 842,000 by *Aila* (Akter, 2009).

A particular study by *Unnayan Onneshan* reported that the number of displaced people in Bangladesh, due to climate change, would continue if necessary actions are

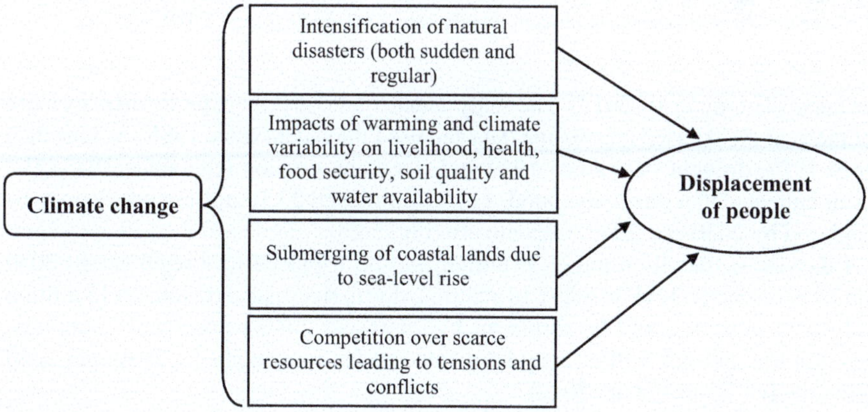

Fig. 5.26 Four ways by which climate change causes displacement of people. *Source* Prepared by the authors

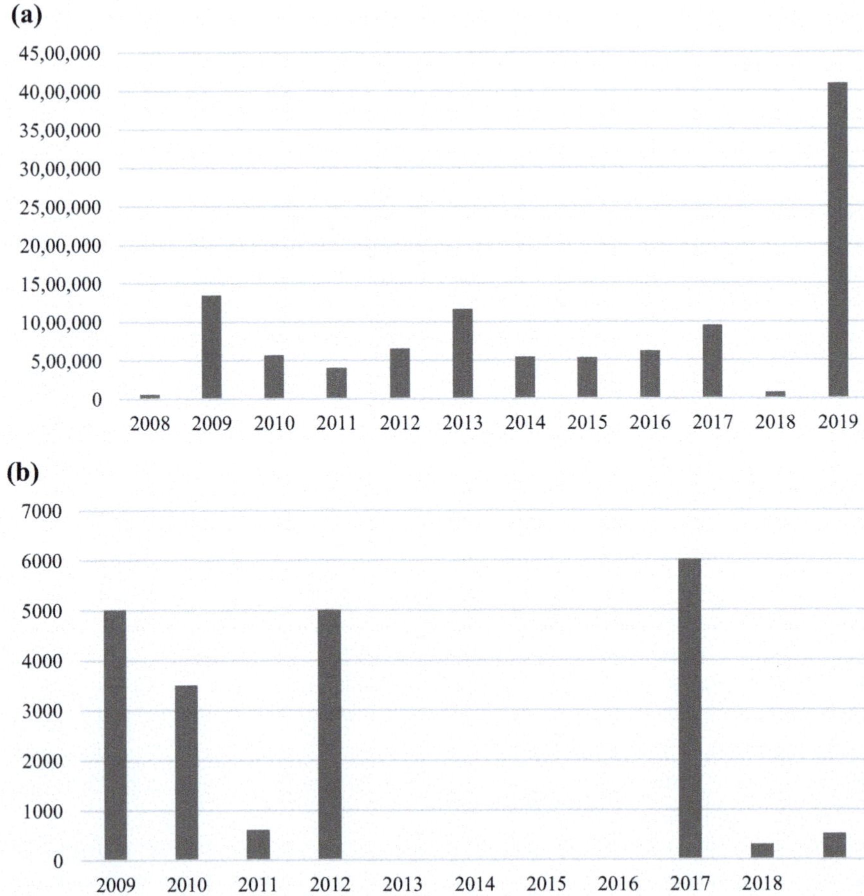

Fig. 5.27 **a** New displacements by natural disasters in Bangladesh, 2008–2019 and **b** new displacements by violence and conflict in Bangladesh, 2008–2019. *Source* IDMC (2021)

not taken (Titumir et al., 2012). The study endeavoured to project future displacement for the year 2040 based on existing data and used flood data since 1986, cyclone data since 1970, drought data since 1949 and riverbank erosion data since 1982. The result indicated that due to the flood, 11.02 million and 23.16 million people may be displaced by 2030 and 2040, respectively (Fig. 5.28).

Likewise, 6.46 million and 26.39 million may be displaced due to cyclones by 2030 and 2040 respectively (Fig. 5.29). In addition, due to riverbank erosion, 26.15 million and 40.67 million could be displaced by 2030 and 2040 respectively. Similarly, 4.64 million and 5.5 million may be displaced due to drought by 2030 and 2040 respectively (Titumir et al., 2012).

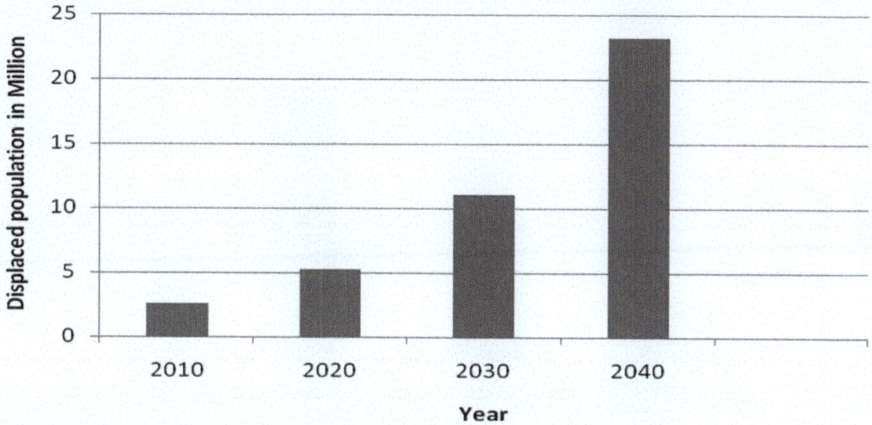

Fig. 5.28 Projection on displacement by flood. *Source* Titumir et al. (2012)

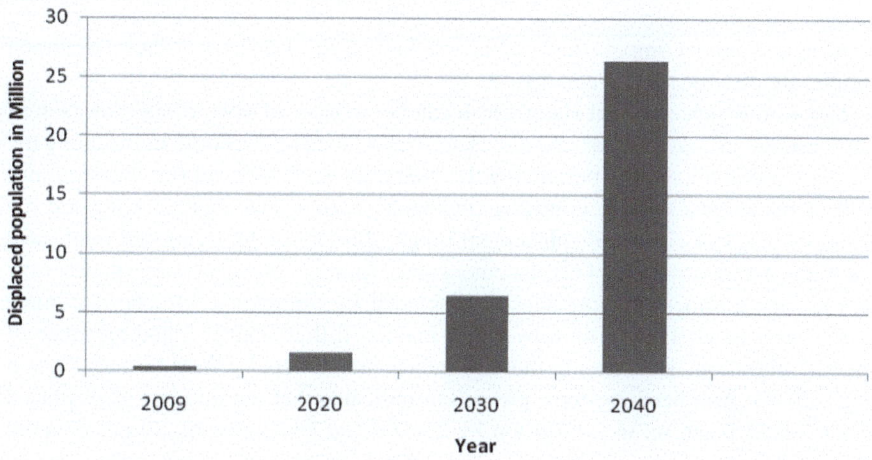

Fig. 5.29 Projection on displacement by cyclones. *Source* Titumir et al. (2012)

Taking into account flood, cyclone, riverbank erosion and drought cumulatively, climate change-induced migration in Bangladesh may reach 22.37 million, 48.28 million and 95.72 million by 2020, 2030 and 2040 respectively (Fig. 5.30).

Apart from natural disasters, the sea-level rising also causes displacement of people, especially those living on the coast of the country. Bangladesh has already shown to have the largest share of land area affected by sea-level rise throughout South Asia (Dasgupta et al., 2007), and by 2050, with a projected 50 cm rise in sea level, the country may lose approximately 11% of its land, affecting an estimated 15 million people who live in the coastal region (EJF, 2021). Sea-level rise also has a negative impact on various ecosystems destroying people's livelihood opportunities. For instance, the biodiversity of the Sundarbans in Bangladesh is under severe threat

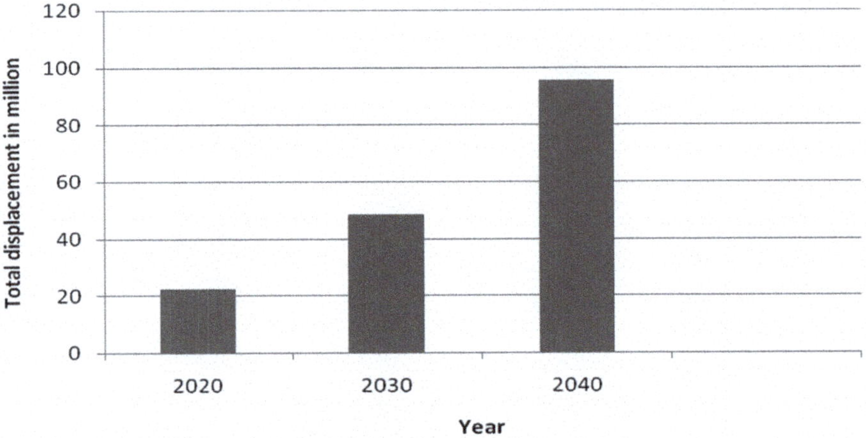

Fig. 5.30 Projection on displacement on the whole. *Source* Titumir et al. (2012)

of extinction and it compels the local people to migrate elsewhere as their livelihood options are disappearing gradually.

Types of displacement in Bangladesh can be divided into two categories such as displacement in coastal areas and displacement in land areas. People living along the coast are predominantly experiencing the dual effects of sea-level rise and cyclones with storm surges. On the contrary, riverbank erosion and river flooding are the primary causes of climate displacement inland. Moreover, droughts and landslides can be the potential causes of displacement in the future as climate change progresses. Overall, the number of Bangladeshis displaced by the diverse effects of climate change may reach 13.3 million by 2050, making it the country's leading cause of internal migration, according to the most recent assessment by World Bank (Rigaud et al., 2018). Another long-term projection indicates that the number of displaced people could reach up to 30 million by 2100 if sea levels rise by 80 cm or more (YPSA, 2014).

5.8 International Co-operation in Financing and Technology Transfer

The first international convention on climate change—the United Nations Framework Convention on Climate Change (UNFCCC)—adopted in Rio de Janeiro, Brazil, in 1992 reached an agreement that each nation should accept its "common but differentiated responsibilities" [Article 3(1)] (United Nations, 1992). It, accordingly, urges the widest possible cooperation and active participation by all countries for tackling climate change. The negotiations taken so far, however, are increasingly fraught with difficulty to bring any effective outcome.

United Nations (UN), under the UNFCCC, organises the 'Conference of the Parties (COPs)' every year since 1995 at which the northern and southern countries discuss to find paths of cooperation in tackling climate change (Table 5.23). The outcomes of these COPs, however, are ineffective overall and can be criticised for their insufficient scale and institutional framework as there has always been a lack of mutuality among the parties. In the most recent context, the discussions at Climate Conference at Bonn in 2022 are criticised in terms of "empty words, hollow promises and false solutions" by recognising the ineffectiveness of the conference (Friends of the Earth International, 2022).

The parties in the negotiation process are the different groups of states (Fig. 5.31). The advanced industrialised nations need to play a key role in meeting the targets, as they are the leaders in climate policy. They keep making pledges but are less inclined to fulfil them. The role of developing countries is also vital and they have shown interest in playing that role from their side. Nevertheless, sometimes they give up and focus on industrialisation to keep pace with the developed countries. Moreover, there are no clear plans for how countries will reach the pledges and virtually no plan is legally binding in most cases. Accordingly, the climate agreements remain hollow and climate changes continue to accelerate perilously.

The fault line, therefore, broadly exists between the northern developed and southern developing countries in terms of equity, adaptation–mitigation and finance. First, southern states acknowledge their responsibilities but at the same time, they want developed countries to take more responsibilities as the developed ones are historically responsible for larger emissions of GHGs. On the contrary, the north has questioned the distinction between Annex and non-Annex countries. Second, the northern countries argue that the developing countries should adapt to the changing situations and they would give financial support in that case. In this way, they try to avoid the mitigation initiatives from their side. Only adaptation by developing countries, however, is not a viable solution for addressing climate change. The continuous emission and further warming of the world would make the result of adaptation measures nothing but a failure in the upcoming future. Third, there is also evidence of transferring finance from the North to the South inappropriately. Developed countries often pay a deaf ear regarding 'fair burden share'. They have agreed to assist in financing developing countries' transition to clean energy and adaptation to climate change, but they have refrained from taking responsibility for the harm that climate change has inflicted elsewhere. The lack of balance of power in the international political economy context actually resulted in the emergence of two opposing groups at the climate negotiation table, indicating the precarious institutional set-up.

The efforts by the developing countries to adopt adaptation and mitigation strategies have not progressed to the expected level because of financial constraints. The Green Climate Fund (GCF) is a mechanism through which the developed countries transfer financial and technological resources to developing countries. The developed countries initially did not promise to provide sufficient finance for the successful implementation of Agenda-21, but later committed to assisting in the future. Practically, their promises are spurious in the form of non-cooperative behaviour at COPs while signing and ratifying several treaties and protocols.

Table 5.23 UN climate Conference of the Parties (COPs)

COP	Duration	Location	Key features	Major drawbacks
COP-1	28 March–7 April, 1995	Berlin, Germany	• First conference with a vision to successfully tackling climate change • See the need to reduce emissions	• Just the beginning of the initiative
COP-2	8–19 July, 1996	Geneva, Switzerland	• Accepted the scientific findings on climate change proffered by the IPCC (1995) • Called for "legally binding mid-term targets"	• Rejected uniform 'harmonised policies' in favour of flexibility
COP-3	1–10 Dec, 1997	Kyoto, Japan	• Adoption of "Kyoto Protocol" initiating legally binding provisions to reduce GHGs emission for all countries including Annex I (developed) countries • Mechanisms such as *emission trading, clean development mechanism* and *joint implementation* came to the forefront	• The protocol could not include USA and China, the world's two largest emitters • Canada also withdrew in 2011 • The protocol demands too much and achieves too little (West, 2016)
COP-4	2–13 Nov, 1998	Buenos Aires, Argentina	• Adoption of Buenos Aires Plan of Action • Some non-Annex developing countries (e.g., Argentina and Kazakhstan) for the first time demonstrated willingness to participate in reducing GHG emissions	• Failed to come to an agreement on the issues that were not resolved in COP-3 • Ended up with adopting only a two-year "Plan of Action"
COP-5	25 Oct–5 Nov, 1999	Bonn, Germany	• Pledge to finalise climate agreement by November 2000	• Primarily a technical meeting that did not arrive at any major conclusion
COP-6	13–25 Nov, 2000	The Hague, Netherlands	• Intended to wrap up three years of negotiations on the implementation of the Kyoto Protocol	• Collapsed as the USA wanted carbon sinks as part of the agreement • The EU was firm in their stance that USA should not get special treatment

(continued)

Table 5.23 (continued)

COP	Duration	Location	Key features	Major drawbacks
	17–27 July, 2001	Bonn, Germany	• The agreements include the issues of: (a) CDM, (b) carbon sinks and (c) financing mechanism	• The resumption of talks resulted in an agreement with little progress
COP-7	29 Oct–10 Nov, 2001	Marrakech, Morocco	• Wrapped up work on the Buenos Aires Plan of Action, finalising most of the operational details and setting the stage for nations to ratify the Kyoto Protocol	• The United States maintained only observer role and declined to take part actively in the negotiations
COP-8	23 Oct–1 Nov, 2002	New Delhi, India	• Called for efforts by developed countries to transfer technologies to minimise the impact of climate change on developing countries	• Marked by hesitation of Russia stating that it needed more time to think over it • Not agreed by European Union, G-77 and China
COP-9	1–12 Dec, 2003	Milan, Italy	• Agreed to use the Adaptation Fund established at COP-7 in 2001 in supporting developing countries	• No new action plan
COP-10	6–17 Dec, 2004	Buenos Aires, Argentina	• Discussed the progress made since the first COP and its future challenges • The Buenos Aires Plan of Action was adopted	• The action plan primarily focused on adaptation • No specific action plan on mitigation measures
COP-11	28 Nov–9 Dec, 2005	Montreal, Canada	• Marked the entry into force of the Kyoto Protocol and strengthening the CDM • Many developing countries showed greater willingness to discuss stronger efforts (e.g., Papua New Guinea, Costa Rica, Brazil)	• USA, again, insisted throughout the negotiations that it opposed any new process

(continued)

Table 5.23 (continued)

COP	Duration	Location	Key features	Major drawbacks
COP-12	6–17 Nov, 2006	Nairobi, Kenya	• Adopted a five-year plan of work to support climate change adaptation by developing countries • Agreed on the procedures and modalities for the Adaptation Fund	• The majority of the discussions avoided any mention of reducing emissions • A disconnect between the political process and the scientific imperative (Black, 2006)
COP-13	3–14 Dec, 2007	Bali, Indonesia	• Adoption of Bali Action Plan (at the end of the first commitment period of Kyoto Protocol)	• Starting from the bottom without proper implementation of Kyoto Protocol
COP-14	1 Dec–12 Dec, 2008	Poznań, Poland	• Agreed on principles for the financing of a fund to help the poorest nations	• Discussion over the former issues; No concrete plan of action
COP-15	7–18 Dec, 2009	Copenhagen, Denmark	• Developed countries agreed to allocate 30 billion USD for the period 2010–2012 under First Start Finance (FSF) mechanism to help developing countries on mitigation and adaptation efforts	• Except few bilateral efforts (mostly diverted money from Official Development Assistance), the world did not see any collective action at the global or even regional scale • Lack of transparency in allocating funds through proper channel • Majority of the disbursement followed multilateral channels (such as the World Bank, UNDP, etc.) and bilateral agencies (such as DFID, GIZ, USAID, etc.), other than developing countries demanded UNFCCC created funds such as the Adaptation Fund and the Least Developed Countries Fund (IIED, 2012) • Did not achieve a binding agreement for long-term action

(continued)

Table 5.23 (continued)

COP	Duration	Location	Key features	Major drawbacks
COP-16	29 Nov–10 Dec, 2010	Cancun, Mexico	• Formalises the commitment set out in Copenhagen • Called for the 100 billion USD per annum Creation of "Green Climate Fund" and "Climate Technology Centre"	• The funding mechanism of Green Climate Fund (GCF) was not agreed upon by the parties
COP-17	28 Nov–9 Dec, 2011	Durban, South Africa	• Annex 1 countries promised to mobilise 100 billion USD additionally by 2020 under GCF	• The basket of commitment further enriched without resolving the previous issues • The deal was not sufficient to avoid global warming beyond 2°C as more urgent action is needed (Harvey & Vidal, 2011)
COP-18	26 Nov–7 Dec, 2012	Doha, Qatar	• Closed with a historic shift in principle but few cuts in GHGs emission • The representatives from the small island states at severe risk from climate change were vociferous • Decision to extend Kyoto Protocol until 2020	• Made little progress towards funding GCF as was promised in COP 17 (Harrabin, 2012) • Russia, Belarus and Ukraine were the states that created turmoil over the negotiations by claiming that they should be allowed extra credit for the emissions cuts • The US, EU, and China accepted the agreement with varying degrees of reservation but US, China and Russia did not support the extension of Kyoto Protocol
COP-19	11–22 Nov, 2013	Warsaw, Poland	• Led to an agreement that all states would start cutting emissions as soon as possible, but preferably by the first quarter of 2015	• Activists and poor countries accused Australian diplomats of not taking the talks seriously • Several countries attending the COP 19 have been criticised for poor performance on stated environmental pollution targets

(continued)

Table 5.23 (continued)

COP	Duration	Location	Key features	Major drawbacks
COP-20	1–12 Dec, 2014	Lima, Peru	• All countries agree to develop and share their commitment to reduce emission	• While this was a conference in the annual series, more attention is directed towards the 2015 Conference in Paris
COP-21	30 Nov–12 Dec, 2015	Paris, France	• Adoption of the historic 'Paris Agreement' governing climate change reduction measures covering the period from 2020 to 2030 • Aimed at limiting rise in temperature 1.5 °C—a more ambitious goal than expected before the summit	• Practically impossible to meet the target in the context of current international political environment • Emissions reductions targets are based on voluntary national pledges, failing to hold the largest emitters accountable • Too weak to help the poor as it excludes poor and fails to put humanity's interest above short-term goals (Harvey, 2015) • States are making promises they are unable to honour (Victor et al., 2017) • Emission rates are falling in industrialised countries but the rate is too low to meet the pledges • USA withdrew from the agreement in 2020, which multiplies the troubles
COP-22	7–18 Nov, 2016	Marrakech, Morocco	• Started turning the Paris Agreement into a detailed blueprint for action • Talked on financing mechanism, balancing on adaptation and mitigation fund	• The agreement on financing was nothing new rather it was just a repetition of previous pledge of contribution of 100 billion USD per year by 2020 • Ignored the problems of the global south (Yousfi, 2016)

(continued)

Table 5.23 (continued)

COP	Duration	Location	Key features	Major drawbacks
COP-23	6–17 Nov, 2017	Bonn, Germany	• Focused primarily on technical details of the Paris Agreement • Finalised the Gender Action Plan and the Local Communities and Indigenous People's Platform	• Developing countries asked for discussing the emission cuts that developed countries are required to make before 2020 under the Kyoto Protocol when the developed countries resisted it arguing that this issue was best discussed in other forums
COP-24	2–15 Dec, 2018	Katowice, Poland	• The objective was to have a full implementation of the Paris Agreement • Welcomed "timely completion" of the Special Report on Global Warming of 1.5 °C	• Failed to adopt key scientific report as USA, Saudi Arabia, Russia and Kuwait objected to the special report on reducing global warming
COP-25	2–13 Dec, 2019	Madrid, Spain	• Focused on some of the rules for implementing the Paris agreement	• How fast the world needs to cut greenhouse gas emissions received little attention • Almost total disconnect between what the science requires and what the climate negotiations are delivering in terms of meaningful action • The decisions about the carbon market and emissions cuts were deferred to the next climate conference • The United States, Russia, India, China, Brazil and Saudi Arabia were the main opponents of the measures

(continued)

Table 5.23 (continued)

COP	Duration	Location	Key features	Major drawbacks
COP-26	31 Oct–13 Nov, 2021	Glasgow, Scotland	• Countries agreed on a new deal named as Glasgow Climate Pact that particularly focused on fossil fuels (more specifically on coal) as the first ever climate deal • An emphasis was given on reducing deforestation • Increase in climate fund to help developing countries to adopt clean energy • Reaffirmed Paris agreement	• Lack of representations from affected developing countries (de Ferrer, 2021) and from many NGOs, which raised the issue of inclusivity. • India and China's last minute move from 'phase-out' to 'phase-down' approach towards coal usage leads to the weakening of the new climate deal (Harvey et al., 2021; Masood & Tollefson, 2021) • Nations failed to agree on creating a 'loss and damage' fund for climate vulnerable countries who are not responsible for climate change (Masood & Tollefson, 2021) • Previous pledges on climate financing by developed countries were not fulfilled (BBC News, 2021); hence, increase in the value of climate fund does not indicate any prospect • New pledges are not again legally binding
COP-27	6–20 November, 2022	Sharm El Sheikh, Egypt	• A 'loss and damage' fund was agreed for the first time • Returned its focus on climate change adaptation issue • Called for reforming financial institutions such as the World Bank • Emphasis was given on nature-based solutions	• Questions remained how the 'loss and damage' fund will be financed and configured and how the claims will be identified as eligible (Carlin, 2022) • The agreement on fossil fuel that was adopted in COP-26 remained unchanged (Rannard, 2022); therefore, it still emphasises on 'phase down' not 'phase out' • Issues relevant to design carbon markets were not settled (Carlin, 2022)

Source Prepared by the authors based on different sources

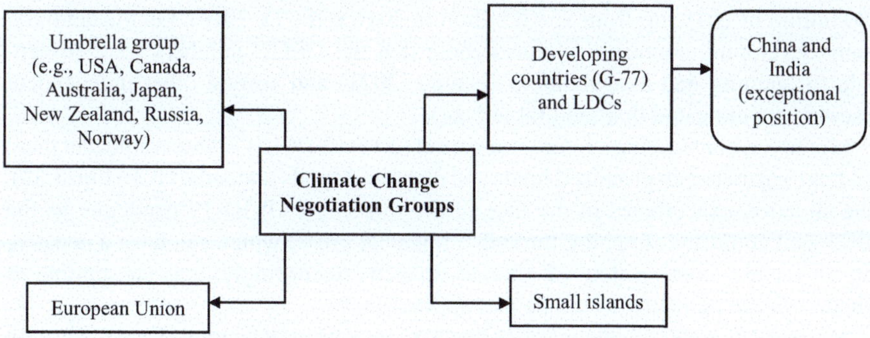

Fig. 5.31 Climate change negotiation groups. *Source* Prepared by the authors

Nevertheless, there has been an increase in the climate finance flow in later years. The annual flow rose to $579 billion, on an average, over the two-year period of 2017–18, representing a $116 (25%) increase from 2015 to 2016 (Fig. 5.32). The finance flow, however, shows a declining trend from 2017 to 2018, which is noteworthy. Moreover, while climate finance has shown a positive trend, action still falls short of what is needed under the 1.5 °C scenario (Averchenkova et al., 2020).

The flow in climate finance shows a rising trend in accordance with the developing countries' escalating demand for finance with the increased impacts of climate change. Moreover, the developing countries have been urged at COP 21 to commit to reducing their GHG emissions in exchange of receiving climate funding totalling $100 billion a year from the developed countries by 2020. Data released by the Organisation for Economic Co-operation and Development (OECD)—a group that represents many of the world's wealthier nations—claimed that the developed countries were close to achieving this target, which is to provide around $50 billion annually to low- and middle-income countries (OECD, 2015). In particular, climate finance provided and mobilised by developed countries reached $71.2 billion in 2017 and

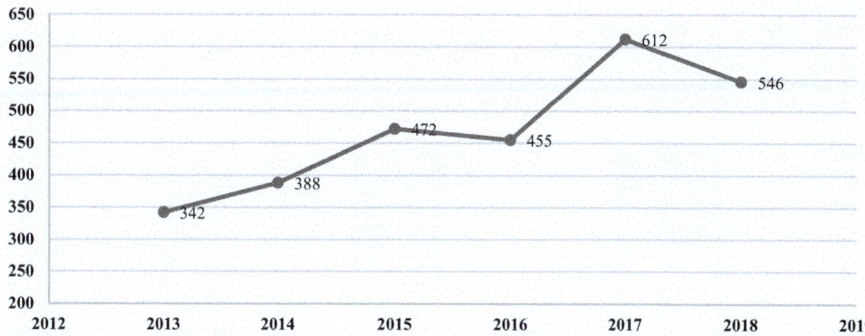

Fig. 5.32 Total global climate finance flows between 2013 and 2018 (in billion $). *Source* CPI (2019)

$80 billion in 2018, up from $58.6 billion in 2016 (OECD, 2019). On the contrary, many developing countries' claims denote that the OECD numbers were susceptible to 'gaming and exaggeration' (Nature, 2021) and several other independent assessments indicated that too (Table 5.24).

Moreover, considering a long-term period (2003–2020), it is observable that there is a huge gap between promised funds and disbursed funds under the GCF (Fig. 5.33). The situation exacerbates in the face of the ongoing COVID-19 pandemic as the developed countries are going through a financial crisis, which can have a negative impact on the existing flow of climate finance. Accordingly, it is susceptible to whether future demands would be met appropriately.

Another key point of contention is that a major part of the climate fund is given in the form of loans and other non-grant instruments. Only $12.5 billion of the estimated $59.5 billion in annual public climate finance reported in 2017–18 was in the form of grants and the rest as loans except $1.5 billion, which was unspecified (Fig. 5.34).

Table 5.24 Differences in assessments regarding climate-specific assistance by developed countries in 2017–18

Assessed by	Findings
OECD (2019)	Public climate finance mobilised by developed countries reached an estimated $59.5 billion (annual average)
Oxfam (2020)	Climate-specific net assistance is $19–22.5 billion (annual average); a small increase compared to 2015–16 when the amount was $15–19.5 billion (annual average)
Averchenkova et al. (2020) (under UN)	Donors were over-reporting climate-funding data by $3 billion to $4 billion

Source Authors' compilation from different sources

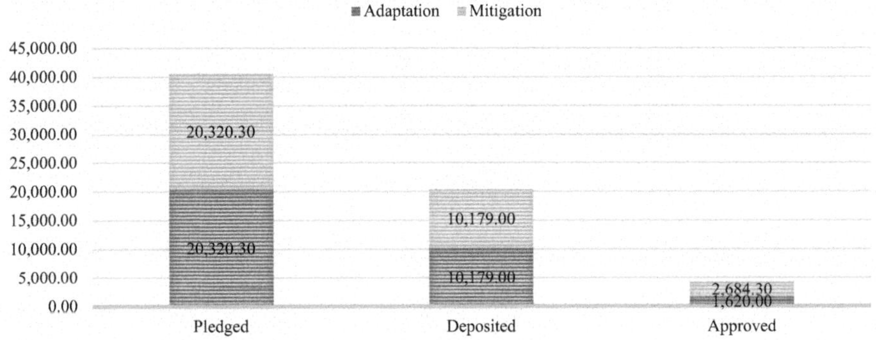

Fig. 5.33 Status of GCF supporting adaptation and mitigation (2003–2020, in million $). *Source* Climate Funds Update 2021 (ODA & HBS, 2021)

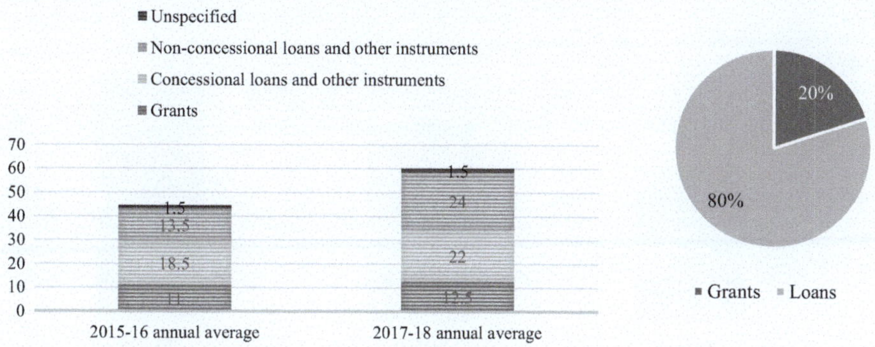

Fig. 5.34 Estimated climate finance by instrument via bilateral and multilateral channels, 2017–18 and 2015–16 (annual averages) (in billion \$). *Source* Oxfam (2020)

Though the total amount of grants has slightly changed from \$11 billion in 2015–16 to \$12.5 billion in 2017–18, the amount of grants compared to loans has decreased. On the other hand, the provision of loans (concessional and non-concessional) has increased significantly during the same period. In total, it was estimated that grants made up about 20% of all public climate finance, with loans and other non-grant instruments accounting for the remaining 80%. Another estimation suggests that the proportion of climate funding given as grants dropped from 27 to 20% between 2013 and 2018 (Nature, 2021). As loans have to be repaid with interest, they keep mounting pressure on the developing countries and they are compelled to take loans to safeguard themselves against the excessive carbon emissions of rich countries. It can be conjectured that the overwhelming provision of climate finance as loans would aggravate the debt burden situation of the developing countries in the post-pandemic era.

A debate over public and private finance is also discernible in relation to the governance of climate finance. It is reasonable to advocate for public finance sources for the committed GCF. The disbursement scenario, however, indicates that a higher portion of the climate finance is flowed by private sectors (Fig. 5.35).

The developing countries favour direct access, arguing that the fund should be managed under UNFCCC, whereas the developed countries proposed disbursing the fund through bilateral aid channels or World Bank-managed climate investment funds. The evidence of such initiatives so far demonstrates the ineffectiveness of the financing mechanism for adaptation programmes in developing countries.

Finally, the distribution of climate finance among the most vulnerable countries at the receiving end is uneven and the scale of funding is not proportionate to estimated needs. Region-wise distribution of climate finance shows that East Asia and the Pacific region receive the highest volume of finance, accounting for \$238 billion per year (41% of all flows) on average during 2017–18 (Fig. 5.36). The second-highest amount goes to Western Europe—an OECD region (\$106 billion), which is quite low compared to the highest one. The distribution is also drastically low for South Asia and Sub-Saharan Africa region.

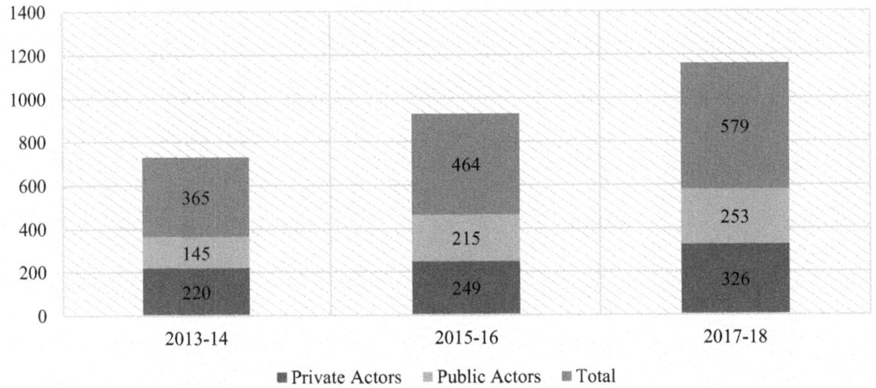

Fig. 5.35 Breakdown of global climate finance flows by public and private actors, 2013–2018 (two-year average, in billion $). *Source* CPI (2019)

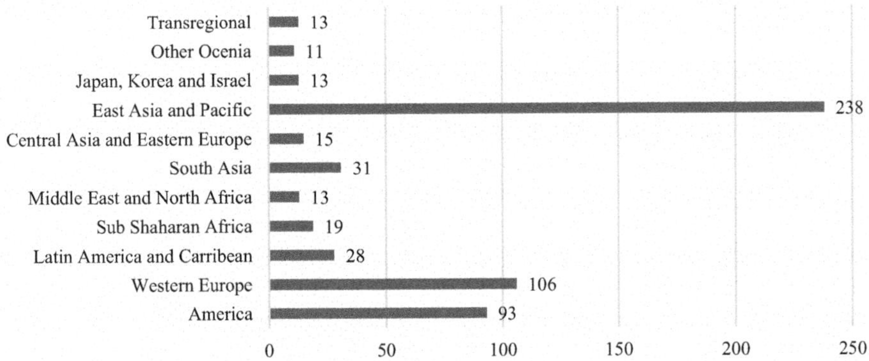

Fig. 5.36 Destination region of climate finance (in billion $, 2017–2018 annual average). *Source* CPI (2019)

Thus, there are evidences of institutional fragility at the international level. The market-centric policy instruments are not well-functioning due to the lack of cooperation among nation states. The policies, primarily backed up by the industrialised countries, favour further exploitation of fossil fuel instead of searching for other solutions. The countries, therefore, actually pursue self-interest while pretending to work together for the common good. Market-based instruments are thus worsening the situation by destroying the mutual belief among the developed and developing countries and considering the atmosphere as a mere commodity.

5.9 Institutional Fragility at the National Level: Methane Emission and Energy Transformation

The historical emission scenario showed in previous discussions that Bangladesh's role in emitting the major GHGs is negligible compared to the developed countries, but the emission of some GHGs exhibits an increasing trend. Such increases in emissions, however, are neither the result of the country's gigantic industrialisation nor the lack of willingness to address the problem of climate change. The reason is an unplanned decision-making process and the lack of effective cooperation at the institutional level, which implies the existence of institutional fragility at the national level.

A recent report claims that Bangladesh is a major contributor to global methane emission as satellite image shows that mysterious plumes of methane gas appear over Bangladesh (Clark et al., 2021). There is a debate, however, regarding the sources of the plumes (The Daily Star, 2021). It can nevertheless be argued that the potential sources obviously emit methane gases to some extent even if the key source does not belong to Bangladesh. The available data signifies that the emission of methane gas has increased over the years though at a moderate rate in Bangladesh (Fig. 5.37). It increased from 68,350 kt of CO_2 equivalent in 1999 to 83,790 kt of CO_2 equivalent in 2018 growing at an average annual rate of 1.08%.

Fig. 5.37 Methane emission in Bangladesh. *Source* World Development Indicators (2021)

The possible sources of emission are paddy production in agricultural land, landfills, leaky natural gas pipelines and coal stockpiles. The contribution of agriculture to methane emission nevertheless is quite negligible, as the scientists have claimed. On the contrary, other sources are the outcomes of a lack of mismanagement at the institutional level in the country. For instance, it has been reported that the methane has primarily originated from the *Matuail* landfill of Dhaka, which emits about 4000 kg of methane every hour (EARTH.ORG, 2021). The landfill site covers 100 acres and takes approximately 2500 kg tonnes of waste per day, and the waste management system is outdated and disorganised, which is primarily the result of the failure of the institutions. In fact, *Matuail* barely meets the basic requirements of a sanitary landfill. The concerned authority simply acquires more and more land instead of enforcing any sustainable landfill use practices (Khan, 2018). The use of landfills by waste management departments over the years appears to be characterised by a lack of a long-term strategy and coordinated effort, resulting in the uncontrolled and unsustainable disposal of waste.

In addition, the country's failure to transform the energy sources into renewable ones is also another indication of institutional failure as well as of technological backwardness. Many countries are now adopting renewable energy sources to lessen their contribution to emissions of GHGs and the top five countries in this regard are China, the USA, Germany, India and Japan (Fig. 5.38a).

Bangladesh is so far successful in enriching its power generation capacity over the years reaching 21,967 MW in 2021 (BPDB, 2021) but largely fails to ensure sustainable energy sources although there are potentialities in the renewable energy sector (e.g., solar, wind, biomass, etc.) (Baky et al., 2017; Nandi et al., 2011). Within 2021, the country has been able to generate only 0.56% of the total electricity from renewable sources (Fig. 5.38b). Still, natural gas accounts for almost 52% of the total electricity generation. It is an urgent need to strengthen the institutional and

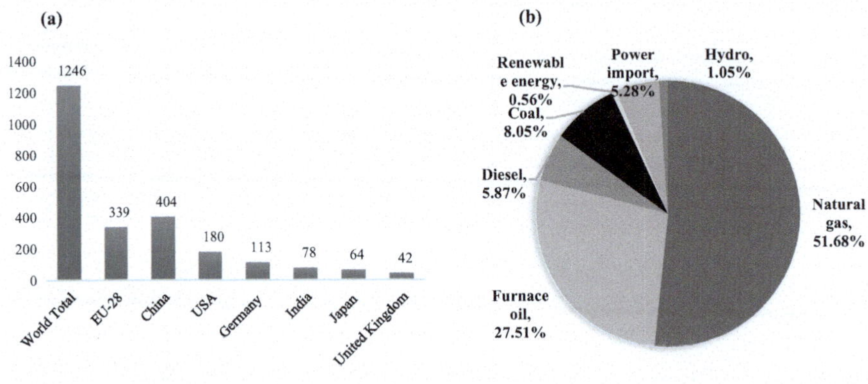

Fig. 5.38 **a** Renewable power capacities in world in 2018 (in gigawatts) **b** renewable power capacities in Bangladesh in 2021 (as a share of total installed capacity). *Source* REN21 (2019), BPDB (2021)

technological capacity of the country to tap the potential of renewable energy sources with a view to addressing the climate change issue.

Overall, the existence of a fragile institutional set-up emerging from the complex dynamics of international political economy and the national level incapacity results in the failure of mitigation and adaptation measures for climate change.

5.10 Material Balance, Resilience and Sustainability

This final section intends to provide a dialectic logical analysis to define the meaning of 'sustainability' in accordance with the key variables of the reformed theoretical framework. In fact, climate change is the most obvious indication of the 'non-sustainability' of nature. This non-sustainability results from the material imbalances between human beings and nature. Therefore, equal distribution of power across regions among different agents could promote the complementary relations between people and nature, which in turn would ensure sustainability. In other words, equitability, fairness and justice are necessary to guarantee the nature's sustainability and at least, to tackle the climate change problem.

It is evident that the agents are in a state of overconsumption with inequality resulting in the reduction of the resilience of nature. This book argues that resilience is external to the market, political settlement and institutions. Resilience can only function if the market works perfectly. The political settlement ensures the balance of power relationships and the institutions provide entitlement to the humans belonging to the nature. The perturbation caused by the imperfect market, capital accumulation and institutional fragility reduces the resilience of the socio-ecological capacity, hence, induces overexploitation. This causes faster accumulation and reduces the time required for innovation to persist in the changing system. Moreover, quicker accumulation results in less reorganisation in the short run and greater catastrophe in the long run.

Therefore, it is plausible to argue that market centrism is at the root of the distortion of the sustainability of nature. The market economy promotes industrialisation to achieve faster economic growth. The powerful developed countries have continued to prioritise economic growth over environmental concerns in order to maintain their leadership positions in the global economy. The market, however, offers several options to protect nature's balance, which are largely faulty in nature. The market-based solutions are just eye-washing tools that intend to spread the message that the market also cares about the nature. Accordingly, rather than relying solely on market-based solutions, the first priority should be to create a balanced nature of power across borders and to enrich the complementary relation between natures and people that existed prior to the market-centric era since the birth of the earth.

The fundamental argument is that defining human beings under *homo economicus* paradigm (human beings are self-interested) is faulty at its root and this is the key problem in the orthodox market-centric economic model in explaining human behaviour. This book argues that such conceptualisation of human beings and concomitant policy initiatives and organisation of economic activity are the major factors creating pressure on the sustainability of nature. It also argues that the behaviour of human beings should be defined as social beings who behave in a reciprocal way and want to lead a life and have a social organisation of production that ensures both the well-being of themselves as well as that of nature.

In this context, they can be defined as *homo sustinens* (Siebenhuner, 2000) in lieu of *homo economicus*. Thus, sustainability necessities to be conceptualised by focusing not on the substitutability aspect but rather by considering the revitalisation of human–nature relations to ensure the material balance and enrich the resilience of the earth's ecosystem. The necessary condition is, therefore, the cooperative management and the sufficient condition is the functioning of the global climate institution through the diffusion of power. Overall, an internationally coordinated response—by recognising and revitalising the inherent relationship between nature and human beings—is required.

5.11 Concluding Remarks

Climate change is an emerging challenge of governing global commons resulting in unequal consequences, which requires 'radical interventions' (Morrison et al., 2022) to bring a meaningful alteration in the trend. The developing countries bear the unequal burdens of climate change despite being the least contributors to GHGs emissions, whereas the primary contributors to climatic disorders are the developed countries. Therefore, the developing countries cannot tackle their problems in their own way based on market reforms. Cooperation from the developed states is also necessary. However, the unequal power structure between these two groups of countries makes the coordinated efforts fail in the past years. The developing countries, having less power in the international arena, could not put pressure on the developed countries to reduce emissions. The non-cooperative behaviour thus creates ecological debt and inequality among the countries. Above all, the competition to concentrate on the rising economic growth by both groups leads to the destruction of nature, signifying the metabolic rift.

Thus, the sustainability of nature and the material balance between nature and people become a function of diffusion of power. The chapter's key argument, thereby, is to define sustainability based on relationships between humans and nature under the concept of human sociality. It reiterates that the nature can be sustained and the problem of climate change tackled only when people would behave collectively with love and care for the nature and when the internationally acknowledged regulatory framework would function effectively.

References

Ackerman, F. (2007). Economics for a warming world. *Post-Autistic Economics Review, 44*, 2–18.

Ackerman, F., Stanton, E. A., & Bueno, R. (2013). CRED: A new model of climate and development. *Ecological Economics, 85*, 166–176. https://doi.org/10.1016/j.ecolecon.2011.04.006

Acosta, A. M., & Pettit, J. (2013) *Practice guide: A combined approach to political economy and power analysis*. Swiss Development Cooperation.

Ahmed, S., & Khan, M. A. (2022). Spatial overview of climate change impacts in Bangladesh: a systematic review. *Climate and Development*, 1–16. https://doi.org/10.1080/17565529.2022.2062284

Ahlstrom, D. (2020). Time for china to accept it is no longer a 'developing country'. *Nikkei Asia*. Retrieved April 02, 2021 from https://asia.nikkei.com/Opinion/Time-for-China-to-accept-it-is-no-longer-a-developing-country

Ahmed, R., & Karmakar, S. (1993). Arrival and withdrawal dates of the summer monsoon in Bangladesh. *International Journal of Climatology, 13*(7), 727–740. https://doi.org/10.1002/joc.3370130703

Ahmed, R., & Kim, I. (2003). Patterns of daily rainfall in Bangladesh during the summer monsoon season: Case studies at three stations. *Physical Geography, 24*(4), 295–318. https://doi.org/10.2747/0272-3646.24.4.295

Akter, T. (2009). *Climate change and flow of environmental displacement in Bangladesh*. Unnayan Onneshan.

Ali, M. S. (n.d.). *Effect of climate change on floods of Bangladesh: Learning from the past*. Unpublished Research Report. Khulna University of Engineering and Technology (KUET), Bangladesh.

Allan, J., Donovan, C., Ekins, P., Gambhir, A., Hepburn, C., Reay, D., Robins, N., Shuckburgh, E., & Zenghelis, D. (2020). A net-zero emissions economic recovery from COVID-19. COP26 Universities Network Briefing.

Atkinson, R. D., & Hackler, D. (2010). Economic doctrines and approaches to climate change policy. *The Information Technology and Innovation Foundation.*

Averchenkova, A., Gonzalez, L. Martinez-Diaz, L., & van Rooij, J. (2020). *Delivering on the $100 billion climate finance commitment and transforming climate finance. The Independent Expert Group on Climate Finance.*

Baky, M. A. H., Rahman, M. M., & Islam, A. K. M. S. (2017). Development of renewable energy sector in Bangladesh: Current status and future potentials. *Renewable and Sustainable Energy Reviews, 73*, 1184–1197. https://doi.org/10.1016/j.rser.2017.02.047

Basak, J. K. (n.d.). *Effects of increasing temperature and population growth on rice production in Bangladesh: Implications for food security*. Unnayan Onneshan.

Basak. J. K. (2009). *Climate change impacts on rice production in Bangladesh: Results from a model. Climate Change and Development Perspective*. Unnayan Onneshan.

Basak, J. K., Ali, M. A., & Islam, M. N. (2009). Assessment of the effect of climate change on 'boro' rice production in Bangladesh using DSSAT model. In *Proceedings of the Training Material on Facing the Challenges of Climate Change: Issues, Impacts and Adaptation Strategies for Bangladesh*, International Training Network (ITN) Centre, BUET, Dhaka.

Basak, J. K., Ali, M. A., Islam, M. N., & Rashid, M. A. (2010). Assessment of the effect of climate change on *boro* rice production in Bangladesh using DSSAT model. *Journal of Civil Engineering (IEB), 38*(2), 95–108.

Basak, J. K., Titumir, R. A. M., & Dey, N. C. (2013). Climate change in Bangladesh: A historical analysis of temperature and rainfall data. *Journal of Environment, 2*(2), 41–46.

Basher, M. A., Islam, A. K. M. S., Stiller-Reeve, M. A., & Chu, P. (2019). Changes in future rainfall extremes over northeast Bangladesh: A Bayesian model averaging approach. *International Journal of Climatology, 40*, 3232–3249. https://doi.org/10.1002/joc.6394

BBC News. (2021, November 15). COP26: What was agreed at the Glasgow climate conference? BBC News. https://www.bbc.co.uk/news/science-environment-56901261

Becker, M., Papa, F., Karpytchev, M., Delebecque, C., Krien, Y., Khan, J. U., Ballu, V., Durand, F., Le Cozannet, G., Islam, A. K. M. S., Calmant, S., & Shum, C. K. (2020). Water level changes, subsidence, and sea level rise in the Ganges–Brahmaputra–Meghna delta. *PNAS, 117*(4), 1867–1876. https://doi.org/10.1073/pnas.1912921117

Behlert, B., Diekjobst, R., Felgentreff, C., Manandhar, T., Mucke, P., Pries, L., Radtke, K. & Weller, D. (2020). World risk report 2020. Bündnis Entwicklung Hilft and Ruhr University Bochum—Institute for International Law of Peace and Armed Conflict (IFHV).

Bellamy, B. R., & Diamanti, J. (2018). *Materialism and critique of energy.* MCM' Publishing.

Berkes, F., Colding, J., & Folke, C. (2003). *Navigating socio-ecological systems: Building resilience for complexity and change.* Cambridge University Press.

Berkhout, F. (2002). Technological regimes, path dependency and the environment. *Global Environmental Change, 12*(1), 1–4. https://doi.org/10.1016/S0959-3780(01)00025-5

Black, R. (2006). Climate talks a tricky business. *BBC News.* http://news.bbc.co.uk/2/hi/science/nature/6161998.stm

BMD. (2017). List of major cyclonic storms from 1960 to 2017. Bangladesh Meteorological Department, Government of the People's Republic of Bangladesh, Dhaka. Retrieved March 23, 2021 from http://live3.bmd.gov.bd/p/Historical-Cyclones/

Böhm, S., & Dabhi, S. (Eds.). (2009). *Upsetting the offset: The political economy of carbon markets.* MayFly.

Bousquet, F., Botta, A., Alinovi, L., Barreteau, O., Bossio, D., Brown, K., Caron, P., Cury, P., D'Errico, M., DeClerck, F., Dessard, H., Enfors Kautsky, E., Fabricius, C., Folke, C., Fortmann, L., Hubert, B., Magda, D., Mathevet, R., Norgaard, R. B., … Staver, C. (2016). Resilience and development: Mobilizing for transformation. *Ecology and Society, 21*(3), 40. https://doi.org/10.5751/ES-08754-210340

Boyle, M. J. (2021). Top 25 developed and developing countries. *Investopedia.* Retrieved September 25, 2021. https://www.investopedia.com/updates/top-developing-countries/

BPDB. (2021). Present installed generation capacity (MW) as on 05 April, 2021. Bangladesh Power Development Board, Government of the People's Republic of Bangladesh.

Burkett, P. (1999). *Marx and nature: A red and green perspective.* St. Martin's Press.

Cabeza-Gutés, M. (1996). The concept of weak sustainability. *Ecological Economics, 17*(3), 17–156. https://doi.org/10.1016/S0921-8009(96)80003-6

Cadham, J. (2020). COVID-19 and climate change. Centre for International Governance Innovation. Retrieved March 26, 2021 from https://www.cigionline.org/articles/covid-19-and-climate-change

Carrington, D. (2021). Children set for more climate disasters than their grandparents, research shows. *The Guardian.* https://www.theguardian.com/environment/2021/sep/27/children-set-for-more-climate-disasters-than-their-grandparents-research-shows

Carlin, D. (2022, December 16). COP 27 Recap: The Good, the bad, and what's next after the climate conference. *Forbes.* https://www.forbes.com/sites/davidcarlin/2022/12/16/cop-27-recap-the-good-the-bad-and-whats-next-after-the-climateconference/?sh=3335097a2c99

CEDMHA. (2020). *Bangladesh: Disaster management reference handbook.* Center for Excellence in Disaster Management and Humanitarian Assistance.

CEGIS. (2011). Final report on programmes containing measures to facilitate adaptation to climate change of second national communication project of Bangladesh. Center for Environmental and Geographic Information Services and Department of Environment (DoE), Government of the People's Republic of Bangladesh, Dhaka.

CEGIS. (2013). In S. K. Saha, M. S. Ali, & S. Shahid (Eds.), *Vulnerability to climate induced drought scenario and impacts. Study Report Under Comprehensive Disaster Management Programme II-CDMP II.* Center for Environmental and Geographic Information Services and Ministry of Disaster Management and Relief, Government of the People's Republic of Bangladesh.

Chaturvedi, A., Pandey, B., Yadav, A. K., & Saroj, S. (2021). An overview of the potential impacts of global climate change on water resources. In B. Thokchom, P. Singh, P. Qiu & P. K. Iyer (Eds.),

Water conservation in the era of global climate change (pp. 99–120). Elsevier. https://doi.org/10.1016/B978-0-12-820200-5.00012-9

Childs, M. (2012). Privatising the atmosphere: A solution or dangerous con? *Ephemera: Theory & Politics in Organization, 12*(1/2), 12–18.

Choudhury, A. M., Haque, M. A., & Quadir, D. A. (1997). Consequences of global warming and sea level rise in Bangladesh. *Marine Geodesy, 20*(1), 13–31. https://doi.org/10.1080/01490419709388092

Clark, A., Wu, J., & Devnath, A. (2021, April 08). Mysterious plumes of methane gas appear over Bangladesh. *Bloomberg Green*. https://www.bloomberg.com/news/articles/2021-04-07/mysterious-plumes-of-methane-gas-appear-over-bangladesh

Clark, B., Foster, J. B., & York, R. (2009). Capitalism in wonderland. *Monthly Review, 61*(1).

Central, C. (2019). *Flooded future: Global vulnerability to sea level rise worse than previously understood*. Climate Central.

Climate Watch. (2021). Bangladesh: Greenhouse gas emissions and emissions targets. *CAIT Data*. Retrieved July 15, 2021 from https://www.climatewatchdata.org/countries/BGD

CPI. (2019). In B. Buchner, A. Clark, A. Falconer, R. Macquarie, C. Meattle, R. Tolentino, & C. Wetherbee (Eds.), *Global landscape of climate finance 2019*. Climate Policy Initiative.

Cumming, G. S., & Peterson, G. D. (2017). Unifying research on social-ecological resilience and collapse. *Trends in Ecology & Evolution, 32*(9), 695–713. https://doi.org/10.1016/j.tree.2017.06.014

Daily, G. C., Ehrlich, P. R., Mooney, H. A., & Ehrlich, A. H. (1991). Greenhouse economics: Learn before you leap. *Ecological Economics, 4*, 1–10.

Daly, H. E. (1994). Operationalizing sustainable development by investing in natural capital. In A. M. Jansson, M. Hammer, C. Folke, & R. Costanza (Eds.), *Investing in natural capital: The ecological economics approach to sustainability*. Island Press.

Daly, H. E. (2008). *A steady state economy*. Sustainable Development Commission.

Dasgupta, S., Laplante, B., Meisner, C., Wheeler, D., & Yan J. (2007). *The impact of sea level rise on developing countries: A comparative analysis*. World Bank Policy Research Paper 4130. The World Bank Group.

DDM. (2017). *Report on impact of ongoing flood*, August, 2017. Department of Disaster Management, Government of the People's Republic of Bangladesh.

de Ferrer, M. (2021, November 2). COP26: Who is being left out of the climate conversation? *Euronews*. https://www.euronews.com/green/2021/11/01/the-missing-voices-of-cop26-who-is-being-left-out-of-the-climate-conversation

Dietz, S., Hepburn, C., & Stern, N. (2008). Economics, ethics and climate change. In *Climate Change and Global Justice Conference*, Department of Politics and International Relations, September 21–22, 2007, University of Oxford.

DoE. (2019). *Sixth national report for the convention on biological diversity*. Department of Environment, Government of the People's Republic of Bangladesh, Dhaka.

Dominioni, G. (2022). Pricing carbon effectively: a pathway for higher climate change ambition. *Climate Policy*, 22(7), 897–905. https://doi.org/10.1080/14693062.2022.2042177

Donat, M. G., Lowry, A. L., Alexander, L. V., O'Gorman, P. A., & Maher, N. (2016). More extreme precipitation in the world's dry and wet regions. *Nature Climate Change, 6*, 508–513. https://doi.org/10.1038/nclimate2941

EARTH.ORG. (2021). A landfill in Bangladesh is leaking huge quantities of methane. Retrieved August 20, 2021 from https://earth.org/landfill-in-bangladesh-is-leaking-huge-quantities-of-methane/

Eckstein, D., Künzel, V., Schäfer, L., & Winges, M. (2020). *Global climate risk index 2020*. Briefing Paper. Germanwatch, Bonn & Berlin.

EJF. (2021). Climate displacement in Bangladesh. Environmental Justice Foundation. Retrieved March 10, 2021 from https://ejfoundation.org/reports/climate-displacement-in-bangladesh

Elflein, J. (2017). Global capacity of mosquitoes (*Aedes aegypt*) to transmit dengue fever 1950–2014. *Statista*. Retrieved March 27, 2021 from https://www.statista.com/statistics/785341/dengue-fever-transmission-capacity-mosquito-globally/

Empson, M. (2015). In P. Burkett (Ed.), *Marx and nature: A red and green perspective*. Review of the book Marx and nature: A red and green perspective. Haymarket Books.

FAO. (2007). In S. Ramamasy & S. Baas (Eds.), *Climate variability and change: Adaptation to drought in Bangladesh. A resource book and training guide. Institutions for Rural Development* (Vol. 9). Food and Agriculture Organization of the United Nations.

Farmer, J. D., Hepburn, C., Ives, M. C., Hale, T., Wetzer, T., Mealy, P., Rafaty, R., Srivastav, S., & Way, R. (2019). Sensitive intervention points in the post-carbon transition. *Science, 364*(6436), 132–134. https://doi.org/10.1126/science.aaw7287

Farmer, J. D., Hepburn, C., Mealy, P., & Teytelboym, A. (2015). A third wave in the economics of climate Change. *Environmental and Resource Economics, 62*, 329–357. https://doi.org/10.1007/s10640-015-9965-2

Farrell, N. (2017). What factors drive inequalities in carbon tax incidence? Decomposing socioeconomic inequalities in carbon tax incidence in Ireland. *Ecological Economics, 142*, 31–45. https://doi.org/10.1016/j.ecolecon.2017.04.004

Farukh, M. A., Hossen, M. A. M., & Ahmed, S. (2019). Impact of extreme cyclone events on coastal agriculture in Bangladesh. *Progressive Agriculture, 30*(1), 33–41. https://doi.org/10.3329/pa.v30i0.41555

FFWC. (2019). Annual Flood Report 2019. Flood Forecasting and Warning Centre, Bangladesh Water Development Board.

Fickling, D. (2021). This week may turn the tide on two centuries of emissions. *Global Carbon Project, World Bank and Bloomberg Opinion*. Retrieved April 20, 2021 from https://www.bloomberg.com/opinion/articles/2021-02-28/china-s-five-year-plan-will-determine-the-future-of-decarbonization?sref=fDPa8NQZ

Fletcher, R. (2012). Capitalizing on chaos: Climate change and disaster capitalism. *Ephemera: Theory & Politics In Organization, 12*(1/2), 97–112.

Folke, C., Carpenter, S. R., Walker, B., Scheffer, M., Chapin, T., & Rockstrom, J. (2010). Resilience thinking: Integrating resilience, adaptability and transformability. *Ecology and Society, 15*(4), 20.

Folke, C., Hahn, T., Olsson, P., & Norberg, J. (2005). Adaptive governance of social-ecological systems. *Annual Review of Environment and Resources, 30*, 441–473. https://doi.org/10.1146/annurev.energy.30.050504.144511

Ford, D. J., Smit, B., & Wandel, J. (2006). Vulnerability to climate change in the Arctic: A case study from Arctic Bay, Canada. *Global Environmental Change, 16*(2), 145–160.

Foster, J. B. (1999). Marx's theory of metabolic rift: Classical foundations for environmental sociology. *American Journal of Sociology, 105*(2), 366–405. https://doi.org/10.1086/210315

Foster, J. B. (2009). *The ecological revolution: Making peace with the planet*. Monthly Review Press.

Foster, J. B., & Burkett, P. (2016). *Marx and the earth: An anti-critique*. Brill.

Foster, J. B., & Clark, B. (2004). Ecological imperialism: The curse of capitalism. In L. Panitch & C. Leys (Eds.), *Socialist register* (pp. 186–201). Monthly Review Press.

Foster, J. D. (2010). *Clean technology manufacturing competitiveness: The role of tax incentives*. Subcommittee Hearing. Subcommittee on Energy, Natural Resources and Infrastructure, United States Senate Committee on Finance.

Friends of the Earth International. (2022). Empty words, hollow promises, and false solutions ring loud at Bonn conference on climate change. Retrieved June 25, 2022 from https://www.foei.org/empty-words-hollow-promises-and-false-solutions-ring-loud-at-bonn-conference-on-climate-change/

Fuentes, R., Galeotti, M., Lanza, A., & Manzano, B. (2020). COVID-19 and climate change: A tale of two global problems. *Sustainability, 12*, 8560. https://doi.org/10.3390/su12208560

Georgescu-Roegen, N. (1975). Selections from "energy and economic myths." *Southern Economic Journal, 41*(3), 347–381.

Ghotge, S. (2018). Climate change and Marx in the 21st century, Part I. *Capitalism Nature Socialism, 29*(2), 29–42. https://doi.org/10.1080/10455752.2018.1467366

Gilbertson, T., & Reyes, O. (2009). *Carbon trading: How it works and why it fails*. DagHammarskjöld Foundation.

Gowdy, J., & Erickson, J. D. (2005). The approach of ecological economics. *Cambridge Journal of Economics, 29*(2), 207–222. https://doi.org/10.1093/cje/bei033

Greenfield, P. (2023, January 20). Revealed: more than 90% of rainforest carbon offsets by biggest provider are worthless, analysis shows. *The Guardian*. https://www.theguardian.com/environment/2023/jan/18/revealed-forest-carbon-offsets-biggestprovider-worthless-verra-aoe

Gunderson, L. H., & Holling, C. S. (2002). *Panarchy: Understanding transformations in human and natural systems*. Island Press.

Gunderson, L. H., Holling, C. S., & Allen, C. R. (2010). The evolution of an idea—The past, present, and future of ecological resilience. In L. H. Gunderson, C. R. Allen, & C. S. Holling (Eds.), *Foundations of ecological resilience* (pp. 423–444). Island Press.

Guoju, X., Qiang, Z., Runyuan, W., Yubi, Y., Yu, L., Jie, S., Fengju, Z., Juying, H, Fei, M., Chengke, L., Jing, W., Hong, Z., Helin, W., & Kai, Z. (2016). Climate warming: Does northwest China face a stark food security challenge? *Applied Ecology and Environmental Research, 14*(3), 613–636. https://doi.org/10.15666/aeer/1403_613636

Harrabin, R. (2012, December 8). UN climate talks extend Kyoto protocol, promise compensation. *BBC News*. http://www.bbc.com/news/science-environment-20653018

Harris, J. M., Roach, B., & Codur, A. (2017). *The economics of global climate change*. Tufts University.

Harvey, F. (2015, December 14). Paris climate change deal too weak to help poor, critics warn. *The Guardian*. https://www.theguardian.com/environment/2015/dec/14/paris-climate-change-deal-cop21-oxfam-actionaid

Harvey, F. (2021, May 14). Third of global food production at risk from climate crisis. *The Guardian*. https://www.theguardian.com/environment/2021/may/14/third-of-global-food-production-at-risk-from-climate-crisis

Harvey, F., Carrington, D., & Brooks, L. (2021, November 14). Cop 26 ends in climate agreement despite India watering down coal resolution. *The Guardian*. https://www.theguardian.com/environment/2021/nov/13/cop26-countries-agree-to-acceptimperfect-climate-agreement

Harvey, F., & Vidal, J. (2011, December 11). Durban deal will not avert catastrophic climate change, say scientists. *The Guardian*. https://www.theguardian.com/environment/2011/dec/11/durban-climate-change-deal?intcmp=239

Hasan, A. B. M. S., & Rahman, M. Z. (2013). Change in temperature over Bangladesh associated with degrees of global warming. *Asian Journal of Applied Science and Engineering, 2*(2).

Hausfather, Z. (2018). Explainer: What climate models tell us about future rainfall. *Carbon Brief*. Retrieved March 17, 2021 from https://www.carbonbrief.org/explainer-what-climate-models-tell-us-about-future-rainfall

Hazra, S., Ghosh, T., DasGupta, R., & Sen, G. (2002). Sea level and associated changes in the Sundarbans. *Science and Culture, 68*(9–12), 309–321.

Hepburn, C., O'Callaghan, B., Stern, N., Stiglitz, J., & Zenghelis, D. (2020). Will COVID-19 fiscal recovery packages accelerate or retard progress on climate change? *Oxford Review of Economic Policy, 36*(S1), 359–381. https://doi.org/10.1093/oxrep/graa015

Holland, S. P., & Moore, M. R. (2013). Market design in cap and trade programs: Permit validity and compliance timing. *Journal of Environmental Economics and Management, 66*(3), 671–687. https://doi.org/10.1016/j.jeem.2013.05.004

Hossain, M. S., Roy, K., & Datta, D. K. (2014). Spatial and temporal variability of rainfall over the south-west coast of Bangladesh. *Climate, 2*, 28–46. https://doi.org/10.3390/cli2020028

Huq, S., Karim, Z., Asaduzzaman, M., & Mahtab, F. (Eds.). (1999). *Vulnerability and adaptation to climate change for Bangladesh*. Springer.

IDMC. (2021). Displacement data: Bangladesh. Internal Displacement Monitoring Centre. Retrieved February 15, 2021 from https://www.internal-displacement.org/countries/bangladesh

IEA. (2020). Global energy review: CO_2 emissions in 2020: Understanding the impacts of Covid-19 on global CO_2 emissions. International Energy Agency. Retrieved April 02, 2021 from https://www.iea.org/articles/global-energy-review-co2-emissions-in-2020

IIED. (2012). *The eight unmet promises of fast-start climate finance*. IIED Briefing. International Institute for Environment and Development. Available at https://pubs.iied.org/sites/default/files/pdfs/migrate/17141IIED.pdf

IPCC. (1990). In J. T. Houghton, G. J. Jenkins, & J. J. Ephraums (Eds.), *Climate change: The IPCC scientific assessment*. Contribution of working group I to the first assessment report of the Intergovernmental Panel on Climate Change. Cambridge University Press.

IPCC. (1995). In R. T. Watson, M. C. Zinyowera, & R. H. Moss (Eds.), *Climate change 1995: Impacts, adaptations and mitigation of climate change: Scientific-technical analyses*. Contribution of working group II to the second assessment report of the Intergovernmental Panel on Climate Change. Cambridge University Press.

IPCC. (1998). Potential impact and coastal zone management. Working group II: Impacts, adaptation and vulnerability. Intergovernmental Panel on Climate Change, Geneva.

IPCC. (2001). In J. J. McCarthy, O. F. Canziani, N. A. Leary, D. J. Dokken, & K. S. White (Eds.), *Climate change 2001: Impacts, adaptation, and vulnerability*. Contribution of working group II to the third assessment report of the Intergovernmental Panel on Climate Change. Cambridge University Press.

IPCC. (2007). In R. K. Pachauri & A. Reisinger (Eds.), *Climate change 2007: Synthesis report*. Contribution of working groups I, II and III to the fourth assessment report of the Intergovernmental Panel on Climate Change. IPCC.

IPCC. (2014). In R. K. Pachauri & L. Meyer (Eds.), *Climate change 2014: Synthesis report*. Contribution of working groups I, II and III to the fifth assessment report of the Intergovernmental Panel on Climate Change. IPCC.

IPCC. (2018). Summary for policymakers. In V. Masson-Delmotte, P. Zhai, H.-O. Pörtner, D. Roberts, J. Skea, P. R. Shukla, A. Pirani, W. Moufouma-Okia, C. Péan, R. Pidcock, S. Connors, J. B. R. Matthews, Y. Chen, X. Zhou, M. I. Gomis, E. Lonnoy, T. Maycock, M. Tignor, & T. Waterfield (Eds.), *Global warming of 1.5 °C*. An IPCC special report on the impacts of global warming of 1.5 °C above pre-industrial levels and related global greenhouse gas emission pathways, in the context of strengthening the global response to the threat of climate change, sustainable development, and efforts to eradicate poverty. Cambridge University Press.

IPCC. (2021). Summary for policymakers. In V. Masson-Delmotte, P. Zhai, A. Pirani, S. L. Connors, C. Péan, S. Berger, N. Caud, Y. Chen, L. Goldfarb, M. I. Gomis, M. Huang, K. Leitzell, E. Lonnoy, J. B. R. Matthews, T. K. Maycock, T. Waterfield, O. Yelekçi, R. Yu, & B. Zhou (Eds.), *Climate change 2021: The physical science basis*. Contribution of working group I to the sixth assessment report of the Intergovernmental Panel on Climate Change. Cambridge University Press.

IPCC. (2022). Summary for policymakers. In: Climate change 2022: Impacts, adaptation and Vulnerability. In H.-O. Pörtner, D.C. Roberts, E.S. Poloczanska, K. Mintenbeck, M. Tignor, A. Alegría, M. Craig, S. Langsdorf, S. Löschke, V. Möller, A. Okem (Eds.), Contribution of working group II to the sixth assessment report of the intergovernmental panel on climate change. Cambridge University Press. https://doi.org/10.1017/9781009325844.001

Islam, A. K. M. S., Murshed, S. B., Khan, M. S. A., & Hasan, M. A. (2014). Impact of climate change on rainfall intensity in Bangladesh. Institute of Water and Flood Management (IWFM) & Bangladesh University of Engineering and Technology (BUET), Dhaka.

Kaplow, L. (2010). *Taxes, permits, and climate change*. Working Paper 16268. National Bureau of Economic Research, Cambridge.

Kapp, K. W. (1978). *The social costs of business enterprise* (3rd ed.). Spokesman.

Karmakar, S., & Nessa, J. (1997). Climate change and its impacts on natural disasters and SW-monsoon in Bangladesh and the Bay of Bengal. *Journal of Bangladesh Academy of Sciences, 21*(2), 127–136.

Karmakar, S., & Shrestha, M. L. (2000). Recent climate change in Bangladesh. SMRC Series No. 4. SAARC Meteorological Research Centre, Dhaka.

Khan, M. (2018, June 10). Where does all our waste end up? *The Daily Star*. https://www.thedailystar.net/star-weekend/environment/where-does-all-our-waste-end-1587943

Khan, M. H. R., Rahman, A., Luo, C., Kumar, S., Islam, G. M. A., & Hossain, M. A. (2019). Detection of changes and trends in climatic variables in Bangladesh during 1988–2017. *Heliyon, 5*, e01268. https://doi.org/10.1016/j.heliyon.2019.e01268

Khan, T. M. A., Quadir, D. A., Akter, F., & Sarker, M. A. (1999). Sea level changes and coastal erosion problems along Bangladesh Coast. *Journal of Remote Sensing and Environment, 3*, 87–103.

King, D., Ganesan, A., Wadhams, P., et al. (2021). Scientists' statement on lowering atmospheric methane concentrations. *Methane Action*. Retrieved June 10, 2021 from https://methaneaction.org/expert-statement-oxidation-methane/

Klenert, D., Funke, F., Mattauch, L., & O'Callaghan, B. (2020). Five lessons from COVID-19 for advancing climate change mitigation. *Environmental and Resource Economics, 76*, 751–778. https://doi.org/10.1007/s10640-020-00453-w

Kovel, J. (2002). *The enemy of nature*. Zed Books.

Lakhani, N. (2022, October 14). Egypt silenced climate experts' voices before hosting Cop27, HRW says. *The Guardian*. https://www.theguardian.com/environment/2022/oct/14/cop27-egypt-host-climate-talks?CMP=share_btn_tw

Leach, M., & Scoones, I. (2006). *The slow race: Making technology work for the poor*. Demos.

Leach, M., & Scoones, I. (Eds.). (2015). *Carbon conflicts and forest landscapes in Africa*. Routledge.

Lebel, L., Anderies, J. M., Campbell, B., Folke, C., Hatfield-Dodds, S., & Hughes, T. P. (2006). Governance and the capacity to manage resilience in regional social-ecological systems. *Ecology and Society, 11*(1), 19.

Leonard, J. (2020). How does climate change affect human health? *Medical News Today*. Retrieved March 19, 2021 from https://www.medicalnewstoday.com/articles/climate-change-and-health

Lindsey, R. (2021). Climate change: Global sea level. *NOAA, Climate.gov*. Retrieved February 15, 2021 from https://www.climate.gov/news-features/understanding-climate/climate-change-global-sea-level

Lohman, L. (Eds.). (2006). *Carbon trading: A critical conversation on climate change, privatisation and power*. Dag Hammarskjöld Foundation.

Lohman, L. (2012). Financialization, commodification and carbon: The contradictions of neoliberal climate policy. *Socialist Register, 48*, 85–107.

Luetz, J. (2008). *Planet prepare: Preparing coastal communities in Asia for future catastrophes*. World Vision Asia Pacific Regional Office.

Magdoff, F., & Foster, J. B. (2011). What every environmentalist needs to know about capitalism. *Monthly Review Press, 61*(10).

Mahtab, F. (1989). *Effect of climate change and sea level rise on Bangladesh*. A report submitted to the Commonwealth Group of Experts. Commonwealth Secretariat, London.

Malhi, Y., Franklin, J., Seddon, N., Solan, M., Turner, M. G., Field, C. B., & Knowlton, N. (2020). Climate change and ecosystems: Threats, opportunities and solutions. *Philosophical Transactions of the Royal Society B, 375*, 20190104. https://doi.org/10.1098/rstb.2019.0104

Maréchal, K. (2012). *The economics of climate change and the change of climate in economics*. Routledge.

Marino, E., White, D., Schweitzer, P., Chambers, M., & Wisniewski, J. (2009). *Drinking water in north-western Alaska: Using or not using centralized water systems in two rural communities*. University of Calgary.

Markandya, A. (2011). Equity and distributional implications of climate change. *World Development, 39*(6), 1051–1060. https://doi.org/10.1016/j.worlddev.2010.01.005

Marx, K. (1887). *Capital: A critique of political Economy* (Vol. 1). Progress Publishers.

Masood, E., & Tollefson, J. (2021). 'COP26 hasn't solved the problem': scientists react to UN climate deal. *Nature*, 599(7885), 355–356. https://doi.org/10.1038/d41586-021-03431-4

Mattauch, L., & Hepburn, C. (2016). Climate policy when preferences are endogenous—And sometimes they are. *Midwest Studies in Philosophy, XL*.

McKenna, J. (2017). The deadly diseases being released as ice thaws. *World Economic Forum*. Retrieved March 19, 2021, from https://www.weforum.org/agenda/2017/05/the-deadly-diseases-being-released-by-climate-change/

Mengistu, D., Bewket, W., Dosio, A., & Panitz, H. (2021). Climate change impacts on water resources in the Upper Blue Nile (Abay) River Basin, Ethiopia. *Journal of Hydrology, 592*, 125614, https://doi.org/10.1016/j.jhydrol.2020.125614

Metcalf, G. E. (2009). Market-based policy options to control U.S. greenhouse gas emissions. *Journal of Economic Perspectives, 23*(2), 5–27.

Miller, F., Osbahr, H., Boyd, E., Thomalla, F., Bharwani, S., & Ziervogel, G. (2010). Resilience and vulnerability: Complementary or conflicting concepts. *Ecology and Society, 15*(3), 11.

MoEF. (2005). National adaptation programme of action (NAPA) 2005. Ministry of Environment and Forest, Government of the People's Republic of Bangladesh, Dhaka.

MoEF. (2009). Bangladesh climate change strategy and action plan (BCCSAP) 2009. Ministry of Environment and Forests, Government of the People's Republic of Bangladesh, Dhaka.

MoEF. (2012). Second national communication of Bangladesh to the United Nations Framework Convention on Climate Change. Ministry of Environment and Forest, Government of the People's Republic of Bangladesh.

MoFA. (2018). Climate change profile: Bangladesh. Ministry of Foreign Affairs, Government of Netherlands.

MoFDM. (2008). Super cyclone *Sidr* 2007: Impacts and strategies for interventions. Ministry of Food and Disaster Management, Government of the People's Republic of Bangladesh, Dhaka.

Morrison, T. H., Adger, W. N., Agrawal, A., Brown, K., Hornsey, M. J., Hughes, T. P., Jain, M., Lemos, M. C., McHugh, L. H., O'Neill, S., & Van Berkel, D. (2022). Radical interventions for climate-impacted systems. *Nature Climate Change*, 12(12), 1100–1106. https://doi.org/10.1038/s41558-022-01542-y

Mondol, M. A. H., Mamun, A., Iqbal, M., & Jang, D. (2018). Precipitation concentration in Bangladesh over different temporal periods. *Advances in Meteorology, 28*. https://doi.org/10.1155/2018/1849050

Montgomery, W. D. (1972). Markets in licenses and efficient pollution control programs. *Journal of Economic Theory, 5*(3), 395–418. https://doi.org/10.1016/0022-0531(72)90049-X

Murshed, S. B., Islam, A. K. M. S., & Khan, M. S. A. (2011). *Impact of climate change on rainfall intensity in Bangladesh.* Paper presented at the 3rd International Conference on Water & Flood Management (ICWFM), January 8–10, Dhaka.

Nandi, S. K., Hoque, M. N., & Ghosh, H. R., & Roy, S. (2011). Potential of wind and solar electricity generation in Bangladesh. *International Scholarly Research Network, 6*. https://doi.org/10.5402/2012/401761

Natamba, L., Zhang, W., Zhang, J., & Zhao, X. (2018). Climate change causing food insecurity in East Africa: Traditional and non-traditional strategies to solve the problem. *Applied Ecology and Environmental Research, 16*(3), 2233–2254. https://doi.org/10.15666/aeer/1603_22332254

Nature. (2021). Global climate action needs trusted finance data. *The International Journal of Science, 589.*

Newell, P., Boykoff, M., & Boyd, E. (2012). *The 'new' carbon economy: Constitution, governance and contestation.* Wiley Blackwell.

Newell, P., & Paterson, M. (2010). *Climate capitalism: Global warming and the transformation of the global economy.* Cambridge University Press.

Newsome, T., & Ripple, W. (2019). 11,000 scientists warn: Climate change isn't just about temperature. *The Conversation.* Retrieved January 15, 2021 from https://theconversation.com/11-000-scientists-warn-climate-change-isnt-just-about-temperature-126261

NOAA. (2021a). Climate at a glance: Global time series. National Centers for Environmental Information, published September 2021a. Retrieved September 24, 2021a from https://www.ncdc.noaa.gov/cag/

NOAA. (2021b). Despite pandemic shutdowns, carbon dioxide and methane surged in 2020. *NOAA Research News*. Retrieved July 15, 2021b from https://research.noaa.gov/article/ArtMID/587/ArticleID/2742/Despite-pandemic-shutdowns-carbon-dioxide-and-methane-surged-in-2020

Nordhaus, W. D. (1992). An optimal transition path for controlling greenhouse gases. *Science, 258*(5086), 1315–1319. https://doi.org/10.1126/science.258.5086.1315

Nordhaus, W. D. (2001). Global warming economics. *Science, 294*, 1283–1284. https://doi.org/10.1126/science.1065007

O'Connor, M. (Ed.). (1994). *Is capitalism sustainable? Political economy and the politics of ecology*. Guilford Press.

ODA & HBS. (2021). In C. Watson & L. Schalatek (Eds.), *Climate finance thematic briefing: Adaptation and mitigation finance. Climate Finance Fundamentals* (Vols. 3 & 4). Climate Funds Update, Official Development Assistance and Heinrich Boll Stiftung.

OECD. (2015). *Climate finance in 2013–14 and the USD 100 billion goal*. A report by the Organisation for Economic Co-operation and Development (OECD) in collaboration with Climate Policy Initiative (CPI).

OECD. (2019). *Climate finance provided and mobilised by developed countries in 2013-17*. OECD Publishing. Organisation for Economic Co-operation and Development, Paris. https://doi.org/10.1787/39faf4a7-en

Oxfam. (2020). *Climate finance shadow report 2020: Assessing progress towards the $100 billion commitment*. Oxfam International.

Pappas, S. (2016). 5 deadly diseases emerging from global warming. Retrieved March 19, 2021 from https://www.livescience.com/55632-deadly-diseases-emerge-from-global-warming.html

Parson, E. A., & Zeckhauser, R. J. (1995). Equal measures or fair burdens: Negotiating environmental treaties in an unequal world. In H. Lee (Ed.), *Shaping national responses to climate change: A post-Rio guide* (pp. 81–114). Island Press.

Posner, E. A., & Sunstein, C. R. (2009). Justice and climate change. In J. E. Aldy & R. N. Stavins (Eds.), *Post-Kyoto international climate policy: Summary for policymakers* (pp. 93–95). Cambridge University Press.

Quadir, D. A., Shrestha, M. L., Khan, T. M. A., Ferdousi, N., Rahman, M., & Mannan, A. (2004). Variations of surface air temperature over the land areas in and around the Bay of Bengal. *Natural Hazards, 31*, 561–584. https://doi.org/10.1023/B:NHAZ.0000023368.81668.e3

Rahman, M. R., Salehin, M., & Matsumoto, J. (1997). Trends of monsoon rainfall pattern in Bangladesh. *Bangladesh Journal of Water Resources Research, 14*(18), 121–138.

Rainforest Action Network (RAN), BankTrack, Indigenous Environmental Network (IEN), Oil Change International (OCI), Reclaim Finance and the Sierra Club. (2021). *Banking on climate chaos: Fossil fuel finance report 2021*.

Rannard, B. G. (2022, November 20). COP27: Climate costs deal struck but no fossil fuel progress. *BBC News*. https://www.bbc.co.uk/news/science-environment-63677466

Rarieya, M., & Fortun, K. (2010). Food security and seasonal climate information: Kenyan challenges. *Sustainability Science, 5*, 99–114. https://doi.org/10.1007/s11625-009-0099-8

Rawat, N., Babu, M. S. U., & Nautiyal, S. (2016). *Climate change and sea-level rise: A review of studies on low-lying and island countries*. Working Paper, No. 359. The Institute for Social and Economic Change, Bangalore, India.

REN21. (2019). *Renewables 2019 global status report*. REN21 Secretariat, Paris.

Rigaud, K., de Sherbinin, A., Jones, B., Bergmann, J., Clement, V., Ober, K., Schewe, J., Brent, S., McCusker, A., Heuser, S., & Midgley, A. (2018). *Groundswell: Preparing for internal climate migration*. The World Bank.

Ritche, H. (2019). Who has contributed most to global CO_2 emissions? *Our World in Data*. Oxford Martin School, University of Oxford. Retrieved March 12, 2021 from: https://ourworldindata.org/contributed-most-global-co2

Ritchie, H., & Roser, M. (2020). CO_2 and greenhouse gas emissions. *Our World in Data*. Oxford Martin School, University of Oxford. Retrieved March 20, 2021 from https://ourworldindata.org/co2-and-other-greenhouse-gas-emissions

Ropke, I. (2004). The early history of modern ecological economics. *Ecological Economics, 50*(3–4), 293–314. https://doi.org/10.1016/j.ecolecon.2004.02.012
Sattar, R. S., Wang, S., Tahir, M. N., & Caldwell, C. (2017). Assessment of smallholder farmer's vulnerability due to climate change in arid areas of Pakistan. *Applied Ecology and Environmental Research, 15*(4), 291–312. https://doi.org/10.15666/aeer/1504_291312
Schmitz, H., & Scoones, I. (2015). *Accelerating sustainability: Why political economy matters.* Evidence Report-Policy Anticipation, Response & Evaluation, No. 152. Institute of Development Studies, England.
Shahid, S. (2010). Recent trends in the climate of Bangladesh. *Climate Research, 42*, 185–193. https://doi.org/10.3354/cr00889
Shamseddin, A. M., & Chaibi, T. (2019). Scanning climate change impacts on water resources of the largest African river basins. *International Journal of River Basin Management, 18*(1), 33–38. https://doi.org/10.1080/15715124.2019.1576699
Siddiqui, T. (2019, November 02). Battling climate induced displacement. *The Daily Star*. https://www.thedailystar.net/opinion/environment/news/battling-climate-induced-displacement-1821784
Siebenhuner, H. (2000). *Homo sustinens*—Towards a new conception of humans for the science of sustainability. *Ecological Economics, 32*(1), 15–25. https://doi.org/10.1016/S0921-8009(99)00111-1
Smith, R. (2010). Beyond growth or beyond capitalism? *Real-World Economics Review* (53).
SMRC. (2003). *The vulnerability assessment of the SAARC coastal region due to sea level rise: Bangladesh case.* SMRC Report No 3. SAARC Meteorological Research Centre, Dhaka.
Söderbaum, P. (1990). Neoclassical and institutional approaches to environmental economics. *Journal of Economic Issues, 24*(2), 481–492. https://doi.org/10.1080/00213624.1990.11505047
Solow, R. M. (1993). An almost practical step towards sustainability. *Resources Policy, 16*(3), 162–172. https://doi.org/10.1016/0301-4207(93)90001-4
Spash, C. L. (2007). The economics of climate change impacts a la Stern: Novel and nuanced or rhetorically restricted? *Ecological Economics, 63*(4), 706–713. https://doi.org/10.1016/j.ecolecon.2007.05.017
SRDI. (2010). Saline soils of Bangladesh. Soil Resource Development Institute, Ministry of Agriculture, Government of the People's Republic of Bangladesh.
Stavins, R. N. (2001). *Experience with market-based environmental policy instruments.* Discussion Paper 01-58. Resources for the Future, Washington, D.C.
Subhani, R., & Ahmad, M. M. (2019). Socio-economic impacts of cyclone *Aila* on migrant and non-migrant households in the south-western coastal areas of Bangladesh. *Geosciences, 9*(482). https://doi.org/10.3390/geosciences9110482
Suh, S. C. (2016). *The failure of climate change negotiations: Irrational countries exclude the poor and the future generations.* Department of Economics Working Paper Series, Working Paper No. 16-07. University of Windsor, Ontario.
The Daily Star. (2019, January 17). Climate change-induced salinity affecting soil across coastal Bangladesh. https://www.thedailystar.net/country/impact-of-climate-change-across-coastal-bangladesh-induced-salinity-affecting-soil-1688872
The Daily Star. (2021, April 10). Huge plumes of methane spotted over Bangladesh. https://www.thedailystar.net/backpage/news/huge-plumes-methane-spotted-over-bangladesh-2075061
The Daily Star. (2022, June 25). Flood death toll now 82 as 9 more die in 24 hours: DGHS. https://www.thedailystar.net/environment/climate-crisis/natural-disaster/news/flood-death-toll-now-82-9-more-die-24-hours-dghs-3056111
The Guardian. (2021, March 24). *Big banks' trillion-dollar finance for fossil fuels 'shocking', says report.* Banking on Climate Chaos—Fossil Fuel Finance Report 2021. https://www.theguardian.com/environment/2021/mar/24/big-banks-trillion-dollar-finance-for-fossil-fuels-shocking-says-report

Tiseo, I. (2021). Largest producers of fossil fuel CO_2 emissions worldwide in 2019, by share of emissions. *US Energy Information Administration 2019*. Retrieved July 15, 2021 from https://www.statista.com/statistics/271748/the-largest-emitters-of-co2-in-the-world/

Titumir, R. A. M., & Basak, J. K. (n.d.). *Climate change and crop production: Impacts on food security of Bangladesh*. Unpublished Research Report. Unnayan Onneshan, Dhaka.

Titumir, R. A. M., & Basak, J. K. (2013). Climate change and food security in Bangladesh: An application of DSSAT model. *Asia Pacific Journal of Environment and Development, 20*(21), 1–18.

Titumir, R. A. M., Kabir, M. H., & Baten, A. B. (2012). *The tempest of exodus: The case of climate change-induced displacement in Bangladesh and international negotiations*. Unnayan Onneshan.

Tollefson, J. (2016). Global warming already driving increases in rainfall extremes. *Nature*. https://doi.org/10.1038/nature.2016.19508

UNDRR. (2019). Global assessment report on disaster risk reduction. United Nations Office for Disaster Risk Reduction (UNDRR), Geneva.

UNEP. (2020). Emission gap report 2020. United Nations Environment Programme, Nairobi.

UNEP. (2021). Adaptation gap report 2020. United Nations Environment Programme, Nairobi.

United Nations. (1992). United Nations framework convention on climate change. FCCC/INFORMAL/84, GE.05-62220 (E) 200705.

United Nations. (2021). Climate change 'biggest threat modern humans have ever faced', world-renowned naturalist tells Security Council, calls for greater global cooperation. United Nations Press Release SC/14445. Retrieved March 24, 2021, from https://www.un.org/press/en/2021/sc14445.doc.htm

Unnayan Onneshan. (2008). *Climate change as if development mattered: A perspective on developing Bangladesh strategy*. Unnayan Onneshan and Oxfam.

Unruh, G. C. (2000). Understanding carbon lock-in. *Energy Policy, 28*(12), 817–830.

van den Bergh, J. C. J. M. (2004). Optimal climate policy is a utopia: From quantitative to qualitative cost-benefit analysis. *Ecological Economics, 48*(4), 385–393. https://doi.org/10.1016/j.ecolecon.2003.10.011

van Wijk, M., Naing, S., Franchy, S. D., Heslop, R. T., Lozano, I. N., Vila, J., & Ballesté-Delpierre, C. (2020). Perception and knowledge of the effect of climate change on infectious diseases within the general public: A multinational cross-sectional survey based study. *PLoS ONE, 15*(11), e0241579. https://doi.org/10.1371/journal.pone.0241579

Victor, D. G., Akimoto, K., Kaya, Y., Yamaguchi, M., Cullenward, D., & Hepburn, C. (2017). Prove Paris was more than paper promises. *Nature, 548*, 25–27. https://doi.org/10.1038/548025a

Warren, P. D. (2021). Forced migration after Paris COP21: Evaluating the "climate change displacement coordination facility". *Columbia Law Review, 116*(8). Retrieved July 27, 2021 from http://columbialawreview.org/content/forced-migration-after-paris-cop21-evaluating-the-climate-change-displacement-coordination-facility/

Warrick, R. A., Bhuiya, A. H., & Mirza, M. Q. (1994). *The greenhouse effect and climate change* (pp. 17–20). Briefing Document No. 1. Bangladesh Unnayan Parishod (BUP), Dhaka.

Weitzman, M. (2009). On modeling and interpreting the economics of catastrophic climate change. *Review of Economics and Statistics, 91*(1), 1–19.

West, L. (2016). What is Kyoto Protocol? Retrieved August 13, 2017, from https://www.thoughtco.com/what-is-the-kyoto-protocol-1204061

Westra, S., Alexander, L. V., & Zwiers, F. W. (2013). Global increasing trends in annual maximum daily precipitation. *Journal of Climate, 26*, 3904–3918. https://doi.org/10.1175/JCLI-D-12-00502.1

World Bank. (2000). *Bangladesh: Climate change and sustainable development*. Report No. 21104-BD. The World Bank, Rural Development Unit, South Asia Region.

World Bank. (2007). *Floods 2007 damage and needs assessment report*. Dhaka, Bangladesh.

World Bank. (2010). *Economics of adaptation to climate change: Bangladesh*. The World Bank Group.

World Bank. (2020). *SDG 13 Climate action: Floods, droughts and heat waves herald a changing climate*. Atlas of Sustainable Development Goals 2020 from World Development Indicators. Retrieved December 30, 2020, from https://datatopics.worldbank.org/sdgatlas/goal-13-climate-action/

World Development Indicators. (2021). Bangladesh—Methane emissions (kt of CO_2 equivalent). Retrieved August 25, 2021, from https://knoema.com/WBWDI2019Jan/world-development-indicators-wdi

WHO. (n.d.). Climate change and infectious diseases. World Health Organization. Available at https://www.who.int/globalchange/climate/en/chapter6.pdf

WHO. (2018). Climate change and health. World Health Organization. Retrieved March 19, 2021, from https://www.who.int/news-room/fact-sheets/detail/climate-change-and-health

WHO. (2020). Coronavirus disease (COVID-19): Climate change. World Health Organization. Retrieved March 27, 2021 from https://www.who.int/news-room/q-a-detail/coronavirus-disease-covid-19-climate-change

WMO. (2020). September was the warmest on record. World Meteorological Organization. Retrieved January 05, 2021 from https://public.wmo.int/en/media/news/september-was-warmest-record

WRI. (2008). In L. Parker & J. Blodgett (Eds.), *Greenhouse gas emissions: Perspectives on the top 20 emitters and developed versus developing Nations*. CRS Report for Congress. World Resources Institute, Washington D.C.

Yousfi, F. (2016, November 16). COP22 in Morocco: Between greenwashing and environmental injustice. *Middle East Eye*. http://www.middleeasteye.net/columns/cop22-morocco-between-greenwashing-and-environmental-injustice-391898113

YPSA. (2014). Bangladesh housing, land and property (HLP) rights initiative: Climate displacement in Bangladesh. Displacement Solutions and Young Power in Social Action, Bangladesh.

Zhang, B. (2013). Market-based solutions: An appropriate approach to resolve environmental problems. *Chinese Journal of Population Resources and Environment, 11*(1), 87–91. https://doi.org/10.1080/10042857.2013.777526

Zhang, J., Feng, M. Q., & Wang, Y. (2019). Effects of climate change on hydrology and water resources. *Applied Ecology and Environmental Research, 17*(4), 9411–9423. https://doi.org/10.15666/aeer/1704_94119423

Chapter 6
Conclusions: Sustainable Transformative Pathways

6.1 Introduction

The final chapter accumulates and integrates the outcomes undertaken throughout the book as well as contains suggestions regarding sustainable resource governance mechanisms. It argues, overall, that in order to proceed towards a sustainable transformative pathway, it is essential to maintain a reciprocal relationship between human beings and nature. It is a fallacy to consider 'nature' and 'human beings' as two different entities of the earth's ecosystem. It is only possible to ensure the well-being of both human beings and nature if they are considered mutually inclusive to the system.

The existing perspectives in resource governance regimes assert that the human beings are the destroyers of nature and suggest tackling the challenge of nature's degradation either through technological innovation or with market-centric policy options both of which regard human beings as exogenous to nature's sustainability framework. Opposing that position, this book contends that it is factual that the anthropogenic drivers are the main factors behind natural resource degradation; however, human beings are the only ones who can also alter the situation and move towards the sustainable transformative pathways because humans and nature cannot be separated in any way. What is imperative, therefore, is to revitalise the symbiotic relationship between the nature and human beings.

6.2 Sustainable Transformative Pathways: Necessary and Sufficient Conditions

The book, throughout the chapters, sets out the context to specify the necessary and sufficient conditions for progressing towards a sustainable transformative pathway that can warrant the well-being of both human beings and nature. The necessary condition entails reviving the historic mutual ties between human beings and nature,

R. A. M. Titumir et al., *Natural Resource Degradation and Human-Nature Wellbeing*,
https://doi.org/10.1007/978-981-19-8661-1_6

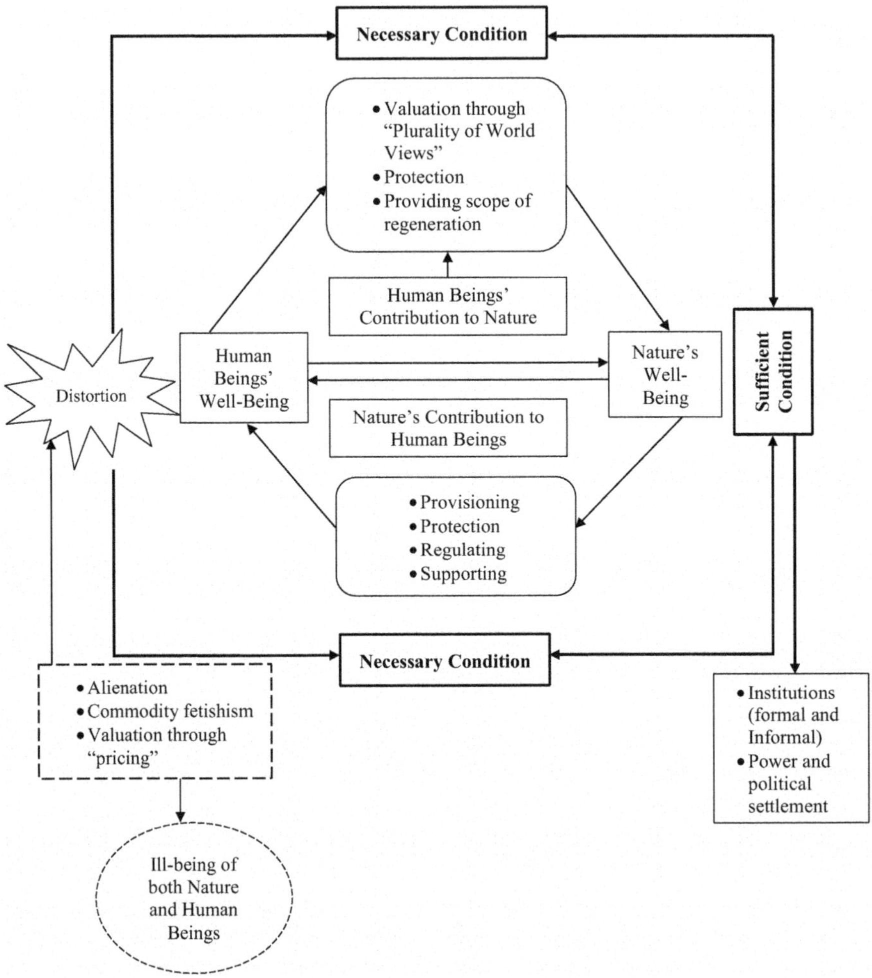

Fig. 6.1 Sustainable transformative pathways. *Source* Prepared by the authors

whereas the sufficient condition is exemplified here by the nature of institutions and political settlement. The conditions, on the contrary, when distorted under the market economy, result in the ill-being of both of the entities (Fig. 6.1).

At present, with the advent of capitalism, human beings consider themselves as independent and masters of nature, despite the fact that they are part of nature. They treat nature as a 'mere matter' that exists just to be extracted for human benefits, destroying existing resources in a variety of ways. Thus, human beings alienate themselves from nature (Titumir et al., 2019). People become estranged from the world of nature when they are unable to recognise its humanity, when they fail to see the world as their world and to ponder themselves as part of it. Hence, alienation stems from people's failure to recognise the sociality or interconnectedness that exists between humans and nature. This alienated relation provokes the commodification of natural resources via market pricing. The commodification here denotes the transformation of commodities into fetishes as human belief in commodities has produced an obfuscated hierarchy of natural resource value on which demands of commodity rely. Accordingly, the most precious commodities are those with a high price tag.

In such a way, valuation becomes equal to the market pricing, and the intrinsic values of natural resources as well as the sociality between nature and humans are ignored, leading to overextraction and thus destruction of natural resources. Because in such conditions, the powerful agents expropriate the resources by using the power of institutions which is more rampant in a transitional economy. In this process, human beings, though apparently seem to gain, also lose in the long run. Overall, if the transformation of nature to sustainability is not ascertained, there would be an imbalance in nature, resulting in resource degradation and loss of endowments for the human beings.

The book argues, however, that humans are a part of the ecology, not just the exclusive agents who only extract resources. There is a long-standing embeddedness between human beings and the ecology, which remains unexplored. Furthermore, humans are often identified as external to the ecological system. As a part of this system, as the book emphasises on the contrary, human beings have maintained an intertwined, intimate and reciprocal relationship with nature. This nexus can be studied from the 'human sociality' perspective (Titumir et al., 2020). The revitalisation of this relationship is at the core to move towards sustainable transformative pathways, as the book contends. Importantly, the revival of the relationship requires that human beings must recognise the intrinsic value of nature. This can also be related to the emerging ideas of 'plurality of worldviews' or 'plurality perspective' or 'value pluralism', which recognises that nature needs to be valued from multiple angles going beyond the monetary valuation (IPBES, n.d; Pascual et al., 2017, 2021).

Some of the important components of such pluralistic valuation include addressing power relations, acknowledging the role of institutions and conflict resolution, among others (IPBES, n.d). However, ideas of pluralistic valuation have only recently emerged in relation to biodiversity conservation. This book makes the argument that the plurality perspective must be internalised into the overall nature governance framework. The understanding can be connected to the recently developed concept of 'societal boundaries' rather than 'planetary boundaries' (Brand et al., 2021). Collective autonomy and self-limitation politics are two fundamental parts of societal boundaries that, when combined, can provide guidance for a just socio-ecological transformation.

Transformative pathways also require the stabilisation of nature that is conservation of natural resources through damage limitation. For the stabilisation of nature, therefore, it is vital to identify both the natural and anthropogenic drivers of ecological degradation. This is where the sufficient conditions—nature of institutions, political settlement, power and class relations, governance structures—are crucial to take into consideration. Importantly, sustainable transformation in a resource governance regime necessitates favourable political settlement and institutional capacity along with the necessary conditions. Countries that fail to adhere to the sufficient conditions appropriately have, thus, been proven to be sabotaging their sustainable development.

This book considers the particular context of the developing or transitional economy of Bangladesh where such failures can be observed. The particular form of materialist incentives of expropriation of natural resources by means of power and coercion existing in the economy (Titumir, 2021) causes unabated loss from the viewpoints of the destruction of resources and overall socio-ecological imbalance. The more money the powerful class accumulates, the greater its influence in extracting resources from nature. A clientelistic network emerges because of this ubiquitous motivation, in which members of the network thrive through primitive accumulation (Titumir et al., 2019). Only the material provision of nature comes forward in this way, condoning the intrinsic and underlying values of mutual existence. Therefore, actions are required to alter the existing governance mechanisms through the reformation of formal institutions and the recognition of informal institutions because the norms and values existing in society are shaped by the nature of institutions. Overall, if the sustainable transformation would be possible to ensure, it would build a conducive ambience for ensuring the well-being of both humans and nature.

6.3 Bending the Curve of Degradation of Biodiversity Resources

Despite about a century and a half of action by policymakers and conservation organisations, global biodiversity is in peril (Pascual et al., 2021). It is more apparent through the failure of the countries in meeting the targets sufficiently envisioned under the Convention on Biological Diversity (CBD). The Convention's 196 signatory countries agreed in 2010 to achieve20 ambitious biodiversity targets for 2020, none of which has been fully met, though six targets have been partially attained (SCBD, 2020). Recently, countries have again adopted a new Global Biodiversity Framework (GBF) with 4 goals and 23 targets under the CBD mainly to ensure the conservation of 30% land and sea areas by 2030 (SCBD, 2022). Although this a comprehensive plan, success of it requires systematic change in public policy (Nature, 2022). Moreover, the accountability of the individual countries, particularly the developed ones, remains a critical issue.

Notably, Bangladesh, as a contracting party to CBD, is also committed to implementing conservation and sustainable management of its biological diversity (Titumir et al., 2020). However, empirical analysis based on findings of Chap. 3 reveals that the most important biodiversity hotspot of Bangladesh—the Sundarbans—is under threat of continuous degradation because of the fragile nature of institutions, unequal power-sharing arrangements and commodification of biodiversity resources under the concrete social–historical context of the country's transitional economy. This process has also a negative impact on the lives and livelihoods of Indigenous Peoples

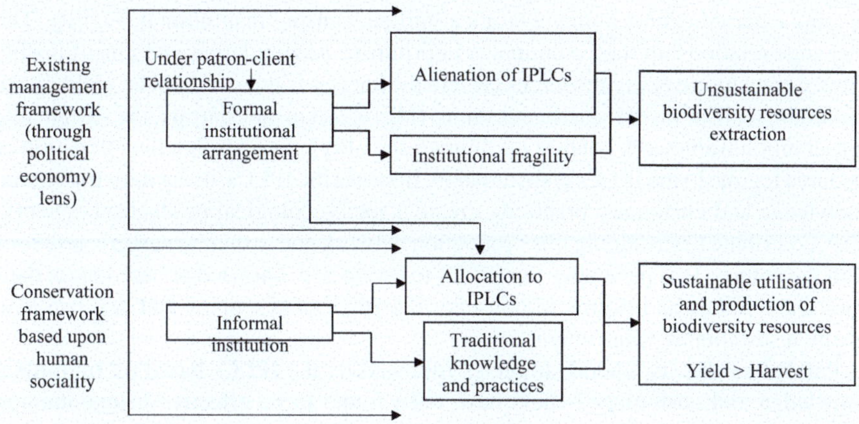

Fig. 6.2 Bending the curve of biodiversity resource degradation. *Source* Prepared by the authors

and Local Communities (IPLCs). Overall, it can be summarised that the Sundarbans' legal and governance system has been weak and contradictory in terms of the implementation process. As a result, the conservation efforts are hampered, and the forest's sustainability is jeopardised due to constant exploitation by powerful groups and the ineffective role of the administrative authority. Moreover, the state also loses a significant share of revenue in the long run.

With a view to altering the existing scenario, the chapter, in contrast to the market-centric perspective, constructs a conservation framework that demonstrates how the interinstitutional pitfall caused by the exclusion of informal institutions and community ownership degrades biodiversity resources. In addition, it states that conservation necessitates recognising a diverse set of values, knowledge and framings of biodiversity landscape that foster cooperation and incentivise conservation for long-term sustainable use of those resources (Fig. 6.2).

The conflicting decision in the face of the degradation of the biodiversity resources is whether the human beings should stop using the resources or continue to use them to meet their growing needs. The market-centric approach argues that biodiversity resources are valuable (in price) in the sense that they are useful for the human beings, and therefore, the resources should be used. While those belonging to the conservationist camp, vehemently oppose this position and argue that the resources should be conserved. Neither of these two strands takes into consideration political economic and institutional factors in comprehending the gain (or loss) of biological resources. Particularly in a transitional economy, the political-economic lens is the key to identifying the drivers of biodiversity resource degradation while the nature of institutions (formal and informal) matters in general to draw viable policy measures for the conservation of those resources.

In the presence of neoliberal policy measures, the exchange process involves a patron-client relationship (Titumir et al., 2020). Under such a scenario, because of unequal power distribution between the political elites and the IPLCs, the IPLCs become external agents in the ecological milieu, causing institutional fragility. The exchange relationship then culminates in primitive accumulation and unsustainable extraction of biodiversity resources (where the harvest is greater than the yield due to maximum realisation of the resources rent) (Fig. 6.2). On the contrary, the sustainable conservation framework centred on 'human sociality' suggests that the allocation of resource regime to the IPLCs is sustainable. Because the IPLCs, using their traditional knowledge and customary practices, create a socio-ecological production network. They contribute to sustaining this production network due to their synergetic relations with the nature. This persuades the IPLCs to invent new knowledge based upon their traditional values and norms to conserve the stock of resources and practise it to ensure a sustainable value chain.

Put differently, livelihood strategies followed by the IPLCs based on traditional knowledge and customary beliefs have been found to be effective in maintaining sustainable resource utilisation and conservation. The rules and practices they follow help conserve the forest by maintaining ecological harmony and a thriving socio-ecological lifecycle. Overall, the IPLCs and their practice of traditional knowledge make the biodiversity resources more resilient (where yield is greater than harvesting) and sustainable (Fig. 6.2). Thus, their livelihood pattern promotes sufficiency rather than infinite growth, redistribution of wealth rather than accumulation and of course, equity (Pascual et al., 2021; Swiderska, 2021). Nevertheless, the inclusion of traditional knowledge is not only a matter of equity but also a source of knowledge (Díaz et al., 2015). Yet, their knowledge has often been neglected by identifying them as 'backward' (Swiderska, 2021), under the formal institutional management system.

The alternative conservation measures suggest that the recognition and inclusion of such informal institutions are at the root of the revitalisation of the human–nature nexus in the context of biodiversity resource governance and accordingly, it is necessary to integrate indigenous knowledge into the contemporary biodiversity governance framework. It directly complements the IPBES's plurality view as it reasons that multiple values can be best understood based on experience-based knowledge and hence, it calls for integrating local level knowledge platform into the mainstream policy track (González-Jimenez et al., 2018; Pascual et al., 2017). There is, therefore, no denying the inevitability to revise laws, regulations and policies governing the use of biodiversity resources and to protect the rights of the IPLCs. Overall, the underlying implication of this alternative understanding is that it does not imply that the use of biodiversity resources should be prohibited (as argued by conservationists). Rather, the resources can be used and managed in a sustainable manner if certain factors are taken into consideration (e.g., mutuality of humans and nature, nexus between formal and informal institutions) in the case of usage and management process of the resources.

6.4 Equalising the Curve for Water Resources

The world is getting thirstier for water in tandem with economic development and population growth (Liou, 2021). Simultaneously, climate change threatens the availability and quality of water, exacerbating the problem of water security (Felter & Robinson, 2021). However, different regions of the world are experiencing varying degrees of difficulty in gaining access to basic water services or managing water resources sustainably. While some regions are geographically located in arid regions, others are blessed with an abundance of freshwater. Bangladesh is one of those countries where water is available, but there are issues in guaranteeing equitable access to water and water resources for all and in ensuring proper utilisation and sustainable management of those resources. This clearly indicates that the problem lies within the governance mechanisms.

The findings from the preceding analyses in Chap. 4 show that there is an acute problem in the provision of water at the household level in urban Bangladesh. Considering the context of the capital city of the country, the analysis exhibits that the market for water is not operating efficiently due to segregated markets. Some of the buyers are obtaining access to water through the formal channel while others are purchasing water from the informal market. It creates an unequal distribution in terms of the provisioning of water. Consecutively, utilising its inelastic demand, a group of intermediaries is securing the rent, making an oligopoly market by putting a high price on the demand for water. It is thus leading to the crisis of water, which at its core is a non-natural crisis originating from the problem of the market. Moreover, the competition over the accessibility to water poses threat to increase the unstable exploitation of water resulting in the declination at the groundwater level.

Secondly, the case of transboundary water governance provides strong evidence of non-cooperation resulting from the imbalance of power among the negotiating parties in the allocation of water rights. The flow was supposed to remain in its natural course but the crossing of political boundaries has made the transboundary water a political commodity. The flow is now subject to national interest and the allocation of water is helping to streamline the policy debates. It is, however, also the cause of conflicts as the mutually agreed upon allocation system is not stable. The absence of a regulatory regime, unequal power-sharing arrangements and inalienable national interest, rather than keeping the flow as it is, have now become the primary reasons for water shortage, especially in the lower riparian countries.

The third case on wetlands provides the proof of institutional fragility in the governance of aquatic resources. The existing institutional frameworks, particularly the policies, for the governance of the wetlands are not performing adequately to realise the maximum revenue along with the preservation of the species or resources. Considering the most vital resource of the *haors* in Bangladesh—the fisheries, the discussion reveals that the stock of capture fish is decreasing and the market-centric policies are excluding the local fishermen (claimants) as the recipient of the resources.

As a result, the claimants are being denied access to rent, which causes disruption in the human–nature nexus, ultimately inducing the degradation of the wetland ecosystem.

Finally, the problems of marine resources governance stem from technological deficiency and unstable property rights regimes. In such a scenario, the revenue earnings from the marine resources are dissipated either by suboptimal resource utilisation or through illegal capture by other powerful neighbouring countries. Thus, the pressure on marine resources and the inefficiencies in governance is increasing. Overall, the water resources in Bangladesh are being threatened by multiple challenges such as shortage, inequitable distribution and degradation. Concurrently, the complexities concerning water resource governance can be broadly understood as the problems of provisioning, the rights and proprietorship paradox and the inequality in benefit-sharing stemming from the political settlements.

In this backdrop, the chapter develops an alternative framework that can ensure the sustainability of water governance by challenging the current framework (Fig. 6.3). The current governance regime is predominantly centred on scarcity escaping from the root causes that stem from the failures of distribution and the price. Accordingly, it promotes the commodification of water and emphasises formal institutions. It fails, however, to explain the complexities of water governance and results in the inequitable provision, dissipation of rent, political contestation creating hostile situations and degradation of water resources. Moreover, resources remain underutilised in some cases. On the whole, the problems of water resources have been merely focused in light of equitable governance and the outcome is the unsustainable utilisation of resources.

Therefore, the alternative understanding necessitates that water is not a scarce resource rather the problem lies within the distribution system. Simultaneously, it emphasises recognising the multiple values of nature going beyond monetary valuation. The United Nations has also called for valuing water from multiple perspectives in recent times (UNESCO/UN Water, 2021; Liou, 2021). However, it is also necessary to strengthen the institutional arrangements (both formal and informal) under a strong regulatory regime to maximise the outcome through the multiple valuation procedure by creating a balance in power-sharing arrangements. In particular, informal institutions often play a significant role in managing some kind of water resources (e.g., wetlands).

Above all, the necessity lies in revitalising the nexus between human beings and water because water security is inextricably linked to the coexistence of human beings, the production system and the environment. This is a prime condition to enhance cooperation in managing water and its resources to alter the existing scenario. Following these principles would result in sustainable water governance indicated by universal provisioning of water, minimisation of conflicts over water, appropriation of underutilised water resources through technological advancement and termination of degradation of those resources.

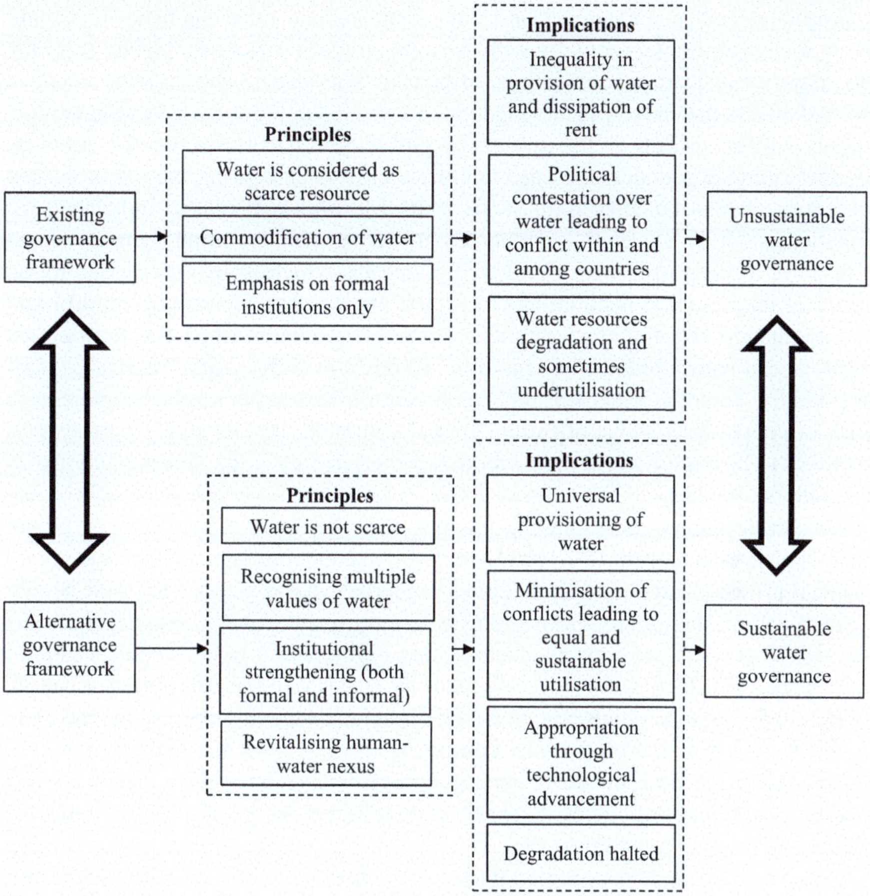

Fig. 6.3 Equalising the curve for water resources. *Source* Prepared by the authors

6.5 Flattening the Curve of Climate Crisis

Climate change is currently one of the most pressing issues. It has in fact emerged as the greatest obstacle to ending poverty and realising rights, threatening the strides made and the journey to lifting people out of poverty, as the poor, particularly the marginalised sections, are hit hard. The severity has reached the point where IPCC suggests that only rapid and drastic reductions in greenhouse gases (GHGs) emissions in this decade can prevent the climate breakdown and save the life of the earth, with every fraction of a degree of additional warming likely to exacerbate the accelerating effects (IPCC, 2021). It signals that ever-stronger efforts are required to arrest global warming which are, however, challenging to make fruitful within a very short span of time.

Empirical evidence examined in Chap. 5 substantiates that the historic preponderant responsibility lies actually with the industrialised nations of the North for the vast majority of historical emissions of GHGs. They have been creating negative external effects through rapid and gigantic industrialisation for the sake of economic growth and for residing in the hub of the global power platform. On the contrary, the developing and underdeveloped countries have been bearing the major burden of climate change but they have so far played a minimum role to imbalance the climatic system. However, these countries are also gradually opting for rapid industrialisation to keep pace with developed countries, which is also partly due to the failure of international negotiations to protect their rights. Discussion on different climate change negotiations, contracts and treaties demonstrates that the devised solutions can hardly reduce the emission of GHGs over the years. There are insufficient enforcement mechanisms and significant imbalances of opposing interests in terms of cooperation and application. Global commons are essentially portrayed as resources to be exploited for the profit of selected parties in the drafting and text of international treaties.

The developing countries having less power in the international arena could not put enough pressure on the developed countries to reduce emissions. Since states with financial prowess can control how global resources would be used for their benefit (Clancy, 1998), there is no motivation for less powerful states to participate in the negotiation process. As a result, there is less participation in preservation treaties and hence, less effective effort to safeguard the balance of the atmosphere. Overall, the political-economic landscape in which climate change discussions are currently taking place contains a plethora of competing interests and contradictions (Böhm et al., 2012). Moreover, the developing countries have technological incapacity and fragile institutional arrangements at the national level, which also deteriorates the climate vulnerability condition.

Bangladesh, as a developing country, also contributes little to increasing atmospheric concentrations of GHGs and yet suffers disproportionately from the impacts of climate change due to its geographical location and its low capacity to cope with the adversity of climate change. Different climate change effects such as recurring natural disasters, food insecurity and forced displacement, among others, have been contributing to augmenting people's vulnerabilities—primarily the poor—in the country. Overall, despite being the least contributor to the cause of climatic vulnerability, Bangladesh is one of the worst victims.

Considering such a scenario, this book reframes the understanding of solution to climate change as well as of sustainability. The alternative understanding posits that the crisis of climate change can be resolved over time by increasing the earth's resilience on the conditions of stabilisation, transformation and sustainability (Fig. 6.4).

Human impact on the environment is principally determined by the specific social structure, the social relations among people and the material or social relations between people and nature (Liodakis, 2010). The social structure, influenced by a

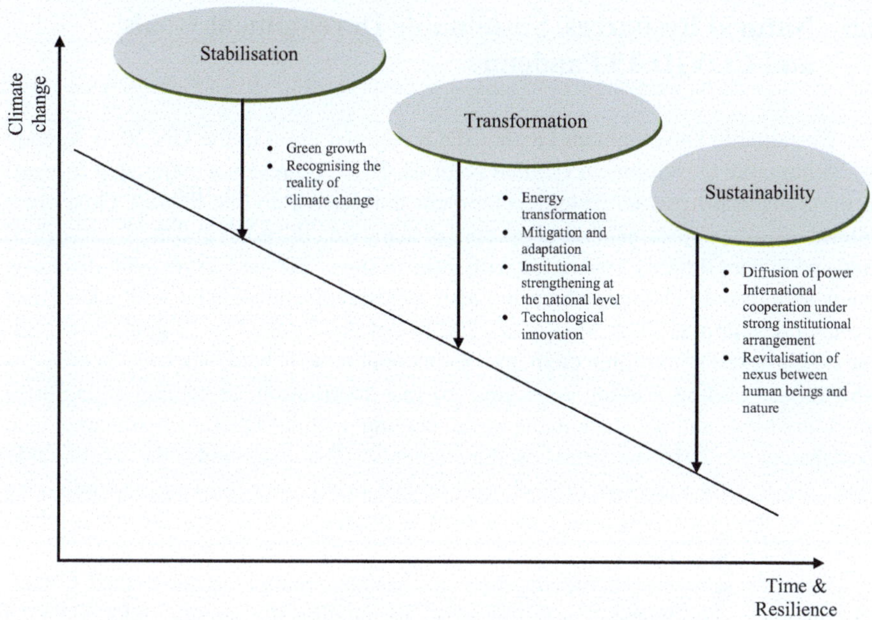

Fig. 6.4 Flattening the curve of climate crisis. *Source* Prepared by the authors

neoliberal regime, sees currency as the source of sustenance. Corporate-driven globalisation, accordingly, aggravates global warming due to the promotion of unsustainable production and privatisation of the global commons. It necessitates to be acknowledged at the first instance to address the climate change issue along with emphasising green growth to stabilise the condition.

Under transformation conditions, the focus would be primarily on mitigation and adaptation measures. In accordance with that, nation states should move towards energy transformation and technological innovation. However, the stabilisation and transformation conditions are necessary but not sufficient to halt the climatic vulnerabilities. Indeed, individual countries are currently considering such conditions and devising policies accordingly but there is no effective outcome. As climate change is a problem in the realm of global governance and sustainability is a holistic issue, the issues need to be problematized considering the whole world. Hence, the material balance of the earth system requires revitalising through the recognition of the complementary relations between nature and human beings, by diffusing the power relations among nation states and through the formation of strong institutional arrangements simultaneously.

6.6 Natural Resources, Sustainable Development Goals and COVID-19 Pandemic

The Sustainable Development Goals (SDGs) agreed upon in the UN 2030 Agenda for Sustainable Development (United Nations, 2015) represent a major step forward in the transition to sustainable development that leaves no one behind. Governing natural resources sustainably is at the core of achieving the goals and there are several goals which are directly associated with three of the concerned issues—biodiversity, water and climate change, which are also inextricably interlinked with each other (Pandya & Sharma, 2022; Shin et al., 2022)—of this book. The 17 goals, however, can be clustered under three elements of sustainable well-being such as (a) efficient allocation: building a living economy, (b) fair distribution: protecting capabilities for flourishing, and (c) sustainable scale: staying within planetary boundaries (see Costanza et al., 2016 for details on the elements). The goals under the 'sustainable scale' are directly addressed in this book (Table 6.1) but other goals are also indirectly related to the concerned issues and which is obvious as the goals are themselves interrelated to each other.

The progress towards achieving the goals has been hampered due to the COVID-19 pandemic. The world was falling short in meeting the targets under SDG 15 even before the pandemic and only one-third of the 113 countries were on track to achieve their national targets for integrating biodiversity into national planning (United Nations, 2020). The pandemic has revealed the importance of protecting biodiversity to mitigate the possibility of future pandemics but for the time being, it has put up barriers to realise the target to protect biodiversity as the financing in this sector might be reduced due to the economic downturn across the world.

The number of people using safely managed water rose by 10% from 2000 to 2017 (Felter & Robinson, 2021) although billions still lack access to water. Due to the increased exploitation of water resources in 2020 during the pandemic, it has been projected that 700 million people might be displaced by water scarcity by 2030 and some countries are experiencing a funding gap of 61% to achieve targets under SDG 5 (United Nations, 2020). It is unlikely, therefore, that countries would achieve the target of ensuring safe and affordable drinking water for all. On the contrary, the drastic reduction in human activity due to the pandemic can have some positive impacts on attaining the targets under SDG 14 (United Nations, 2021). Nevertheless, such positive impacts of the pandemic on water would probably sustain on a short-term basis, and accordingly, it cannot be concluded that the goal would be successfully met by 2030.

Finally, progress towards SDG 13 is also not satisfactory as is evident from the recent assessment of the IPCC, which claims that it is only conceivable to escape warming of 1.5 or 2.0 °C if massive and immediate cuts in GHGs are made (IPCC, 2021; McGrath, 2021). The pandemic has brought about a slight benefit in terms of reduction in carbon emissions but its long-term implication is negative. The pandemic has shifted the global attention towards itself while climate change is an equally important—in some instances more important—issue. Moreover, funding

Table 6.1 SDGs that are directly addressed in the book. *Source* Prepared by the authors

Goal	Description	Specific targets directly addressed in the book	Relevant chapter(s)
Goal 6: clean water and sanitation	Ensure availability and sustainable management of water and sanitation for all	Target 6.1: safe and affordable drinking water Target 6.5: implement IWRM Target 6.6: protect and restore water-related ecosystems (e.g., wetlands)	Chapter 4
Goal 13: climate action	Take urgent action to combat climate change and its impacts	Target 13.1: strengthen resilience and adaptive capacity to climate-related disasters Target 13.a: implement the UNFCCC	Chapter 5
Goal 14: life below water	Conserve and sustainably use the oceans, seas and marine resources for sustainable development	Target 14.2: protect and restore ecosystems Target 14.4: sustainable fishing Target 14.5: conserve coastal and marine areas	Chapters 4 and 3
Goal 15: life on land	Protect, restore and promote sustainable use of terrestrial ecosystems, sustainably manage forests, combat desertification and halt and reverse land degradation and halt biodiversity loss	Target 15.1: conserve and restore terrestrial and freshwater ecosystems Target 15.2: end deforestation and restore degraded forests Target 15.5: protect biodiversity and natural habitats Target 15.6: eliminate poaching and trafficking of protected species	Chapters 3 and 4

for new economic policies may divert the emergency funds usually afforded to climate funding (Markard & Rosenbloom, 2020). Therefore, it could be challenging to address climate change as envisioned under Agenda 2030. Overall, it distracted both the government and the public from environmental problems, which were the burning issues at the global platform before 2020.

Against this backdrop, there is an urgent need to rethink the governance mechanisms of natural resources. Post-pandemic recovery efforts should therefore prioritise plummeting environmental vulnerabilities over the medium and long terms. Overall, there is still a long way to go in achieving the desired ultimate goal to 'leave no one behind'—either to conserve biodiversity resources or to guarantee availability and sustainable management of water or to combat climate change. It is expected that the alternative understandings developed in this book about reversing the process of natural resource degradation can aid to trace the transformative pathway necessary to meet the goals in the near future, even if not possible by 2030. Policy initiatives grounded on the understandings can steer the course of socio-economic development onto a pathway that can ensure the well-being of both human beings and nature concurrently.

References

Böhm, S., Murtola, A., & Spoelstra, S. (2012). The atmosphere business. *Ephemera: Theory & Politics in Organization, 12*(1/2), 1–11.

Brand, U., Muraca, B., Pineault, É., Sahakian, M., Schaffartzik, A., Novy, A., Streissler, C., Haberl, H., Asara, V., Dietz, K., Lang, M., Kothari, A., Smith, T., Spash, C., Brad, A., Pichler, M., Plank, C., Velegrakis, G., Jahn, T., ... Görg, C. (2021). From planetary to societal boundaries: An argument for collectively defined self-limitation. *Sustainability: Science Practice and Policy, 17*(1), 265–292. https://doi.org/10.1080/15487733.2021.1940754

Clancy, E. A. (1998). The tragedy of the global commons. *Indiana Journal of Global Legal Studies, 5*(2), 601–619.

Costanza, R., Daly, L., Fioramonti, L., Giovannini, E., Kubiszewski, I., Mortensen, L. F., Pickett, K. E., Ragnarsdottir, K. V., De Vogil, R., & Wilkinson, R. (2016). Modelling and measuring sustainable wellbeing in connection with the UN Sustainable Development Goals. *Ecological Economics, 130*, 350–355.https://doi.org/10.1016/j.ecolecon.2016.07.009

Díaz, S., Demissew, S., Carabias, J., Joly, C., Lonsdale, M., et al. (2015). The IPBES conceptual framework-connecting nature and people. *Current Opinion in Environmental Sustainability, 14*, 1–16.https://doi.org/10.1016/j.cosust.2014.11.002

Felter, C., & Robinson, K. (2021). *Water stress: A global problem that's getting worse*. Council on Foreign Relations. Retrieved July 15, 2021 from https://www.cfr.org/backgrounder/water-stress-global-problem-thats-getting-worse

González-Jimenez, D. G., Berghofer, U., Berghofer, A., Heubach, K., Kosmus, M., & von Bertrab-Tamm, A. (2018). *Beyond measurement: Multiple values of nature and their diverse conceptualization*. Intergovernmental Science-Policy Platform on Biodiversity and Ecosystem Services (IPBES).

IPBES. (n.d). *Contrasting approaches to values and valuation: Value monism versus value pluralism in policy*. Retrieved August 25, 2021 from https://ipbes.net/ru/node/15967

IPCC. (2021). Summary for policymakers. In V. Masson-Delmotte, P. Zhai, A. Pirani, S. L. Connors, C. Péan, S. Berger, N. Caud, Y. Chen, L. Goldfarb, M. I. Gomis, M. Huang, K. Leitzell, E. Lonnoy, J. B. R. Matthews, T. K. Maycock, T. Waterfield, O. Yelekçi, R. Yu & B. Zhou (Eds.) *Climate change 2021: The physical science basis. Contribution of working group I to the sixth assessment report of the intergovernmental panel on climate change*. Cambridge University Press. In Press.

Liodakis, G. (2010). Political economy, capitalism and sustainable development. *Sustainability, 2*, 2601–2616. https://doi.org/10.3390/su2082601

Liou, J. (2021). *World water day 2021: Optimizing water resources with nuclear and isotopic techniques*. Retrieved July 15, 2021 from https://www.iaea.org/newscenter/news/world-water-day-2021-optimizing-water-resources-with-nuclear-and-isotopic-techniques

Markard, J., & Rosenbloom, D. (2020). A tale of two crises: COVID-19 and climate. *Sustainability: Science Practice and Policy, 16*(1), 53–60. https://doi.org/10.1080/15487733.2020.1765679

McGrath, M. (2021). *Climate change: IPCC report is 'code red for humanity'*. BBC News. https://www.bbc.com/news/science-environment-58130705

Nature. (2022). Biodiversity faces its make-or-break year, and research will be key. *Nature Editorials, 601*, 298. https://doi.org/10.1038/d41586-022-00110-w

Pandya, A. B., & Sharma, P. (2022). Water governance—Role for assuring water security. *Irrigation and Drainage, 71*(5), 1417–1419. https://doi.org/10.1002/ird.2778

Pascual, U., Balvanera, P., Díaz, S., Pataki, G., Roth, E., Stenseke, M., et al. (2017). Valuing nature's contributions to people: The IPBES approach. *Current Opinion in Environmental Sustainability, 26*, 7–16.https://doi.org/10.1016/j.cosust.2016.12.006

Pascual, U., Adams, W. M., Diaz, S., Lele, S., Mace, G., & Turnhout, E. (2021). Biodiversity and the challenge of pluralism. *Nature Sustainability*. https://doi.org/10.1038/s41893-021-00694-7

SCBD. (2020). *Global biodiversity outlook 5–Summary and conclusions*. Secretariat of the Conventional on Biological Diversity.

SCBD. (2022). Decision adopted by the Conference of the Parties to the Convention on Biological Diversity: Kunming-Montreal Global Biodiversity Framework. Secretariat of the Convention on Biological Diversity. https://www.cbd.int/doc/decisions/cop-15/cop-15-dec-04-en.pdf

Shin, Y., Midgley, G. F., Archer, E. R. M., Arneth, A., Barnes, D. K. A., Chan, L., Hashimoto, S., Hoegh-Guldberg, O., Insarov, G., Leadley, P., Levin, L. A., Ngo, H. T., Pandit, R., Pires, A. P. F., Pörtner, H., Rogers, A. D., Scholes, R. J., Settele, J., & Smith, P. (2022). Actions to halt biodiversity loss generally benefit the climate. *Global Change Biology, 28*(9), 2846–2874. https://doi.org/10.1111/gcb.16109

Swiderska, K. (2021). *Here's why indigenous economics is the key to saving nature*. International Institute for Environment and Development. Retrieved August 20, 2021 from https://www.iied.org/heres-why-indigenous-economics-key-saving-nature

Titumir, R. A. M. (2021). *Numbers and narratives in Bangladesh's economic development*. Palgrave Macmillan. https://doi.org/10.1007/978-981-16-0658-8

Titumir, R. A. M., Paran, M. S., & Pash, M. W. (2019). "The Sundarbans is our mind": An exploration into multiple values of nature in conversation with traditional resource users. In *Satoyama initiative thematic review (SITR)* (Vol. 5). United Nations University Institute for the Advanced Study of Sustainability (UNU-IAS).

Titumir, R. A. M., Afrin, T., & Islam, M. S. (2020). Traditional knowledge, institutions and human sociality in sustainable use and conservation of biodiversity of the Sundarbans of Bangladesh. In O. Saito, S. M. Subramanian, S. Hashimoto & K. Takeuchi (Eds.), *Managing socio-ecological production landscapes and seascapes for sustainable communities in Asia: Mapping and navigating stakeholders, policy and action* (pp. 67–92). Springer. https://doi.org/10.1007/978-981-15-1133-2_5

UNESCO/UN Water. (2021). *The United Nations world water development report 2021: Valuing water*. World Water Assessment Programme. United Nations Educational, Scientific and Cultural Organisation.

United Nations. (2015). *Transforming our world: The 2030 agenda for sustainable development*. Outcome Document for the UN Summit to Adopt the Post-2015 Development Agenda: Draft for Adoption. New York.

United Nations. (2020). The sustainable development goals report 2020. New York. Available at https://unstats.un.org/sdgs/report/2020/The-Sustainable-Development-Goals-Report-2020.pdf

United Nations. (2021). *Conserve and sustainably use the oceans, seas and marine resources for sustainable development*. United Nations Statistics Division (UNSD), Department of Economic and Social Affairs (DESA). Retrieved July 27, 2021 from https://unstats.un.org/sdgs/report/2020/goal-14/

The manufacturer's authorised representative in the EU is Springer
Nature Customer Service Centre GmbH, Europaplatz 3, 69115 Heidelberg,
Germany. If you have any concerns regarding our products, please
contact ProductSafety@springernature.com

Printed and bound by CPI Group (UK) Ltd, Croydon, CR0 4YY
30/09/2026
02225317-0013